NUCLEAR POWER
ISSUES
AND
CHOICES

REPORT OF THE NUCLEAR ENERGY POLICY STUDY GROUP

NUCLEAR POWER ISSUES AND CHOICES

Foreword by McGeorge Bundy

THE NUCLEAR ENERGY POLICY STUDY GROUP

Spurgeon M. Keeny, Jr., Chairman

Seymour Abrahamson	Carl Kaysen
Kenneth J. Arrow	Hans H. Landsberg
Harold Brown	Gordon J. MacDonald
Albert Carnesale	Joseph S. Nye
Abram Chayes	Wolfgang K. H. Panofsky
Hollis B. Chenery	Howard Raiffa
Paul Doty	George W. Rathjens
Philip J. Farley	John C. Sawhill
Richard L. Garwin	Thomas C. Schelling
Marvin L. Goldberger	Arthur Upton

Sponsored by the Ford Foundation
Administered by The MITRE Corporation

Ballinger Publishing Company · Cambridge, Massachusetts
A Subsidiary of J. B. Lippincott Company

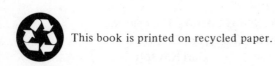

International Standard Book Number: 0-88410-065-0

Library of Congress Catalog Card Number: 77-693

Printed in the United States of America

Library of Congress Cataloging in Publication Data

Nuclear Energy Policy Study Group.
  Nuclear power issues and choices

  1. Atomic power industry—United States. 2. Power resources—United States.
I. Keeny, Spurgeon M. Jr. II. Title.
HD9698.U52N79 1977   333.7   77-693
ISBN 0-88410-065-0

Contents

Foreword

Ours is a time of contests over issues incompletely understood. To those with a sense of history there is no novelty in that, but what does seem new in our age are the number and variety of the issues of this sort, and also the fact that a few of them really do have a visible connection to the ultimate question of human survival.

Policy toward nuclear energy is such an issue. Where optimists see a front door to unlimited low-cost energy, pessimists see a back door to worldwide nuclear calamity, and neither perception is wholly unfounded. For more than thirty years both promise and menace have been strongly argued, and in the early 1970s the urgency of the debate was intensified as a consequence of heightened concern for energy supply, environmental hazard, and nuclear proliferation.

The present study originated from the Ford Foundation's exposure, in 1974 and 1975, to the widespread belief that this debate was suffering from a shortage of disinterested analysis. Because of our concern with general questions of energy policy, demonstrated in a number of grants and most conspicuously in our Energy Policy Project, we were approached for help by a number of advocates and opponents of nuclear power. Each side believed that any fair and objective study would substantiate its position and each sought support for such a study—although it is fair to add that at the outset there were wide differences as to the kind of study that would be "fair."

We pursued the matter through a series of meetings with individuals representing essentially every serious position in the nuclear debate and then brought a number of these people together for a roundtable discussion. We found a high measure of agreement that the public debate on nuclear power issues was poorly structured and undisciplined. The various actors were talking past each other to the crowd; irresponsible statements were going unchallenged; and implicit value judgments were unacknowledged. Moreover the debate was confused because the real evidence was complex, diffuse, and technical, while attempts to condense and explain it were suspect because so many of those making such efforts were seen as having already chosen sides.

So partisans of sharply different positions united in urging us to sponsor a fresh and independent study of both facts and issues. Our Board of Trustees endorsed this recommendation, with the addition of three very important ground rules:

1. The participants in the study must be recognized as highly qualified in their own fields of investigation and analysis.
2. As a group they must be—and must be recognized as being—essentially open-minded on the general debate raging around nuclear power.
3. The participants themselves, as a group, must have sole and complete responsibility for their findings.

This is the way it has worked out. We were fortunate in finding that the trustees and management of The MITRE Corporation fully shared our views of the right ground rules; MITRE has given the study skillful and flexible administrative support at every stage, without a trace of intervention on matters of substantive judgment. We were still more fortunate in finding twenty-one people of the first caliber who were willing to join in this sustained collective endeavor. And we were perhaps most fortunate of all in that Spurgeon M. Keeny, Jr., was willing to serve as Chairman of the group. His fairness and firmness, and his insistence on hard work and clear expression, are what have made this volume possible.

In the two years since we began to plan this study, the debate on nuclear energy has made real progress, both here and abroad. Decisions in public referenda and by public authorities have tended to go against the more extreme positions of both supporters and opponents. The relation between certain forms of nuclear energy and the spread of nuclear weapons is receiving more attention than it did a few years ago, and there seems to be a new determination to approach these questions with a balanced concern for future needs and dangers. Two major and encouraging findings of this study are that on the most dangerous matters we can afford to take the time that is needed to learn more about what should be done and that there is plenty that we can sensibly do in the meantime.

This study will certainly not be the last word on the subject. Another major inquiry is currently proceeding under the sponsorship of the National Academy of Sciences and the National Academy of Engineering, and as new issues emerge new work will be necessary. But I believe that the authors of this study have succeeded in moving the discussion to a new level of accuracy in evidence, trenchancy in analysis, and fairness in judgment. It is an honor to commend this work to the attention of all who care about these questions.

McGeorge Bundy

Preface

This report has been prepared in the hope that it will help clarify the issues underlying the debate on nuclear power at home and abroad. It is the result of a year-long study, made possible by a Ford Foundation grant to The MITRE Corporation. The report is the sole responsibility of the members of the Nuclear Energy Policy Study Group, which operated with complete intellectual independence of both the Ford Foundation and The MITRE Corporation.

Our study and report can best be characterized as an inquiry. The Study Group brought together individuals with a wide variety of expertise, experience, and viewpoints. None of us had taken a strong position for or against nuclear power. We did share a conviction that the questions involved were of great importance, the answers not simple, and the subject one for which a collegial inquiry was particularly appropriate.

Working in this spirit, we became profoundly conscious of the uncertainties that remain to be resolved by further research and by the unfolding of events. We therefore took as our primary task the development of a framework for decision-making. At the same time, we have been impressed with the far-reaching nature of the specific decisions that the United States has to face immediately in the nuclear power field. For this reason, we have gone beyond the analysis of alternatives and have offered our best judgment on a number of these decisions.

In the Overview, which summarizes our assessment of the issues and our major conclusions and recommendations, we have endeavored to put nuclear power in perspective relative to broader economic, social, and security objectives. These findings, which are the product of a long and challenging process, are the agreed views of the entire group.[a]

[a]During the course of this study, Albert Carnesale was appointed to membership on the Nuclear Regulatory Commission Hearing Board looking into the matter of wide-scale use

The matters dealt with in the Overview are developed at length in the body of the report. During the study, one or more members of the group examined specific aspects of the problem and, in consultation with outside experts, prepared working papers for consideration by the entire group. These working papers were discussed at meetings of the entire group and were further developed and modified as a result. The body of the report is based on these working papers, as variations in style may suggest.

In preparing the report, we have not aimed at complete coverage of the subject or the general education of the reader. Rather, our emphasis has been on the critical examination of selected problems to illuminate the central policy issues. While the whole group has reviewed and is in general agreement on the entire report, not every member necessarily agrees with every statement, and a member may not have a position on some of the more detailed material outside his own field of expertise.

In the course of the study, the group met thirteen times, including a week-and-a-half summer session, for a total of thirty-one days. In early sessions, aspects of the problem were discussed with government officials, scientific and industrial experts, and members of public interest groups. Among those consulted were Dean Abrahamson, William Anders, Hans Bethe, Thomas Cochran, Gordon Corey, Donham Crawford, George Kistiakowsky, James MacKenzie, Louis Roddis, Anthony Roisman, Robert Seamans, John Simpson, J. Gustave Speth, Arthur Tamplin, and Eric Zausner.

In addition to meetings of the entire group, several subgroups examined specific areas such as safety, resources, health, demand and costs, and so on. For example, an economics subcommittee considered economic models currently used in government and industry and examined in detail the consequences of various economic assumptions. Other members drew on electric utilities and architectural engineering firms in assessing the comparative costs of coal and nuclear power. Many meetings were held with experts in such fields as nuclear safety and the supply of resources.

We focused much of our attention on the critical review of major reports previously prepared on nuclear power. While we could only sample the extensive literature in the field, we did attempt to examine a broadly representative spectrum of views. We did not intentionally exclude any viewpoint or source of information.

We take this opportunity to thank the following consultants who assisted us in specific aspects of our study: Joan Aron, Samuel Berman, David Coward, Milton Klein, William Lowrance, Alan Manne, Harry Perry, Malvin Ruderman, Daniel Schiff, and Alan Strout. They and the many other individuals who helped us do not of course share responsibility for our findings.

---

of recycle plutonium in mixed oxide fuel in light-water reactors. Accordingly, he has refrained from expressing personal views on this matter and he should not be associated with the group's findings related to plutonium reprocessing and recycle.

We express our special appreciation to Tom Neff, senior staff member, whose many contributions and tireless efforts were vital to the success of the study. We also thank Larry Ruff, Ford Foundation Program Officer, and Deborah Elcock, Dorothy Mathers, and Peter Zimmerman, whose support contributed substantially to the project. We also are indebted to Millie Forrest, project secretary, who with the able assistance of Jewell Freeman, Claudia Johnson, Betsy McKelvain, and Pamela Miller, prepared and somehow kept all our paper work organized. Finally, we want to thank the Ford Foundation for making our study possible and The MITRE Corporation for excellent support in the conduct of our activities.

Before undertaking this study, we had varied views on the importance and proper role of nuclear power. None of us had systematically considered the full range of problems reviewed in this report. During our year-long exposure to these problems, we have all learned a great deal from one another and from sources outside the group. We hope that our investigation of various bodies of data and contrasting points of view, our analysis and reflection, and our intensive discussion among ourselves give value and relevance to the consensus we have reached.

Spurgeon M. Keeny, Jr., Chairman

Seymour Abrahamson

Kenneth J. Arrow

Harold Brown

Albert Carnesale

Abram Chayes

Hollis B. Chenery

Paul Doty

Philip J. Farley

Richard L. Garwin

Marvin L. Goldberger

Carl Kaysen

Hans H. Landsberg

Gordon J. MacDonald

Joseph S. Nye

Wolfgang K.H. Panofsky

Howard Raiffa

George W. Rathjens

John C. Sawhill

Thomas C. Schelling

Arthur C. Upton

Jan. 5, 1977

NUCLEAR POWER
ISSUES
AND
CHOICES

Overview

The debate over the future of nuclear power has become increasingly dominated by dedicated advocates and opponents of this source of energy. While polemics may focus attention on the problem, they do not provide a sound basis for public understanding of the issues or for national decision-making. In this study we have tried to take a fresh and independent look at the role that nuclear power should play in the United States and the rest of the world in this century. We have sought to develop a framework for assessing the difficult problems relating to nuclear power now before the U.S. government. Imminent decisions with far-reaching domestic and international consequences must be made on the following issues: (1) the reprocessing and recycle of plutonium, (2) the breeder reactor program, (3) the management of nuclear wastes, (4) the expansion of uranium enrichment capacity, and (5) the export of nuclear technology and materials.

In assessing the role of nuclear power, we have examined a broad range of difficult and controversial questions, including:

- How important is nuclear power to the economic growth and prosperity of the United States, of our major allies and other advanced countries, and of developing countries?
- How do the economics of nuclear power compare with those of coal and other alternatives?
- What are the extent and distribution of the world's uranium resources, and how do they affect the future of nuclear power?
- What are the economic prospects and developmental time scales of energy sources such as solar and fusion energy that may be alternatives to nuclear and fossil energy?

- What are the effects of nuclear power on the environment and human health as compared with those of coal and other alternatives?
- How safe is nuclear power, and how does the possibility of accidents affect the comparisons between coal and nuclear power?
- Can nuclear wastes be disposed of in an acceptable manner?
- How serious are the possibilities of sabotage to nuclear facilities, or of the diversion of materials to make nuclear weapons?
- What is the relationship between the worldwide growth of nuclear power and the proliferation of nuclear weapons, and how can the decisions of the United States affect the likelihood of proliferation?

While such questions are now being debated most actively in the United States, they are relevant to the rest of the world as well. Even in this country these questions have until recently been largely ignored by the public and, to a disturbing extent, have been treated complacently by the government itself. For more than twenty years, it has been the clear and almost unchallenged policy of the government to promote the development of peaceful nuclear energy at home and abroad.

The intense debate of recent years between those who emphasize the promise of nuclear energy and those who fear its consequences has identified but generally failed to clarify the underlying issues. Many critics attack nuclear power as an unacceptably dangerous source of energy that is being forced on the public despite unfavorable economic prospects. They question its economic benefits by pointing to escalating costs of nuclear construction and fuel, poor reactor performance, and hidden subsidies. They emphasize the gravity of the health hazard inherent in the nuclear fuel cycle, pointing to the possibility of catastrophic accidents and to a persistent threat from nuclear wastes, which may endanger civilizations thousands of years in the future. Finally, they assert that nuclear power will lead inevitably to the proliferation of nuclear weapons throughout the world and that every reactor and fuel cycle facility is a potential target for terrorists interested in sabotage or materials for bombs. Some critics conclude that the only solution to these dangers is a moratorium on further construction of nuclear plants.

Its proponents advocate nuclear power as a safe, clean source of energy that is indispensable to the future U.S. and world economies. They assert that it can generate base-load electricity at significantly lower cost than any fossil fuel alternative and that without it the rising demand for electricity cannot be met. They argue that it is demonstrably less dangerous to the environment and to human health than fossil fuel alternatives. They point to the excellent safety record of reactors and calculate that, while an accident could be serious, the probability of its occurrence is vanishingly small. Nuclear wastes, they assert, can be handled in ways that essentially eliminate the possibility of future accidents. They emphasize that nuclear power is an essential component of energy

independence. Finally, they maintain that the hazard of nuclear weapons proliferation exists independently of American nuclear power and warn that restrictions on programs or exports would simply turn potential markets over to
foreign competitors and reduce U.S. influence over nuclear power developments
abroad. Some advocates conclude that future energy demands can only be met
by a massive government-supported program to accelerate nuclear power, including early introduction of plutonium reprocessing and recycle and the plutonium breeder.

The "energy crisis" arising from the oil embargo of 1973 and the subsequent
large increase in OPEC oil prices introduced a new and confusing element into
the debate. Although these events did call attention to the limits of oil resources,
they did not themselves define—much less determine—the much larger and more
complex issues of long-range energy policy.

In our study, we have found merit in many of the points raised by both the advocates and the critics of nuclear power, but we have not been persuaded by their
conclusions as to the future role of nuclear power. We have also been concerned
about broader economic and security issues raised by the current "energy crisis,"
but we believe that great care must be taken to relate these issues properly to the
question of nuclear power.

To put nuclear power in some perspective, it must be recognized that the
world is not running out of energy. Although the relative contribution that oil
and gas make to the world's energy supplies will diminish before the end of the
century, substantial amounts of these fossil fuels remain. Coal resources are vast,
and uranium resources are probably much larger than currently estimated.
Further in the future, solar energy, probably fusion energy, and possibly geothermal energy can provide essentially unlimited sources of power. If these options are successfully pursued, the world can have plenty of energy in the future,
although probably at costs significantly higher than those of 1976. Thus, the
long-range energy problem is one of higher costs rather than one of absolute
limitations on energy availability.

Over a reasonable period of time, the impact of increased energy costs on the
world's economy in general, and the U.S. economy in particular, will not be as
great as is often assumed. Sudden sharp increases in the price of energy can cause
serious temporary hardships and dislocations, as was demonstrated by the post-
1973 quadrupling of oil prices. Permanently higher real energy costs will reduce
the economic resources available for other needs—an effect of particular importance in developing countries. The cost of energy, however, is a small enough
factor in the overall economy that long-term cost increases of the magnitude we
foresee will not cause major changes in the economic or social future. Economic
growth can be sustained even with large increases in the price of energy. In any
case, higher future energy costs, which are probably inevitable, are largely independent of the rate at which nuclear power is developed and deployed over the
next twenty-five years.

Our analysis indicates that nuclear power has and will probably continue to have a small economic advantage on the average over coal, the closest alternative for the generation of electricity in the United States. Regional and other variations in the cost of electricity from nuclear power and coal, however, are sufficiently large that coal is and will continue to be a competitive source of energy in many areas. Moreover, the ranges of possible social costs, such as health and environmental impacts, associated with coal and nuclear power also overlap to such an extent that neither has a clear advantage. We find such large uncertainties and unknowns in both the economic and social costs that the average comparative advantage could shift either way in the future.

In these circumstances, we believe that there is time for a broad and sustained approach to energy problems. While nuclear power is one of the options that should be pursued, it is not as critical to future economic development as its advocates claim. There is time therefore to assess carefully the potential risks as well as the benefits of nuclear power and to avoid hasty and uncritical decisions.

At present, the range of uncertainties in the comparative costs of coal and nuclear power is such that a mix provides a useful hedge against uncertainties. Since there is considerable overlap in the present costs of coal and nuclear power, economic forces may be expected to produce such a mix of coal and nuclear plants. For the longer term, a balanced research and development program should develop additional options based on improved use of coal and nuclear power as well as solar, geothermal, and fusion energy. These energy sources should play a role in the future when and where they can compete economically.

In general, this analysis also applies to the role of nuclear power in other advanced nations. However, countries that lack domestic supplies of coal, as well as oil and gas, face a more difficult problem than the United States in obtaining assured, diversified energy sources and in dealing with balance of payments problems. While complete dependence on nuclear power is not a solution for such countries, they may have, or may think they have, a special interest in nuclear power.

Nuclear power, with its large, complex, capital-intensive plants poses special problems for developing economies. The higher capital costs per unit of generating capacity of small nuclear plants, suitable for small power grids and limited demands, would tend to eliminate the economic advantage of nuclear power even against imported fossil fuels. Nevertheless, some fifteen to twenty developing countries may find economic justification for nuclear power in this century.

By far the most serious danger associated with nuclear power is that it provides additional countries a path for access to equipment, materials, and technology necessary for the manufacture of nuclear weapons. We believe the consequences of the proliferation of nuclear weapons are so serious compared to the limited economic benefits of nuclear energy that we would be prepared to recommend stopping nuclear power in the United States if we thought this would prevent further proliferation. However, there are direct routes to nuclear weapons in the absence of nuclear power, and the future of nuclear power is not under the unilateral control of the United States. Most advanced countries are now actively developing and utilizing nuclear power, while less developed countries count it among their expectations. In fact, abandonment of nuclear power by the United States could increase the likelihood of proliferation, since the United States would lose influence over the nature of nuclear power development abroad. With continued nuclear power development, however, the U.S. government must give greater weight to the proliferation problem in its decisions on nuclear matters and its relations with other nations.

Nuclear energy is now a fact of international life and will provide a significant portion of the world's electricity by the end of the century. At the same time, nuclear power is only one of several energy options, and decisions about it should be made on the basis of sound national and international economic considerations, realistic accounting of social costs, and a paramount concern to avoid further proliferation of nuclear weapons.

THE CURRENT STATUS OF NUCLEAR POWER

Nuclear power is a present reality, not a future prospect. In mid-1976, nuclear power plants in the United States accounted for about 40,000 megawatts of electric capacity (MWe), or about 8 percent of total national capacity. Abroad, nuclear power totaled about 35,000 MWe. In the United States, some 170,000 MWe of additional nuclear capacity are under construction or on order and scheduled to begin operation by the mid-1980s; abroad, construction and orders are reported to amount to about 130,000 MWe. While there has been a substantial cutback in plans for both U.S. and foreign nuclear power plants in the past two years, this reduction appears to be primarily the result of economic recession and reduced projections of electricity demand rather than a rejection of nuclear power.

Projections of U.S. nuclear power capacity for the year 2000 vary greatly, and have been sharply reduced recently. In 1975 the Energy Research and Development Administration (ERDA) estimated that nuclear power capacity could be as high as 1,250,000 MWe by the year 2000. These estimates were revised downward by ERDA in July 1976, to range from 450,000 to 800,000 MWe, and in September 1976, were further reduced to range from 380,000 to 620,000 MWe.

Measured against the total use of energy, the present contribution of nuclear power is still small, amounting to about 2 percent in the United States. For the foreseeable future, nuclear power will only be used for generation of electric power, which currently consumes some 28 percent of the primary energy used in this country. In contrast, fossil fuels can be used not only for electric power but also for transportation, industry, residential heating, and petrochemicals. It is generally expected, however, that the use of electric power will grow more rapidly than overall energy consumption. Moreover, in producing electricity, nuclear power releases some oil for other purposes. In the future, it could also release coal for other uses such as the production of synthetic oil and gas if there are constraints on the utilization of coal.

Nuclear power has become international. Thirty countries, in addition to the five nuclear weapon states, have nuclear power plants in operation, under construction, or on order. There is now highly competitive international commerce in reactors, uranium (natural and enriched), and supporting equipment. The principal suppliers in addition to the United States are West Germany, France, the United Kingdom, Canada, Japan, South Africa, and the Soviet Union.

The principal motive for this interest in nuclear power is the desire for a cheaper source of energy. Although the original promise of abundant nuclear electric power at a fraction of the cost of other sources has faded, the more modest belief persists that economic benefits will accrue over the lifetime of current plants. Buying into nuclear technology is widely believed necessary to share in future economic payoff. This economic rationale, which is central to the nuclear power debate, is examined in detail in the following section.

Considerations other than economic have also undoubtedly influenced some nations' interest in nuclear power. One of these is the desire to develop a diversified energy base to increase the security of supply. While a reliable energy supply is an objective to which most nations must give attention, only a few countries have the resources for total independence or self-sufficiency. Since much of the present world economy relies on oil and gas and since nuclear power cannot directly substitute for oil in such areas as transportation and petrochemicals, an assured supply of nuclear power would not eliminate the need for other fuels. Substitution of electricity for other energy uses would be a lengthy and gradual process involving major changes in capital equipment. In the next few decades, nuclear power can do little to reduce the impacts of sudden changes in energy supplies such as oil. Principal reliance during this period must be placed on stockpiles, resource sharing, and other measures. However, concern about the long-term security of energy supplies, particularly of oil, will probably lead to a preference for nuclear power, particularly in Japan and Western Europe.

In the longer term, nuclear power may present advantages in guarding against interruption in supply since uranium stockpiles are more manageable and less costly than comparable stockpiles of oil or coal. Some countries may see breeder reactors as providing greater independence from external fuel supplies for electricity. Countries seeking independence through the breeder, however, would have to be able to reprocess plutonium and fabricate new fuel within their own borders, and, aside from Japan and Western Europe, would still be dependent on a few outside suppliers for reactors and critical equipment. The level of economic and political cost individual nations may be prepared to bear in pursuit of energy independence is a hard choice each will have to make for itself.

Another motive for acquiring nuclear power is prestige. The status accorded nuclear power as a high technology industry was initially stimulated in large part by U.S. promotion of the Atoms for Peace program, beginning in 1955. Today, even in less developed nations for whom the investment may be disproportionately large and the benefits questionable, nuclear power may be politically attractive for its perceived glamour.

A final motive for the development of nuclear power is to acquire a technical base for a nuclear weapons option. Even if a country has no present plans to manufacture weapons, it may want to be in a better position to acquire a nuclear weapons capability in response to a future threat. That a shortened lead time toward a potential nuclear weapons capability may be a motive for acquiring nuclear power indicates the close tie between energy policy and foreign policy.

ECONOMIC CONSIDERATIONS

The principal argument for nuclear power has been that it would produce cheaper electricity than alternative energy sources. The present worldwide level of investment in nuclear power may suggest that nuclear power has achieved a clear economic advantage. The charge is frequently made, however, that if costs were properly accounted for and subsidies stripped away, nuclear power would not be competitive. To clarify this fundamental issue, we have examined the economics of nuclear power as compared with coal, the closest competitor, and with other alternative energy sources. We have also considered the broader significance of the differences in cost between nuclear power and alternative sources to general economic growth and prosperity.

Comparison of Coal and Nuclear Power

Like so much else in the nuclear debate, the comparative economics of nuclear power and other energy sources for electric power has become shrouded in controversy. The comparative economics of coal and nuclear power is a genuinely complex problem about which there can be honest differences of opinion. Plants committed today will not begin operation until 1986 and are intended to have a useful life of thirty years. On such a time scale, the projection of lifetime costs is a very speculative business. Not only have construction costs of both coal and nuclear power plants escalated substantially in recent years but so too have the prices for uranium and coal. Moreover, stricter environmental controls on nuclear power and fossil fuels could have far-reaching economic effects. Finally, new scientific information on long-range environmental effects or events relating to safety or nuclear proliferation could lead to decisions that would have major economic effects. In such an uncertain environment, projections must be made with considerable caution.

Despite these large uncertainties, our analysis leads us to the conclusion that nuclear power will on the average probably be somewhat less costly than coal-generated power in the United States. However, coal will continue to be competitive or preferable in many regions since there are large regional cost differences and wide variations even within a region. The advantage for nuclear power is likely to be most significant in New England and in parts of the South. In large areas of the West, containing a small fraction of the country's population, coal-generated power is likely to be less costly than nuclear power. In much of the country, however, the choice is so close and the uncertainties sufficiently large that the balance could easily shift either to increase or to eliminate the small average advantage that nuclear power presently enjoys.

Conclusions on the comparative economics of coal and nuclear power depend on estimates of future capital charges, which include construction costs and interest, and the cost of uranium and coal. Each of these cost factors involves special problems that contribute to the uncertainty and possibility of different interpretations.

Capital Charges. The cost of electricity from nuclear plants is dominated by the capital charges for the plant.[a] At present, more than 70 percent of the cost of electricity from a light-water reactor (LWR) is attributable to capital charges. Over the past decade, construction costs of nuclear power plants have risen markedly faster than the rate of inflation. In addition, as a result of outages and reduced operation, the average capacity factor[b] for nuclear plants has been less than expected, particularly for the large new 1,000 MWe plants. Reductions in the capacity factor have the same effect on power cost as increases in capital charges.

Analysis of capital charges of nuclear power is complicated by the surprisingly wide range in the construction costs of plants of similar design, depending on the location, the builder, and the method of doing business. If one starts with high-cost examples and assumes that costs will continue to rise as they have in the past, it is indeed possible to conclude that nuclear power will become economically noncompetitive. Our analysis indicates, however, that construction costs (in constant dollars) will tend to level off in the future. Costs of labor and materials have escalated more rapidly than the general rate of inflation. The regulatory process has lengthened the construction period and necessitated numerous design changes and retrofits. The rate of escalation in costs should not continue to diverge from the regular rate of inflation, and the construction and licensing period (now typically ten years) should not continue to lengthen and may be shortened. With experience and standardization, design changes and retrofits should be brought under progressively better control. Finally, although aging factors not yet encountered might further reduce the capacity factor, it seems more likely that the somewhat disappointing operating experience to date

[a]Costs are for electricity at the point of generation; distribution costs add 50 to 200 percent to the price to the consumer.

[b]The "capacity factor" of a plant is the ratio of electric energy actually delivered during the year to that which would have been delivered if the plant operated 100 percent of the time at full capacity.

is part of the shakedown experience typical of a new technology and that capacity factors will improve in the future.

Capital charges are a smaller fraction of the total cost of electricity from coal than from nuclear power plants. Nevertheless, these charges still represent a substantial portion of the cost of coal-generated electricity (35 to 65 percent depending on location). Construction costs of coal plants have also increased in recent years for some of the same reasons as for nuclear plants. In addition, the use of new equipment (scrubbers) to reduce sulfur emissions may increase construction costs as much as 20 to 25 percent. In view of evolving pollution standards and technical responses, the uncertainties in construction costs of coal plants are probably as great as for nuclear plants. Nevertheless, with experience, the construction costs of coal plants should also level off and capacity factors should improve.

Uranium. At present, uranium accounts for only 5 to 10 percent of the cost of electricity at the power plant. The recent rapid rise in the price of uranium from around $6 per pound in the early 1970s to spot prices reportedly as high as $40 per pound has further clouded the economic picture. Although the increase from $8 to $40 per pound (if sustained) would only increase the cost of electricity by around 20 percent, such a rapid advance in prices raises basic questions about the future price and availability of uranium. If uranium is indeed in short supply and becomes very expensive, the current generation of light-water reactors (LWRs) will have difficulty competing with coal plants. This asserted future shortage of uranium is the principal argument advanced for early introduction of breeder reactors, which in theory might expand the power produced by a given supply of uranium by a factor of as much as 100.

Our review convinces us that current official estimates of uranium reserves and resources substantially underestimate the amounts of uranium that will be available at competitive costs. We believe that there will be enough uranium at costs of $40 (1976 dollars) per pound to fuel light-water reactors through this century and, at costs of $40 to $70 per pound, well into the next century.

While mineral reserves are commonly understood to be identified resources (i.e., known in location, quantity, characteristics, and economically recoverable with current technology), uranium reserves are specified over a range of estimated costs, which at the higher end will only become of commercial interest in the future. Indeed, early estimates of uranium reserves reflected industry judgment on how much uranium could be produced profitably at the government-set price of $8 per pound. Consequently, there was little commercial interest in more expensive uranium, and presently identified higher cost deposits represent principally the by-product of exploration for low-cost reserves. In the past two years, uranium prices have risen sharply in response to an anticipated expansion in demand, present low production levels and doubts about the pace of expansion, and the increase in oil prices. It is difficult to estimate the increased supplies that will become available at higher prices. This is not unusual, since mineral reserves estimates almost always lag when prices rise and markets expand.

It is not enough to wait for market forces to press firming up of potential resource estimates and the discovery of these "new" reserves, since more reliable estimates are needed for long-range decisions on future energy policy and technical programs. Although the National Uranium Resource Evaluation (NURE) program has been established for this purpose, we do not believe it will produce better estimates soon enough, since it is almost entirely dependent on private efforts to locate, define, and report reserves. This program should be reoriented with higher priority assigned to improving estimates of uranium reserves and resources.

Coal. In a coal-fired electric power plant, the cost of mining and transporting the coal accounts for 30 to 60 percent of the total cost of the electricity. This wide spread in fuel charges reflects differences in costs of transportation, of the sulfur content of coal, and of mining methods. The price of coal increased in response to OPEC's increase in the price of oil and is now about twice the pre-embargo level. The real cost of coal mining, however, has not increased very much in spite of increased labor and environmental charges.

Unlike uranium, U.S. coal reserves are definitely known to be very large. Reserves are estimated at more than 400 billion tons in place with 50 to 80 percent recoverable, compared with an annual production in 1975 of just over 600 million tons. Total coal resources are estimated at some 4 trillion tons, most of which would be available only at higher costs since it is contained in seams too thin or deep to qualify as reserves. In view of the enormous coal reserve, the real cost of mining coal should not increase substantially until well into the next century. Cost increases due to increased mining safety standards, environmental controls, or more difficult deposits should be at least partly offset by technological improvements in mining and transportation.

Recent official policy has projected a doubling of coal production by 1985. Although production has not expanded significantly in the past three years, we believe that it should be possible to achieve this production by the 1985-90 period and to double it again by the turn of the century if the demand appears. Doing so would require an increase in production of some 6 percent annually, a rate we judge achievable if basic problems are resolved. Large investments will be required to open new mines and to improve transportation. Public conflicts will have to be resolved on environmental and land use issues before adequate commitments will be made by private investors.

Three years after the Arab embargo, the coal industry is still not operating at full capacity; and, in the absence of new demand, coal prices have fallen from their peak. Nevertheless, the prospects for coal should not be underestimated since coal will be generally competitive with nuclear power for a long time to come and will in all probability become the material from which synthetic gas and oil will be manufactured. A larger role for coal exports in the international energy trade is also foreseeable and desirable. In these circumstances, the priority and expectations for coal should not be less than for nuclear energy. Both government and the private sector should demonstrate this priority by increased efforts to develop more efficient mining technology, improved transportation systems, better control of sulfur oxides and other emissions, radically improved long-distance electrical transmission, and development of more efficient, lower cost synthetic gas and oil conversion. Equally important is a climate of confidence with regard to land use regulation, public energy policy, and federal-state regulatory relations.

Comparative Economics in Other Countries

The foregoing comparison of coal and nuclear electric power is based on expectations in the United States. Most other countries have fewer resources and less access to technology and investment capital. Nevertheless, the arguments concerning the comparative costs are similar in all countries, as long as fossil fuels and uranium can be relied on as items of international commerce.

Considerable emphasis on nuclear power has developed in Western Europe and Japan. These countries have experienced a gradual reduction of coal reserves and persistent difficulties with their coal industries. As a result, following World War II they turned to low-cost Middle East oil, which, together with natural gas, became their principal energy source. These countries are now planning extensive nuclear power programs to reduce their reliance on oil, which is no longer cheap and constitutes a far larger share of imports for them than for the United States. For most of them, a shift to heavy reliance on coal would require increasing dependence on imports from the United States and Eastern Europe. The political acceptability of such dependence is not clear. There is also a question as to how large a foreign market this country could supply and still meet its own growing domestic demand. For these reasons, a greater preference for nuclear power should be expected in these countries than in the United States. While official pronouncements affirm this preference, there are also indications of increasing public opposition to nuclear power in these countries. On balance, a diversified energy program appears to be the best choice for Western Europe and Japan both economically and politically.

In less developed countries, the share of future energy demands that will be supplied by nuclear power is very uncertain. Nuclear power may be competitive in some twenty developing countries by the year 2000, and others may install it for noneconomic reasons. As a practical matter, the large 1,000 MWe nuclear power plants now being built to achieve economies of scale are not matched to the small power grids of most developing countries. More suitable, smaller plants (less than 600 MWe) would have significantly higher capital costs per kilowatt and, in the absence of demand, are no longer being built. For these reasons, nuclear power may be ruled out as an economic energy option for many developing nations.

Alternative Sources of Energy

It is frequently argued that solar, geothermal, or fusion energy would be viable alternatives to nuclear power if they received a fair share of the research and development funds. It is our judgment that these forms of energy cannot compete with nuclear, coal, or other fossil fuels as major sources of electric power until well into the next century. We believe, however, that vigorous research and development should be carried out in these fields to develop the long-range options and to provide a hedge against possible unforeseen problems with fossil or nuclear power.

For the long run, solar energy is especially interesting, since it is essentially unlimited. It takes many forms, from direct radiation to energy stored in the ocean, waves, winds, and vegetation. Applications such as house heating will be practical in the near future in favorable situations and, in conjunction with improved insulation and other conservation measures, can reduce the growth in demand for electricity. However, solar electric power will become competitive only after considerable research and development and a large increase in the cost of electricity. While we favor a continuing effort to develop solar electric power, we see little prospect, given the state of the technology and the high capital costs, that solar electricity can compete with nuclear and coal plants in this century.

Geothermal energy, which is being exploited on a very limited scale at a few unique locations, constitutes a huge potential resource. Most of this energy, however, will be very difficult to exploit. We see little prospect therefore that geothermal energy will prove competitive for electric power on a large scale in this century.

Fusion, like solar energy, offers the promise of practically unlimited energy. Important scientific progress has been made recently in this extremely sophisticated technology. Although it is still premature to predict success, we believe that fusion reactors will probably demonstrate a useful energy output by the year 2000. There is little prospect, however, that fusion will supply electricity on a competitive basis in the next fifty years. Fusion reactors will involve large capital costs and complex systems with unknown capacity factors, and it remains for future generations to see when they will become competitive.

Despite our pessimistic assessment of the near-term prospects of these alternative energy sources, they support our optimistic assessment that in the longer view of human affairs, adequate energy will be available—at a price.

In the search for new sources of energy, a variety of proposals have been advanced that make use of the energy in nuclear explosives. The proposed applications of Peaceful Nuclear Explosives (PNEs) range from the stimulation of gas and oil and "in-situ" retorting of oil shales to the direct heating of steam for power plants. The economic merit of these proposals ranges from highly dubious to clearly noncompetitive. All of the proposals that could have any serious impact on the energy situation would require a frequency of nuclear explosions and the production and transportation of nuclear devices on a scale that would dwarf nuclear weapons activities. For example, two 50 kiloton explosions per day would be required for a single hypothetical 1,000 MWe power plant. The security implications are staggering since PNEs are fundamentally indistinguishable from military explosives. We believe, therefore, that the now dormant U.S. PNE program should not be revived even as a research and development effort and that the United States should discourage interest in PNEs abroad, since they provide a convenient cover for weapons development.

Long and expensive research and development efforts will be required before advanced technologies can be commercially competitive. With such time horizons, there is little prospect that the private sector will support these activities. Energy research and development is currently funded by the government at over $3 billion per year, about 4 percent of the current value of the energy output. Further development of nuclear power, particularly breeders, and improved utilization of fossil fuels, compete for these funds. While these more developed technologies deserve a large share of the government effort, we believe that a serious effort should continue on advanced technologies to provide assurance of long-range energy supplies and to provide a hedge against the possibility that limitations may have to be placed on either fossil or nuclear energy for some reason.

The research and development chain includes the following steps: fundamental research, basic applicable research, applied research, development, and finally commercialization. Costs increase dramatically as one moves from fundamental research to commercialization. Since the advanced technologies aim at targets as far as fifty years away, the decision to go forward with full-scale development should be taken with great care. The transition from federally supported programs to unsubsidized commercial use can be achieved only if the private sector finds investment in plants economically attractive. Thus there is little value in demonstrating clearly noncompetitive technology unless the demonstration substantially advances the engineering of the technology at a cost commensurate with the value of the advance. If the demonstration takes place before it is economically justified, the government may have to subsidize the program at a high level for a long time after demonstration, and the ultimate product may also be inferior to that which would have resulted from continued development. In addition, premature commitment to expensive demonstration programs can distort the balance of the federal energy program. We believe that the government must exercise greater care in the future before moving into the very costly phases of the development chain.

The Economic Significance of Nuclear Power

Whatever is done about nuclear power over the next few decades, real energy costs will continue to increase into the next century. We have considered the likely effects of this overall cost increase on growth, income and employment, and have estimated the differences that would result from possible variations in the timing and character of the U.S. nuclear future. We find that nuclear power choices have limited bearing on these larger social and economic conditions.

Energy is an important factor in an economy, and any unexpected interruption in supply, such as an oil embargo, will have serious disruptive effects. But the cost of energy is less critical than assured supply. Even after OPEC quadrupled oil prices, primary energy costs are only about 5 percent of the U.S. GNP. Another doubling in real energy costs would result in the shifting of up to 5 percent of the economy from the production of final products to the production of energy or of goods to pay for imports. If such a shift occurred rapidly, it could be accompanied by serious, short-term dislocations in the economy, including increased unemployment, loss in output, and a reduction in growth. The rate of investment might have to be increased permanently from 10 percent to 12 or 13 percent of GNP to provide the larger amounts of capital needed, and current consumption expenditures would be reduced by 2 or 3 percent. In the long run, however, the economy should be able to absorb higher energy costs with little effect on growth or employment.

In actual practice, the impact of higher energy costs on income and capital needs will probably be reduced significantly by the market response to the higher prices and by other energy conservation actions. Energy-saving industrial processes such as the cogeneration of electricity with process heat, energy-efficient building designs and transportation systems, and less energy-intensive consumption habits are widespread in other developed economies where higher energy prices have long prevailed. Now that even higher energy costs are expected everywhere, new conservation opportunities will be developed and exploited as an alternative to expansion of supply. Market forces will produce most of these conservation measures, but the government must play an important role by helping market forces to operate, by providing information and leadership, and by modifying policies which may uneconomically encourage greater energy use.

Although we anticipate that the market response will significantly reduce the growth of energy use, there are hazards in formulating long-range energy policy on specific predictions based on such inherently uncertain factors. Projections of U.S. energy demand in the year 2000 differ by a factor of 2. For example, the Institute for Energy Analysis has estimated that U.S. energy demand, which was about 70 quads (one quad is 10^{15} BTU) in 1975, would grow to 101–124 quads by the year 2000, while the Edison Electric Institute has estimated a demand of as much as 194 quads by then. Because of the large uncertainties in demand determination, we attach little credence to long-term projections that rely on extrapolations from historical experience. Although there has been historical correlation between energy and economic growth, there is no reason to believe that the same relationship will hold under conditions of rising rather than falling energy prices. Fundamental energy policy decisions should therefore be designed to meet a broad range of possible future conditions.

Whatever the income loss due to higher energy costs, nuclear power can do little to reduce it in this century since nuclear power will at best have only a small cost advantage over coal. To understand the ranges of the economic benefits associated with nuclear power, we have used a simple computer model of the U.S. economy to explore the effects of various economic assumptions and program decisions. Even with assumptions favorable to nuclear power, the benefits from the continued growth of light-water reactors (LWRs) and the early introduction of the breeder are very small in this century (a small fraction of 1 percent of GNP), and only 1 or 2 percent in the next century. Relatively conservative assumptions for the reduction in demand in response to higher energy prices were used in the calculations. If, as we anticipate, demand reduction turns out to be easier than assumed in this formal analysis, the income effect of higher energy costs will be even less. Even though it will not affect the long-range economic growth of the country, 1 or 2 percent of the large GNP anticipated for the next century is a large absolute dollar amount of income and should be given up only if there are strong noneconomic reasons for doing so.

The desirability of maintaining or changing any particular style of life has not entered into our analysis. Some critics of nuclear power include among their arguments disapproval of industrial society and of continued economic growth. The broad range of issues relating to the style of life of our society is not, however, central to nuclear power. These issues should be addressed directly on their merits. The style of life that evolves in the future will depend on many factors other than the existence of nuclear power or central power stations or the price of energy. Increases in the price of energy may gradually modify attitudes toward specific energy-intensive activities relative to other activities. But in themselves, higher prices for energy need have relatively little effect on the evolution of the basic style of life of the future.

We have analyzed the impact of energy on the economy from a long-range perspective that smooths out short-term effects. While substantial changes in energy prices can be accommodated in the long run, sudden stoppages or sharp increases can indeed force severe temporary cutbacks in industrial operations with attendant unemployment and hardship. The level of economic activity will be affected since individual and institutional plans and attitudes take time to adjust. The amount of nuclear power available will have little to do with the cause, severity, or duration of such events. The choice made between coal and nuclear power will have little or no effect in insulating the United States from the short-term effects of sudden changes in oil prices and availability. The response to these situations must be by other means.

Although our analysis has focused on the U.S. economy, the same conclusions are broadly applicable in other industrialized countries willing to rely on world markets for fuel supplies and to assume the associated foreign exchange costs. The situation in less developed countries is more serious. In these countries, economic growth is more dependent on expansion of the industrial base, which requires capital and energy, particularly for industry and transportation. Higher energy costs may therefore have a more serious effect on their economic growth. Moreover, the already stringent balance of payments problem of many less developed countries will be further aggravated by the necessity of importing fuel at higher costs. This constraint immediately affects their prospects for economic growth. Nuclear power, however, is not well suited to the needs of many of these countries, since it is capital-intensive and limited to the production of large amounts of electricity. Moreover, the need for increased energy for transportation, industry, and agriculture implies the use of nonelectrical energy that cannot be supplied by nuclear power. The extent to which nuclear power is an economic response to high world oil prices will therefore depend on the circumstances of individual countries.

This assessment of the economic impact of nuclear power has been dominated by market considerations. Responsible policy decisions must also consider the external social costs of risks to public health and the environment and the implications for national security and world peace. In the following section, we consider the social costs that are not included in ordinary market calculations.

HEALTH, ENVIRONMENT, AND SAFETY

Nuclear power has been widely attacked as a threat to human health. Critics are primarily concerned about the possibility of catastrophic reactor accidents and the health and environmental problems associated with nuclear wastes and plutonium. These risks are real and must be considered in any assessment of nuclear power.

As with market economics, the risks and social costs of nuclear power should be compared with those of coal, which is the principal energy alternative for electric power in this century. This comparison is not an easy task. The possible social costs of coal and nuclear power involve such diverse health effects as prompt and delayed deaths, genetic diseases, illness, and discomfort; the environmental effects range from land use problems to the possible modification of the atmosphere leading to worldwide climatic changes. Some of these social costs (such as the costs of improving the safety of reactors, reducing pollution from coal, and payments to miners with black lung disease) are reflected in the market economic comparisons between coal and nuclear power since they are included in the cost of electricity. However, the general effects of emissions from coal and nuclear power plants are not included in such cost comparisons.

Analysis of social costs raises difficult and controversial methodological problems in valuing human life and health now and in the future. The greatest difficulty, however, is the uncertain state of knowledge regarding the effects on health and the environment of low levels of chemical and radioactive pollution and regarding the probability of nuclear accidents. Since there is little operating experience with nuclear power, it is impossible to estimate accident probabilities with any precision. Some risks may be unknown. In the case of coal, several hundred years of experience have not produced quantitative understanding of the health consequences and even less understanding of the possible effects on the world's climate of the carbon dioxide and particulates released during coal combustion.

The range of uncertainty in social costs is so great that the balance between coal and nuclear power could be tipped in either direction with resolution of the uncertainties. It is unlikely, however, that the principal uncertainties will be resolved in the near future. We do not believe therefore that consideration of social costs provides a basis for overriding our conclusions, based on economic analysis, of the comparative attractiveness of the two technologies and the desirability of maintaining a mix.

Public Health—Normal Operations and Accidents

In principle, one can compare the impact on public health of coal and nuclear power directly in terms of the deaths and illness they cause. In normal operations, a 1,000 MWe nuclear power plant has been estimated to produce roughly one fatality per year from occupational accidents and radiation risks to workers and to the public. A comparable new coal plant, meeting current new source standards, has been estimated to produce from two to twenty-five fatalities per year. Accidents in coal mining and transportation account for roughly two fatalities per year, and the rest of the range is attributed to the health effects of sulfur-related pollutants. This wide range results from the very large uncertainties in the actual effects on human health of the pollution chains resulting from sulfur oxides and from significant differences resulting from plant location with respect to population. The analysis of the risk at specific locations is complicated by uncertainties in meteorology, chemistry, synergistic effects involving other pollutants, and existing backgrounds. In addition to fatalities, pollution from coal plants contributes to large-scale nonfatal illnesses and discomfort for which there is no nuclear counterpart. There may also be significant effects from nitrogen oxides, carcinogenic hydrocarbons, and heavy metals for which a quantitative basis has not yet been established.

Thus, in a comparison of normal operations, nuclear power has smaller adverse health costs than coal. However, in an overall comparison of health effects, the possibility of accidents must also be taken into account. The possibility that nuclear accidents could have very serious consequences for public health has long been recognized as a unique problem associated with nuclear power. It is difficult to compare such rare but extremely severe events with the continuous health burden due to fossil fuels, but some perspective is gained from averaging the consequences of estimated accidents over an extended period. This requires knowledge of the probabilities and consequences of a spectrum of possible nuclear accidents.

To date, the safety record of nuclear power reactors in the United States has been excellent, at least as far as public health is concerned. However, the experience of some 200 reactor years of commercial nuclear power does not provide an adequate statistical basis for risk predictions covering the 5,000 reactor years expected during the rest of this century. Probabilistic judgments must be made on related technical experience and theoretical computations. Such an analysis of the current light-water reactors was undertaken in The Reactor Safety Study (WASH-1400, frequently referred to as the Rasmussen Report), published by the Nuclear Regulatory Commission (NRC) in October 1975. This report examined in a systematic fashion a large number of possible paths that could lead to an accident, estimated the overall probability of a nuclear core meltdown and breach of containment, and developed a probabilistic assessment of the consequences of such an accident, averaged over location and weather. Although WASH-1400 is a valuable resource for the study of the safety problem, we believe that it seriously underestimates uncertainties and has methodological flaws that are discussed in our report.

Without attempting to duplicate the massive analysis of WASH-1400 but taking its uncertainties into account, we have attempted to gain some perspective on the possible social costs of reactor accidents by considering the following questions:

- How does the predicted rate of reactor accidents affect the average rate-of-loss comparison between nuclear power and coal?
- How serious might the consequences of a reactor accident be?
- How likely might an extremely serious nuclear accident be if the associated uncertainties are all viewed pessimistically?

The average rate-of-loss due to reactor accidents calculated in WASH-1400 is only about 0.02 fatalities per year for a 1,000 MWe nuclear power plant. This rate is very low compared with the one fatality per year predicted for normal nuclear operations or the two to twenty-five fatalities per year attributed to a comparable new coal plant. Although we have not made an independent estimate of this average value, for losses due to nuclear accidents, our analysis indicates that with extremely pessimistic assumptions the WASH-1400 estimate might be low by a factor of as much as 500. On the other hand, it could be on the high side as well. In the most pessimistic case, which we consider very unlikely, the average rate-of-loss could be as high as ten fatalities per year for a 1,000 MWe nuclear power plant. However, even in this extremely unlikely situation, the average fatalities would not exceed the pessimistic end of the range of estimated fatalities caused by coal. Thus, on an average rate-of-loss basis, nuclear power compares favorably with coal even when the possibility of accidents is included.

An extremely serious accident under very adverse conditions is estimated by WASH-1400 to kill as many as three or four thousand people over a few weeks, cause tens of thousands of cancer deaths over thirty years, and cause a comparable number of genetic defects in the next generation, as well as more than $10 billion in property losses. Despite large uncertainties in biological effects, this appears to be a reasonable assessment of the potential consequences. While such an accident would clearly be a major disaster, the consequences would not be out of line with other peacetime disasters that our society has been able to meet without long-term social impact. For example, the United States has experienced a number of hurricanes that have taken over a thousand lives, produced physical damage in the billions of dollars, and required massive evacuation. In such a nuclear accident, the delayed deaths from cancer would not be an immediate effect but might result in a 10 percent increase in the incidence of cancer in the exposed population over a period of thirty years. It must be emphasized that a nuclear accident would probably have much less severe consequences than those estimated for this extremely serious accident and that most nuclear accidents would result in few, if any, fatalities.

The most serious accident considered in WASH-1400 is assigned an exceedingly low probability of occurrence (only one chance in 200 million years of reactor operation). This calculation is based on the combination of a number of low probability estimates for a series of events, most of which are extremely uncertain. When the uncertain factors are viewed in the most pessimistic light, there is a significant chance that such an event might occur during this century if the nuclear program grows at the projected rate. While it is very unlikely that this pessimistic assessment correctly describes the probability of such accidents, it does place an upper bound on the problem.

Having examined nuclear accidents from each of the above perspectives with very pessimistic assumptions, we have concluded that, even when the possibility of reactor accidents is included, the adverse health effects of nuclear power are less than or within the range of health effects from coal. At the same time, this analysis underscores the importance of continuing efforts to reduce the probability and consequences of accidents by improved safety designs and siting policies.

A foreign reactor accident would not necessarily be evidence of risk in this country, since some foreign reactors may be less safely constructed or operated than those in the United States. Nevertheless, a foreign nuclear accident could have a major psychological impact in this and other countries. A high premium should be put on reducing the probability and consequences of reactor accidents wherever they might occur.

Nuclear Wastes

The potential health hazards of radioactive wastes and plutonium produced during reactor operation are unique to nuclear power. Plutonium and other waste components present special problems since they decay very slowly and remain dangerous for hundreds of thousands of years. Critics of nuclear power question the morality of creating this threat to future generations or even to future civilizations.

We are convinced that nuclear wastes and plutonium can be disposed of permanently in a safe manner. If properly buried deep underground in geologically stable formations, there is little chance that these materials will reenter the environment in dangerous quantities. Even if material were somehow to escape eventually in larger quantities than seems possible, it would not constitute a major catastrophe, or even a major health risk, for future civilizations.

Despite our confidence in the feasibility of permanent disposal, nuclear wastes remain a very serious potential health problem until isolated from the environment. We are, therefore, more concerned about the current worldwide management of nuclear wastes before they are sequestered permanently than we are about the unlikely prospect that they will affect society subsequently. Inadequate management of wastes from the nuclear weapons and earlier civilian power programs here and abroad has already created potential contamination problems that can only be overcome at considerable cost.

Until very recently, all decisions on waste disposal were deferred apparently in anticipation that the problem would be resolved as a by-product of the assumed early introduction of plutonium reprocessing and recycle. As a consequence, it is widely believed that reprocessing is a necessary stage in waste management and disposal, when in fact it may simply complicate the process. If spent fuel is reprocessed to recover plutonium, the possibilities of waste management failure increase because of the additional steps involved. The risks in permanent waste disposal, however, appear to depend little on whether reprocessing has occurred. The impact of reprocessing on waste management is that it substitutes a larger immediate contamination risk for a small reduction in the long-term hazard from permanent disposal.

Environmental Effects

In addition to direct effects on human health, the generation of electricity by either nuclear power or coal has environmental effects on air, land, and water, and potentially on global climate as well. On balance, however, nuclear power has significantly less adverse environmental impact than coal.

Local thermal pollution is common to both sources, although somewhat more severe for nuclear power. In other respects, however, the coal cycle presents the more serious problems. Coal mining has a more disruptive effect on the land than uranium mining and milling, although this difference will diminish as lower grades of uranium ore are mined. Coal mining results in acid runoff that pollutes waterways, and combustion of coal leads to acid rain that damages land and crops.

The most serious potential environmental impacts from greatly increased power generation are changes in global climate. The thermal output of both coal and nuclear power plants contributes directly to the long-term heating of the atmosphere. A much more immediate atmospheric heating problem, however, results from the carbon dioxide produced when coal is burned. The carbon dioxide, which probably cannot be prevented from entering the atmosphere, heats the earth by the so-called greenhouse effect since it is transparent to incident solar radiation but absorbs some of the heat that the earth reradiates. It has been estimated that the carbon dioxide from fossil fuels burned in the last two centuries has increased the mean temperature about 0.3° C above what it would otherwise have been. However, the extent of the actual heating is complicated by the uncertain, possibly compensating, effects of particulates and other pollutants that are also emitted by coal combustion. More fundamentally, the current unpredictability of natural climate variations, which can be significant over relatively short periods, makes an empirical assessment of the actual impact of increased use of fossil fuels very difficult. Moreover, since short-term fluctuations would generally mask any trend for a number of years, these questions are not likely to be resolved for some time.

Despite these uncertainties, major increases in the use of fossil fuels could have a significant effect on climate. Whether the impacts of carbon dioxide will combine with natural changes in climate to the net disadvantage or advantage of mankind or to particular regions such as the United States cannot be judged at this time. On the basis of greater knowledge, however, this effect could take on overriding significance in a comparison of coal and nuclear power. This factor argues against putting complete reliance on coal power at this time.

Risk Reduction

Much better information is needed for the comparative assessment of the social costs of coal and nuclear power. There is little prospect, however, that this information will be available soon. While it is being developed, much can be done to reduce the risks of both coal and nuclear power.

For nuclear power, we have been particularly impressed by the variability of risk with location in case of accident. The predicted consequences of accidents at different sites can vary a hundredfold depending upon the distribution of population and prevailing weather. A more restrictive siting policy would increase somewhat the costs of nuclear power in some locations, but we believe it is warranted by the uncertainties in the probabilities of accidents and by the large risk reductions that are possible. Special measures such as underground siting should be considered if nuclear power is to be sited at high risk locations.

We believe that in research and development more emphasis should be placed on actually improving safety as compared with proving that reactors are "safe enough." The present government safety program, which is oriented toward the latter confirmatory approach, will ultimately narrow the range of uncertainty, but it is unlikely to reduce the probability of accidents. Steps should also be taken to ensure that the regulatory process does not inadvertently create disincentives to improvements in safety design.

For coal, stricter regulations have greatly reduced occupational hazards in mining and substantially reduced the levels of pollutants. Improvements in current technology such as scrubbers and new combustion technologies such as fluidized-bed combustion can further greatly reduce the health hazards to the public. With scrubbers, for example, the health effects from sulfur-related pollutants could be reduced by as much as a factor of 10 below present levels for new plants if low-sulfur coal were also used. The land damage from strip mining can be largely eliminated with only a small increase in the cost of coal. With these and other measures being taken on their own merit, the comparison between the social costs of nuclear power and coal will present a constantly evolving picture.

NUCLEAR PROLIFERATION

In our view, the most serious risk associated with nuclear power is the attendant increase in the number of countries that have access to technology, materials, and facilities leading to a nuclear weapons capability. The growth and diffusion of nuclear power thus inevitably enhance the potential for the proliferation of nuclear weapons. If widespread proliferation actually occurs, it will prove an extremely serious danger to U.S. security and to world peace and stability in general. By 1985, most advanced and many industrializing countries will have nuclear power plants in operation and be only a few steps from a weapons capability.

Expectations, knowledge, and trade in nuclear facilities and materials are so widespread that the United States is not in a position to stop the expansion of nuclear power. Moreover, advanced countries, and some developing countries, are not dependent on nuclear power to produce nuclear weapons. None of the present nuclear weapon states developed its weapons through nuclear power. Each followed the direct path of producing the fissionable materials for its weapons in facilities designed specifically for the purpose.

Despite this somber appraisal of the technical situation, nuclear weapons are not an inevitable consequence simply because the technical capability has been achieved, as West Germany, Japan, Canada, and Sweden demonstrate. The perils of proliferation are recognized by most countries. For most, nuclear weapons offer little advantage and considerable risk. There is widespread though not complete support for the Treaty on the Non-Proliferation of Nuclear Weapons (NPT) and the international safeguards on peaceful nuclear programs administered by the International Atomic Energy Agency (IAEA). These institutions, which reflect an international consensus against the acquisition of nuclear weapons by additional states, provide a framework for a nonproliferation policy and legitimize bilateral and multilateral restraints on nuclear trade.

Not all nonweapon states have foresworn nuclear weapons, and some may seek a nuclear weapons capability. Their possible motives are understandable if questionable: achieving prestige and status for some; overcoming isolation and insecurity for others. The response to these motivations is to give such countries confidence and standing, internationally and regionally, without recourse to nuclear weapons, and to limit the desires for nuclear weapons by regional settlements and easing of tensions and by security alliances and support for insecure states.

The motivations and responses to states seeking weapons capability underscore the essentially political nature of the nuclear proliferation problem. A strategy to constrain proliferation must be complex and comprehensive. U.S. nuclear power policies and programs can be shaped to support such a strategy, but they can be only partially effective unless they are meshed with broader political actions and international arrangements.

Some of the elements of a U.S. nonproliferation strategy that are broader than nuclear power are: a foreign policy in support of international security, peace, and stability; security commitments to reduce the perceived need for nuclear weapons; use of influence to discourage apparent preparatory moves for a nuclear capability; arms limitation agreements (e.g., a comprehensive test ban) to build additional barriers to proliferation; deemphasis of nuclear weapons in military policy, particularly doctrines that present nuclear weapons as acceptable and necessary armaments for limited application or political pressure; and cooperation in international development of the full range of energy resources.

Other measures more specifically related to nonproliferation include: support of the NPT and encouragement of present nonparties to adhere, and increased financial and technical support of the IAEA as it undertakes the increased burden of applying safeguards to a rapidly expanding, worldwide nuclear power industry. IAEA safeguards play a limited, but important, role in the effort to control proliferation. These safeguards cannot prevent proliferation but can discourage it by providing a system for warning that material has been diverted from proper peaceful uses. Such a system, while not foolproof, is a valuable deterrent. It would not be an attractive choice, for most states, to base a nuclear weapons program on clandestine diversions in violation of a formal international treaty.

There are also actions and policies that relate directly to nuclear power and the nuclear fuel cycle that would help to control nuclear proliferation in important ways. The nonproliferation system will inevitably be flawed and unstable if plutonium and highly enriched uranium, materials suitable for nuclear weapons, and the facilities to produce them become increasingly widespread. The time required for achieving a nuclear weapons capability would be greatly reduced and the temptation to make an irreversible decision to fabricate, and even use, nuclear weapons might be difficult to resist in a crisis. Facilities for plutonium separation and enrichment of uranium are thus particularly sensitive.

Fortunately, current reactors do not require plutonium or highly enriched uranium and instead use slightly enriched or natural uranium that cannot be used directly for weapons. These reactors do generate plutonium as part of normal operation, but it is mixed with highly radioactive fission products in the spent fuel and requires separation before it can be used in weapons. If this plutonium is chemically separated (reprocessing), it can be used again (recycled) in current reactors. The plutonium breeder reactor, which produces more plutonium than it consumes, requires reprocessing and recycle of plutonium for fuel. The potential danger of these technologies to proliferation is abundantly clear.

Current uranium enrichment facilities, located in the five weapon states, use gaseous diffusion technology, which inherently requires large capital investment and consumes large amounts of electricity. New methods that can be applied on a smaller scale are in various stages of development. Centrifuge plants, suitable for smaller programs and easily modified to produce highly enriched material, are beginning operation in Western Europe; most other countries would need to import the main equipment. Laser separation, which could prove successful in the 1980s, may permit production of highly enriched uranium on a small scale for relatively little capital investment and very small amounts of electricity. If not controlled, these new technologies will further complicate the proliferation problem.

International arrangements to control the nuclear fuel cycle are imperative to buttress the NPT and its safeguards system. Continued adherence by the United States to its long-standing policy of banning export of reprocessing and enrichment facilities is one contributing measure. Coordination with other suppliers of nuclear technology and materials is also essential. A few suppliers, however, cannot impose an enduring fuel cycle arrangement on the rest of the world. There must be developed a significant consensus among buyers as well as suppliers that it would be in their common interest to control the nuclear fuel cycle and thus establish physical constraints against a chain reaction of proliferation that could undermine international stability. Within the framework of such a common appreciation, it should be possible to develop effective constraints on the export of sensitive technologies.

The early introduction of plutonium recycle and plutonium breeders has been widely believed to be critical to the economic use of uranium and nuclear power. These beliefs have been encouraged by the emphasis on these programs in the nuclear development activities of the United States and the other principal nuclear suppliers. If the nuclear fuel cycle is to be controlled internationally, other countries will have to be convinced that there are no significant economic penalties in deferring these technologies. This will be hard to do if the United States is proceeding with reprocessing and breeder commercialization. While deferral of these programs would not necessarily convince all others that they should foreswear trading or acquisition of such facilities, it probably would convince some and would seriously influence the thinking of all countries. Deferral would also remove the appearance of discrimination in asking developing countries to forego these technologies and in achieving common supplier action to ban the export of such technology. Conversely, a decision by the United States to proceed with plutonium reprocessing and recycle would probably ensure worldwide movement to incorporate plutonium in the fuel cycle.

Postponing plutonium recycle and the plutonium breeder will increase concern about the availability of slightly enriched fuel for the present generation of light-water reactors. Conversely, assured supplies of such fuel from the United States and other suppliers will reduce the pressure for the new technologies directed at extending fuel supplies. Assured supplies of slightly enriched uranium at reasonable prices will also greatly reduce the economic rationale for other countries to build indigenous enrichment plants. The fact that a number of countries will be able to furnish enrichment services should provide further confidence of assured supplies.

Within such a framework of national and international constraints on the nuclear fuel cycle, we believe that, with concerted efforts by the United States and the international community to meet national security concerns and to reduce international tensions, the risk that nuclear power will lead to proliferation can be substantially reduced. The specific nuclear power decisions that the United States will have to take to accomplish this objective are considered below under "Issues for Decision."

Terrorism

A particularly disturbing aspect of nuclear proliferation is that it could extend to subnational terrorist groups. While a completed nuclear weapon would be a more convenient target, a highly organized terrorist group might have the capability to fabricate a crude nuclear weapon from stolen plutonium or highly enriched uranium. Since neither of these materials is available in the present fuel cycle, this threat will only emerge if plutonium is reprocessed and recycled or if reactors requiring highly enriched uranium are introduced.

The difficulty and danger of designing, planning, and constructing a crude weapon from reactor-grade plutonium should not be underestimated. Although it would not prove as easy as sometimes suggested, it is conceivable that a well-organized group supported by knowledgeable individuals could construct a device that might have a yield equivalent to a few hundred tons of TNT. When one considers that the largest conventional bombs of World War II contained only a few tons of TNT, the destructive potential of such a weapon is apparent.

A terrorist group might sabotage a nuclear facility in an attempt to inflict damage or threaten sabotage of a seized facility to blackmail authorities. The most serious target would be an operating nuclear reactor, where trained and knowledgeable saboteurs could cause a major accident. This threat cannot be quantified, but it clearly adds to the probability of an accident. Whatever that additional probability is, it can be reduced by measures designed to prevent or deter sabotage.

We believe that additional measures should be taken to reduce the possibility of terrorist acts to divert materials or sabotage facilities. Physical and personnel security should be improved at nuclear power plants and fuel cycle facilities. At present, the responsibilities of federal, state, and local agencies with respect to jurisdiction over crimes involving nuclear facilities are ill-defined, a situation that could lead to an inappropriate response to an emergency. The federal government should lead in developing improved security practices, coordinating procedures for law enforcement, and making expertise available.

We are convinced that measures to improve security substantially can be taken without infringing on the civil liberties of employees of the nuclear industry and the general public. Overzealous and ill-conceived measures, however, could endanger civil liberties and set dangerous precedents. The government should, therefore, be particularly sensitive to the broader legal implications of measures undertaken to improve security against, or the ability to respond to, terrorist activities.

Nuclear terrorism is international in scope. Terrorist acts in the United States could result from materials or devices seized abroad and smuggled into this country. The United States thus has a critical interest in the improvement of nuclear security and should encourage the development and implementation of effective physical security measures in all countries.

INSTITUTIONAL FRAMEWORK

Formulation of policy on nuclear power is affected by the institutional framework within which decisions must be made and implemented. Nuclear power interacts with public health and safety, the environment, foreign policy and national security as well as the economy. At the same time, nuclear power is, or should be, part of overall national energy policy, which in turn should be part of broader national economic policy. As a result, the complex issues have increasingly cut across institutional lines within the Executive Branch, regulatory agencies, Congress, state and local government, and the private sector.

In some respects, this institutional complexity is characteristic of energy in general. In other respects, nuclear power presents unique institutional problems arising from the original government monopoly in nuclear energy, the special risks of accidents and theft, the risks of nuclear proliferation, and the complex of treaties and agreements that have developed in the field. Until 1974, the government's role in nuclear power was largely the monopoly of the Atomic Energy Commission (AEC), which served as both promoter and regulator of the industry. This arrangement, while effective in developing nuclear power as an energy alternative, tended to make nuclear energy an end in itself, isolated from broader energy policy and potentially out of balance with other domestic and international considerations.

In a major organizational reform, the AEC was abolished in 1974 and the Nuclear Regulatory Commission (NRC) was created to deal with the regulatory aspects of nuclear power. The Energy Research and Development Administration (ERDA) was given responsibility for all research and development in energy including nuclear power. This action eliminated the anomalous situation where the same institution was both promoter and regulator, and placed the development of nuclear power in a common framework with all energy development activities. Although these institutional reforms have corrected some of the obvious problems with the AEC, the decision process remains fragmented. Despite its new charter, ERDA has to some extent continued to place emphasis on nuclear energy, in part because it inherited the substantial organization and facilities of the AEC.

Despite the extensive reorganization of energy institutions and the intense interest and activity in the field since the 1973 embargo, the government has not formulated a clear national energy policy. We believe that such a policy is necessary to provide a basis on which the various agencies concerned with energy can establish priorities and make and implement specific decisions. Such a policy need not, and we believe should not, be a highly structured long-range plan but rather a strategy for developing choices for the future. The policy should establish consistent and achievable objectives and priorities in areas such as the development of new energy supplies, conservation, energy independence, emergency supplies, and the weight to be given to nonproliferation.

We are convinced, after a year's exposure to the range of problems involved, that the President must be directly involved in the formulation of both overall energy and nuclear energy policy. There is no lower level that can have the authority to resolve the diverse domestic, foreign policy, and security interests. While not attempting to advise the President on how to organize the Executive Office, we believe that some arrangement should be devised to assist him directly in this area. Although this approach can be criticized as establishing a pattern that would break down if extended, as will be suggested by some, to many other complex areas such as resources and food, we believe the energy problem does merit priority attention.

Even where energy policy has been established, there is no clear mechanism for implementing it within the Executive Branch, and agencies are often left to interpret government policy independently. The Executive Office must play a stronger role in seeing that policy is understood and carried out responsibly by each lead agency and coordinated with other agencies having responsibilities for various aspects of the nuclear power problem.

With the demise of the AEC, the Joint Committee on Atomic Energy lost its unique position and has now been abolished. Other Congressional committees have acquired increased responsibilities for separate aspects of nuclear power. This has been a useful process in developing broader Congressional and public understanding of the issues. However, Congressional responsibility for nuclear power has become too diffuse, and there should be some consolidation so that Congress can deal effectively with nuclear power in the broader perspective of overall energy and foreign policy. In the final analysis, the ability of Congress to contribute to the policy-making process and produce constructive legislation will depend on the presentation of an overall energy plan to Congress by the President.

In a narrower sense, there are important institutional problems affecting nuclear power in the relations of the federal government with state and local governments. Regulatory authority over nuclear energy facilities is widely dispersed among federal, state, and local authorities. The licensing of a power plant requires dozens of separate permits and approvals. Each authority considers the question not only in a limited geographical setting, but also from a narrow functional perspective—economic, safety, environmental, or aesthetic. The licensing process can become an obstacle course, resulting in delays and increased costs and frustrating sound decision-making.

Under the Constitution, Congress could pass legislation establishing federal control over the whole licensing process, but the variety and intensity of local and regional interests argue strongly against wholesale displacement of their authority. Short of this, it would be desirable to act at the federal level to simplify and rationalize this process by reducing the number of separate hearings and proceedings required before final approval. Federal statute could provide for a single consolidated proceeding at which all aspects of the problem could be covered with all interested parties having an opportunity to appear and present evidence. This would produce a consolidated record on which local, state, and federal authorities could make decisions which could be reviewable in a single appeal to a Federal Court of Appeals.

From the perspective of a rational overall energy policy, one might argue that provision should be made for federal preemption over decisions by state or local authorities. However, in keeping with our conclusion that economic factors and appraisals of social costs should lead to an appropriate mix of coal and nuclear power plants, we believe that it is desirable to allow considerable leeway for local preferences. If it should develop that the cumulative effect of local preferences would endanger a reasonable national mix of coal and nuclear power plants, the case for federal preemption would be stronger.

Government-industry relations in the nuclear power area are currently in some disarray. The government has historically taken the leading role in nuclear development, originally a government monopoly. Private firms were initially contractors or chosen instruments. Even today, nuclear power is not really a private industry in the normal sense, since the government retains a dominant role in such areas as uranium enrichment, waste management, and research and development.

It is in the interest of a sound U.S. energy economy to let the market establish the rate of nuclear power growth. This does not mean, however, complete private ownership of the fuel cycle in view of the large capital requirements, technical and economic uncertainties, and security sensitivity of facilities such as those for uranium enrichment, plutonium reprocessing, or permanent waste disposal. Siting policy is also an appropriate area of government responsibility in view of the strong dependence of accident risks, affecting large and widely dispersed populations, on the specific location of power plants. Working out a government-industry relationship that is economically and managerially sound and provides a clear basis for planning is an important national objective; doctrinaire assignment of proper roles for government and the private sector is not warranted.

In sum, we do not believe nuclear power can be treated as just another industry. Utility choices between coal and nuclear plants should be based on market considerations, within a regulatory framework that deals adequately with social costs external to the industry. At the same time, the special security implications of nuclear power demand continued close government control and participation in critical stages of the nuclear fuel cycle.

ISSUES FOR DECISION

The United States faces a number of early decisions having an important bearing on the future of nuclear power and on the worldwide risks in the nuclear fuel cycle. These decisions, which are closely interrelated, must be considered in the context of the economic, energy supply, social costs, and international security issues discussed above. From this broader perspective we have examined the pending decisions: whether to proceed with plutonium reprocessing and recycle; how to conduct a breeder program most appropriate to long-term energy needs; how to manage and dispose of nuclear waste; when and how to expand enrichment capacity; and how to develop a nuclear export policy which minimizes threats to international peace and stability.

The significant common thread in these decisions is the question of whether plutonium should be introduced into the nuclear fuel cycle. We have concluded that there is no compelling reason at this time to introduce plutonium or to anticipate its introduction in this century. Plutonium could do little to improve nuclear fuel economics or assurance here or abroad. This conclusion rests on our analysis of uranium supply, the economics of plutonium recycle in current reactors, and the prospects of breeder reactors. In the longer term, beginning in the next century, there is at least a possibility that the world can bypass substantial reliance on plutonium. If this is not the case, the time bought by delay may permit political and technical developments that will reduce the nuclear proliferation risks involved in the introduction of plutonium.

Plutonium Reprocessing and Recycle

The principal immediate issue affecting nuclear power is whether the United States should proceed with the reprocessing and recycle of plutonium. Until recently, it was generally assumed that spent fuel from light-water reactors (LWRs) would be reprocessed to recover the plutonium produced during operation and that the plutonium and any unused uranium-235 would be recycled as fuel in LWRs. The expectation was that this process would take place on a commercial scale as soon as the nuclear power industry had expanded to the point to justify the large facilities needed for economic operation. The decision whether to license this activity is now before the NRC. Statements by both candidates during the 1976 Presidential campaign indicated, however, that these assumptions are being challenged on a bipartisan basis and that a consensus is emerging not to proceed at this time with reprocessing.

In a major statement on nuclear policy on October 28, 1976, President Ford announced that "reprocessing and recycling of plutonium should not proceed unless there is sound reason to conclude that the world community can overcome effectively the associated risks of proliferation." This does not, however, constitute a decision on reprocessing but rather an identification of the issue. Although the Administrator of ERDA was directed not to assume that reprocessing would proceed, he was also directed "to define a reprocessing and recycle program consistent with our international objectives." During his campaign, President Carter stated in San Diego on September 25 that he would "seek to withhold authority for domestic commercial reprocessing until the need for, the economics, and the safety of this technology is clearly demonstrated."

The risks associated with reprocessing and recycle of plutonium weigh strongly against their introduction. The use of plutonium in the commercial fuel cycle would expose to diversion and theft material directly usable for weapons. With widespread adoption of the plutonium fuel cycle, there would be increased pressures for independent national reprocessing facilities. The proliferation of such facilities would reduce the time necessary for a national decision to develop weapons.

Despite these widely recognized problems, it has been argued that the economics of reprocessing and recycle of plutonium in LWRs is so compelling as to make their introduction inevitable. Although plutonium and unburned enriched uranium have substantial value, the recovery of these materials from the highly radioactive wastes in spent fuel has proven to be much more difficult and expensive than anticipated. As reprocessing and recycle have moved closer to commercial practice, cost estimates have escalated rapidly. The first two U.S. commercial reprocessing ventures failed, one for economic and the other for technical reasons. The Allied Chemical plant at Barnwell, South Carolina, the only remaining U.S. commercial venture in this field, is likely to incur substantial losses and is seeking government support. European ventures are not yet operating on a commercial basis and are unwilling to contract except on a cost-plue-fee basis.

The most recent government analysis of reprocessing and recycle shows at best a 1 or 2 percent reduction in the cost of electricity in the latter part of the century. These estimates, however, are based on assumptions that appear to underestimate some elements of plutonium fuel cycle costs. Our own analysis of the costs indicates that any net economic benefit during this century is questionable.

Even if plutonium recycle proves of little economic importance, some countries may consider the plutonium inventory in spent fuel reassuring in view of the uncertainties in future uranium supply. In the case of the U.S. program, however, recovery of plutonium and unburned enriched uranium from spent fuel would only reduce uranium fuel requirements by some 20 percent. The incremental value of recycle would be largely irrelevant if access to reasonably priced supplies of fuel can be assured. Specific measures to accomplish this are discussed below.

It has been argued that early reprocessing of LWR fuels is important to build up plutonium inventories for future breeders. Our analysis indicates that the time when breeders may be economically competitive is sufficiently distant that the present value of establishing plutonium inventories now for future breeders is very small and thus recovery of plutonium is not economically justified for many years. Furthermore, spent fuel can be stored retrievably, so that the plutonium could be recovered if plutonium breeder reactors are actually deployed in the futrue.

An incentive to defer reprocessing and recycle also comes from the complexity it introduces into the waste management problem. Wastes are converted in these operations from relatively easy to manage spent fuel to a number of new forms—high level waste, acidic liquid waste, cladding hulls, process trash contaminated by plutonium, and others. As experience with reprocessed military and civilian wastes has shown, these operations introduce opportunities for waste management failures. While it has been commonly believed, particularly abroad, that reprocessing to remove plutonium decreases the long-term hazards of waste, we have concluded that any reduction in long-term risk is small in comparison with the more immediate risks potentially arising in reprocessing and in the use of plutonium in the active fuel cycle.

On the basis of our analysis of plutonium reprocessing and recycle, we have concluded that the international and social costs far outweigh economic benefits, which are very small even under optimistic assumptions. We believe therefore that a clear-cut decision should be made by the U.S. government to defer indefinitely commercial reprocessing of plutonium. Although the question of plutonium reprocessing and recycle is now before the NRC, we believe that, in view of the important international implications, the President should make the decision to defer plutonium reprocessing. If a decision to postpone this technology indefinitely is articulated and carried out effectively, it can have a major influence on the assessment of costs and benefits of reprocessing and recycle by other countries that are, or soon will be, facing similar decisions. Conversely, a U.S. decision to go ahead with reprocessing or actions that appeared to foreshadow such a decision would accelerate worldwide interest in the plutonium fuel cycle and undercut efforts to limit nuclear weapons proliferation. For this reason, we conclude that the government should not take over or subsidize the completion and operation of the Barnwell facility.

The Breeder Reactor Program

The priority and timing of the plutonium breeder is inevitably a central budget and policy issue since the commitment to this program currently dominates federal energy research and development activities. The plutonium breeder, which produces more plutonium than it consumes in operation, can in principle improve the utilization of uranium by a factor of as much as 100. When used in light-water reactors (LWRs), current estimated uranium reserves would provide only one-tenth the energy of coal reserves; in breeders, these same uranium reserves could in principle provide ten times the energy of coal reserves. The breeder thus opens up a vast additional energy resource and answers the criticism that nuclear power will price itself out of the market as soon as low-cost uranium is exhausted.

The Liquid Metal Fast Breeder Reactor (LMFBR) has become the centerpiece in the U.S. energy research and development program. The LMFBR program is focused on the early commercialization of a power plant to compete with the current generation of LWRs. ERDA has estimated that this program will cost at least $12 billion to complete, assuming utilities will be able and willing to start buying breeders within ten years without government subsidies.

The plutonium breeder involves a full commitment to the plutonium fuel cycle and would introduce tremendous quantities of plutonium into national and international commerce. In these circumstances, the pressure for indigenous plutonium reprocessing facilities would grow rapidly and be difficult to oppose. The breeder would thus greatly complicate the proliferation problem and increase the possibility of theft or diversion of material suitable for weapons. The economics of the breeder have generally been considered so persuasive that this serious disadvantage has until recently been largely dismissed in government planning.

Past government policy on the LMFBR has been predicated on a belief that nuclear power would exhaust reserves of low-priced uranium in a few decades, making breeder introduction economically attractive by the early 1990s. Our analysis, however, indicates that the early economic potential of the breeder has been significantly overstated. The LMFBR, as presently envisaged, will have higher capital costs than the LWR and must therefore operate at a significantly lower fuel cycle cost to be economically competitive. There appears to be little prospect that these fuel cycle costs can be reduced to a point that would give the LMFBR a significant economic advantage over the LWR in this century or the early decades of the next century. The current assessment of uranium reserves probably substantially understates the supplies that will become available; uranium, at prices making light-water reactors competitive with breeders, will be available for a considerably longer time than previously estimated. New enrichment technologies may also extend these supplies. Moreover, coal available at roughly current costs will look increasingly attractive if the costs of nuclear power rise. Finally, demand projections on which breeder economic assessments have been made in the past were unrealistically high and have already been substantially reduced. These considerations lead us to the conclusion that the economic incentive to introduce breeders will develop much more slowly than previously assumed in government planning.

This conclusion applies to other countries, as well, provided that they have access to low-enriched uranium to meet their nuclear fuel requirements. Moreover, the contribution of breeders to energy independence is questionable for most countries since the complexity and scale of the breeder fuel cycle would make an autonomous breeder system too costly for all but the largest industrial economies. Therefore, the prospect of a large export market for breeders in this century is illusory.

Despite this negative assessment, we believe that a breeder program with restructured goals should be pursued as insurance against very high energy costs in the future. This situation could develop if additional uranium reserves do not become available, environmental problems place limits on the utilization of coal, and other alternative energy sources do not become commercially viable at reasonable prices in the first decades of the next century. The present U.S. program, directed at the early commercialization of the LMFBR, is not necessary to the development of the breeder as insurance. The ultimate success of the breeder may even be compromised by telescoping development stages to meet an early deadline, freezing technology prematurely. We believe therefore that the breeder program should deemphasize early commercialization and emphasize a more flexible approach to basic technology. In such a program, with a longer time horizon, the Clinch River project, a prototype demonstration reactor costing $2 billion, is unnecessary and could be canceled without harming the long-term prospects of breeders. In fact, premature demonstration of a clearly noncompetitive breeder could be detrimental to its ultimate prospects.

Although long lead times are required for a project as complex as the breeder, we believe that the decision on commercialization, now set for 1986, can safely be postponed beyond the end of the century. The cost, if any, of such postponement will be small, and there is a strong possibility that postponement will help in restraining large-scale, worldwide commerce in plutonium and buy time to develop institutions to deal with this problem. The option of bypassing the plutonium breeder altogether should not be prematurely foreclosed since there is at least a possibility that the plutonium breeder may never become necessary, or even economically competitive, compared to other energy sources that may become available in the next century.

Nuclear Waste Management

The United States must greatly improve the management of its rapidly growing accumulation of nuclear wastes and decide soon on the strategy for its disposal. This long-deferred action is closely related to decisions on plutonium reprocessing and the timing of the breeder program. The need for action on this problem was recognized by President Ford in his directive on October 28, 1976, calling on ERDA to undertake an accelerated program to demonstrate all components of waste management technology by 1978 and a complete disposal repository by 1985. However, the question as to what strategy should actually be followed in managing and disposing of wastes has yet to be resolved.

As indicated in the earlier discussion, we are persuaded that nuclear wastes can be disposed of permanently with acceptable safety by deep burial in salt and other stable geological formations that are isolated from ground water intrusion. This conclusion holds equally whether the nuclear wastes are contained in spent fuel or in processed form. Until they are securely isolated from the environment, however, nuclear wastes are potentially extremely dangerous.

In the past, the waste disposal problem has generally been approached on the assumption that spent fuel would be reprocessed to recover plutonium for recycle. Decisions on disposal were continually deferred pending successful introduction of plutonium recycle. As a consequence, it is widely believed that reprocessing is a necessary stage in the waste disposal process. However, if plutonium is not recycled in light-water reactors or used eventually in breeders, there is no reason to reprocess spent reactor fuel. In fact, reprocessing potentially increases short-term risks associated with management of nuclear wastes and does not significantly reduce long-term risks after disposal. Spent fuel can be disposed of directly, and probably at costs comparable to those for reprocessed wastes. Therefore, if plutonium reprocessing for recycle in LWRs is deferred indefinitely as we recommend, waste disposal is made no more difficult.

The breeder, which we believe should continue as insurance against the uncertainty of future energy needs, presents a more difficult problem since large inventories of plutonium (or highly enriched uranium) would be required to start a commercial breeder program. The time scale of commercially competitive breeders is sufficiently distant, however, that separation of plutonium for this purpose will not be justified economically for some time. Nevertheless, as part of the breeder insurance program, some portion of the spent fuel should be retrievable and not disposed of permanently. The retrievable fraction should depend on the evolving time schedule of breeder development.

In the immediate future, spent fuel can be kept in the cooling ponds at nuclear power plants where it is presently stored. These facilities can easily be expanded when necessary. While this arrangement is acceptable as a temporary measure, it is not satisfactory for extended storage of large quantities of material that can be anticipated with the growth of the nuclear power industry. Therefore, we believe that the waste management and disposal program should develop both permanent and retrievable and irretrievable storage for spent fuel in stable geological formations. While security of storage will have to be balanced against ease of retrieval, the emphasis should be placed on security since retrieval may be long delayed or perhaps unnecessary.

We believe that liquid wastes accumulated from the military program and the abandoned West Valley commercial plant should be disposed of permanently to eliminate a potential safety hazard and to demonstrate the seriousness of the government's concern about the waste management problem. Experience gained in this activity could be applied to handling and disposal of wastes from future reprocessing plants if a decision is made eventually to use breeder reactors.

Our confidence in the ability to manage spent fuel is sufficiently high that we believe the United States should be willing to take back spent fuel from countries lacking waste facilities for retrievable storage or disposal if this will reduce risks to international health or of proliferation of nuclear weapons.

Expansion of Uranium Enrichment Capacity

The United States must have a clear policy on its long-term role in providing enriched uranium fuel to both domestic and foreign nuclear power programs. If future requirements are to be met, present facilities will eventually have to be expanded. The timing and magnitude of this expansion depend not only on the anticipated growth of domestic demand for enriched uranium fuel but also on the extent to which this country wishes to be able to assure fuel for others. The issue has become entangled in the question whether expansion would best be carried out by the private sector or the government. At present all enrichment facilities are owned by the government but operated under contract by private firms.

An assured supply of uranium fuel is a major factor in limiting worldwide proliferation capabilities. The assured availability of fuel at reasonable prices limits the pressure on other countries to seek indigenous enrichment facilities that would provide a capability leading to weapons. An assured fuel supply also reduces the incentive to recycle plutonium or to develop breeders in an attempt to stretch available fuel supplies.

In 1974, when projected demand appeared to call for more enriched uranium than the United States could supply, this country stopped entering into long-term commitments to supply fuel for new reactors abroad. This unwillingness to guarantee supply (even when the purchaser was willing to supply natural uranium for "toll" enrichment) is reportedly the main reason for Brazil's 1975 agreement with West Germany that provides for Brazil's eventual purchase of both enrichment and plutonium reprocessing facilities.

Present U.S. plans for new enrichment capacity are still based on earlier demand projections that are now being revised sharply downward. Cutbacks in nuclear construction plans here and abroad have delayed the time when additional fuel facilities will be required. Requirements recently estimated for the mid-1980s now seem unlikely to be reached before 1990. The ongoing program to upgrade existing U.S. enrichment plants will increase capacity by more than 50 percent in the mid-1980s. By the mid-1980s, Eurodif, in France, and URENCO, a West German/British/Dutch consortium, plan to have a combined capacity approaching, and possibly greater than, present U.S. capacity. The new private and government facilities proposed in legislation submitted to Congress in 1976 would be equal to or greater than the total present U.S. capacity. This would bring the total free world capacity in the mid-1980s to three or four times present levels, well beyond currently projected needs. The Soviet Union also has excess capacity and is selling toll enrichment services on the world market. In short, no shortage of enriched uranium need occur in the 1980s, and there is ample time to meet the needs beyond 1990.

The rapid pace of technological development further complicates decisions on expansion of uranium enrichment since new separation techniques may significantly reduce the cost of enrichment. While two of the proposed U.S. plants would employ the same gaseous diffusion technology as existing plants, three others would employ the centrifuge technology which may prove less expensive. More significant is the prospect that laser isotope separation will reduce drastically the cost of enrichment. If, as we anticipate, this new technique proves commercially feasible, both the construction and operating costs of an enrichment plant will probably be much less than those of either the new gaseous diffusion or centrifuge plants. Laser separation may also stretch uranium supplies and reduce costs by making the extraction of a larger portion of available uranium-235 economically feasible. In these circumstances, it would appear to be prudent to let the technology of centrifuges and lasers evolve further before making major new commitments unless they are urgently required.

While centrifuge and laser technology may support nonproliferation objectives by making enriched fuel available at lower costs, they also create proliferation problems since they make possible much smaller plants that can be converted to the production of weapons-grade enriched uranium. This could prove to be a particularly serious problem with laser enrichment plants, although it may be somewhat compensated by the very high technology and special lasers associated with some of the approaches to laser isotope separation. The inherent size and capital cost of gaseous diffusion plants have provided something of a natural barrier to the spread of indigenous enrichment facilities.

We believe the United States should maintain adequate uranium enrichment capacity to meet worldwide nuclear power requirements. However, in view of rapidly changing demand projections and the possibility of radical technological developments, decisions in this area should not be taken hastily. Nor should the United States attempt to monopolize the world market, since availability of alternative suppliers of safeguarded fuel will enhance confidence in fuel supply. Decisions on expansion should therefore take account of worldwide enrichment capabilities, which should be looked on as an added resource.

The present argument for private ownership of new capacity is not persuasive. Proposals involve extensive government guarantees and few of the advantages of a competitive market. Although we favor the principle that nuclear energy should be judged by unsubsidized competition with other energy sources, we believe that the government should at this time retain control of both enrichment facilities and new enrichment technologies. If this is done, the U.S. government will be in a stronger position to use this resource in support of its nonproliferation policy and in dealing with the security problems that may be created by new enrichment technologies.

Export Policy

Most countries have access to nuclear facilities and fuel only through imports from a small number of supplier states. The terms of such trade can contribute significantly to nonproliferation objectives. The United States cannot, however, unilaterally determine international nuclear trade policy since a growing number of countries are competing for the nuclear export market. Attempts to restrict trade in nuclear fuel cycle facilities must take into account the U.S. commitment in the NPT to facilitate the peaceful use of nuclear energy in exchange for the commitment of other states to forego nuclear weapons. Moreover, attention must be given to the impact of export policies on the political concerns of the few key countries that must reach their own decisions not to develop nuclear weapons.

Despite these limitations and complications, U.S. export policy can significantly support its nonproliferation objectives. As the long-time leader in the field, the United States has considerable influence with other suppliers, most of whom are allies. Although the ultimate response to proliferation is political, important actions can be taken in the export field in conjunction with a broader diplomatic effort to develop a consensus on the merits of a nuclear fuel cycle without plutonium.

From the nonproliferation standpoint, the focus of concern must be on exports of facilities that can separate plutonium or produce highly enriched uranium. Such facilities provide a capability for all but the final steps in the production of weapons. This capability might significantly influence a nation's political decision to develop nuclear weapons in a time of crisis. Such facilities also greatly increase the threat that material suitable for weapons might be stolen by terrorists.

Fortunately, at the end of 1976, other major nuclear suppliers seemed to be moving in the direction of U.S. policy, which has always refused to license exports of these sensitive facilities. The United States should build on these emerging attitudes to develop a consensus among supplier and consumer nations alike against the spread of national plutonium separation and uranium enrichment facilities. This agreement should be built upon a common recognition that it is more in the interests of both suppliers and consumers to reduce the possibility of nuclear proliferation than to pursue marginal economic gains or status from the sale or acquisition of these sensitive facilities. The success of such an effort will depend largely on the extent to which it is widely recognized as a major U.S. priority with strong Presidential support.

A U.S. proposal for international reexamination of the economics of plutonium recycle and breeders will hardly be credible unless the United States is itself prepared to defer its own plutonium recycle and breeder commercialization programs on valid economic and energy supply grounds. Such action will not necessarily convince all countries but will certainly influence their thinking and will preempt charges of discrimination or of failure to honor NPT commitments.

If plutonium reprocessing and recycle and attempts to commercialize breeders are postponed, there will be increased concern about the future availability of enriched fuel for light-water reactors. U.S. export licensing procedures should permit the United States to make credible guarantees of fuel supply to countries that are in compliance with the NPT or other agreements in the operation of their nuclear programs.

Although safeguards cannot prevent a nation from developing a nuclear weapons capability, they can help to deter it and provide assurances to others that it has not done so. The value of safeguards is obviously greatly weakened if they do not apply to all of the nuclear facilities in a country and, in particular, to any indigenous facilities for reprocessing plutonium or enriching uranium. While the NPT obligates nonnuclear weapon states that are members to place all their nuclear facilities under safeguards, the United States and other suppliers have interpreted their treaty obligations on exports to nonNPT members as requiring that the nonmembers place under safeguards only the exported facilities or materials and not *all* their nuclear facilities. To strengthen safeguards, the United States should seek in future agreements with nonNPT members to have all nuclear facilities placed under IAEA safeguards. An effort should also be made to renegotiate existing agreements to include this provision and to persuade other suppliers to adhere to a similar policy. To emphasize that inspection is not intended to serve a discriminatory purpose, the United States should move promptly to implement its offer made in late 1967 in connection with the negotiation of the NPT to put all U.S. peaceful nuclear facilities under IAEA safeguards. The United States should give generous financial and technical assistance to the IAEA so that it will be in a position to handle its rapidly growing responsibility for safeguards.

Above all, in approaching the energy problem, the United States should discard the promotional approach to nuclear energy that has characterized so much of its program since the initiation of the original "Atoms for Peace" program. In all countries, but particularly in developing countries, the United States should encourage realistic assessments of energy needs and options and not press the nuclear option.

In this overview, we have endeavored to put nuclear power in realistic perspective relative to broader economic, social, and security objectives and to develop a framework for considering the current policy decisions on nuclear power. In the body of the report, we examine the elements of the nuclear power problem in greater detail. The report is divided into four parts: Part I deals with the economics of energy, energy supplies, comparative costs of nuclear power, and potential alternative energy sources; Part II deals with the impact of nuclear power on human health and the environment in normal operations and the potential effects of reactor accidents and nuclear waste; Part III examines the relationship of nuclear power to the proliferation of nuclear weapons and nuclear terrorism; and Part IV presents a more detailed analysis of the specific policy decisions currently facing the U.S. government.

Part I

Energy Economics and Supply

Chapter 1

Energy and the Economic Future

The principal justification for nuclear power is that it can make an important contribution to the U.S. and world economy. If only economics were involved, one might be prepared to leave other issues to be resolved by the play of economic forces. But that is not the case. Nuclear power carries with it noneconomic risks such as the potential impact on the environment or national security which we will examine in detail in subsequent chapters. Society's willingness to accept these risks will depend strongly on the economic costs of avoiding them by limiting reliance on nuclear power or foregoing particular applications. Thus decisions affecting the availability or timing of plutonium recycle, the breeder reactor, or even nuclear power itself cannot be made wisely without a careful analysis of their economic consequences.

Elsewhere in this report the specific economics of energy resources and of various energy production technologies are described and analyzed. In this chapter, we examine the economics of energy in a more general way, looking beyond supply considerations to broader questions. How important is energy to economic welfare? Can the economy adjust to higher energy costs without reduction of economic growth or unemployment in the long run? How might rising real incomes, higher energy costs, and changing values and lifestyles interact to influence energy demand over the next half century? And, finally, if noneconomic considerations suggest foregoing or delaying currently projected nuclear development, what would be the economic consequences?

The chapter begins with a general discussion of the economics of energy, explaining why we regard energy as an economic variable, rather than as something requiring special analysis. Energy markets are imperfect for many reasons, and these imperfections must be recognized in analysis of energy and in the formulation of policy. With due account for the principal market failings, one can use market concepts to understand the role of energy in the economy and to help implement energy policy. Short-run disturbances in energy markets, of the sort experienced in the past few years, can cause temporary economic dislocations; and for many reasons, macroeconomic difficulties will continue to be a fact of life. But primary energy is a sufficiently small cost item in a modern economy that, so long as adequate supplies are available, severalfold increases in unit energy costs, occurring over a number of decades, could be accommodated without significantly aggravating problems of unemployment, income distribution, or growth.

Real energy costs are likely to increase over the next half century at rates, which are only slightly dependent on current decisions relating to nuclear power, and will probably level off near the middle of the next century. Even if this cost increase is greater than expected, the effect could be limited to a slightly lower rate of economic growth during the period of rising energy costs, with income somewhat lower but continuing to grow thereafter. Neither the basic social and economic structure, nor the economic wherewithal to deal with social problems, need be seriously threatened by higher energy costs in the long run.

With rising incomes, energy use would continue to increase rapidly, except for the inhibiting effects of increasing scarcity and of environmental and political constraints, all of which cause higher energy costs. These higher costs will result in personal, institutional, regulatory, economic, and technological changes, which will reduce demand for conventional sources below what it would be if costs remained low, both because market demand will be reduced by higher prices and because policy will become more conservation-minded. There is substantial scope for reduction in energy use—whether strictly price-induced or resulting from government energy conservation policy—in response to higher energy costs. The basic conclusions of this analysis, however, do not depend on "optimistic" assumptions about conservation potential.

In our analysis of these economic issues, we made use of a particular computer model, the Energy Technology Assessment model, using a range of illustrative assumptions. The analysis confirms the general conclusion that, whatever is done about nuclear power today, energy costs are not likely to become high enough to require major social or economic changes in the long run. U.S. national income in the year 2025 might be 10-20 percent less than it would be if energy costs remained at 1975 levels, equivalent to a reduction in the rate of economic growth of a few tenths of a percentage point. Furthermore, because nuclear power has at best a small cost advantage over other energy sources, its availability does not add much to national income before the year 2000. Therefore, the costs of delaying any additional nuclear power until then would be very small in this century, rising to at most 1 or 2 percent of GNP early in the next century. Because future GNP will be large, these costs could eventually reach scores of billions of dollars per year, so that as a pure economic matter such a delay in nuclear power is unwarranted. But if there were overriding noneconomic reasons for such a delay, the economy could bear the costs. Furthermore, since plutonium recycle and early introduction of breeder reactors have a questionable cost advantage over the present generation of power reactors without recycle, the cost of deferring the use of plutonium as an energy source is very small.

The analysis which leads us to these conclusions assumes that U.S. coal consumption increases by a factor of three or four over the next thirty to fifty years, an increase we regard as quite feasible. In fact, much larger increases in coal production appear to be technically possible over the next fifty years in the United States and elsewhere. Even countries without large domestic coal reserves can use coal economically on a large scale if they are willing to be dependent on foreign supplies and to bear the balance of payments costs involved. We have not done detailed economic analysis of coal-fired power as a large-scale alternative to nuclear power, except to verify that the economic value of nuclear power depends critically on the extent to which coal production and use can be expanded. The value of nuclear power or the economic cost of delaying nuclear power would be virtually eliminated if coal could be relied upon as the principal energy source of the early twenty-first century.

In less developed countries, expanded energy use may be more closely linked with economic growth than in richer, industrialized countries. Here, economic growth is more dependent on expansion of the industrial and transportation base of the economy, requiring large amounts of capital and energy which often must be imported. In this situation, higher energy costs will have large and immediate effects on current incomes and on balance of payments, which may significantly affect growth prospects. Whether nuclear power, which is capital-intensive, imported, expensive, and economically attractive only for large increments of electricity, is an economic response to high world oil prices depends on the individual circumstances of specific countries. In general, however, the need for increased energy for transportation, industry, and agriculture implies the use of nonelectrical energy that cannot be supplied by nuclear power.

THE GENERAL ECONOMICS OF ENERGY

In any policy analysis, the broad outlines of the results are determined by the assumptions. If a complex model produces results not understandable in terms of the basic assumptions, the model should not be trusted. Especially when dealing with a subject as broad as energy, where the uncertainties are so large, it is important to be explicit about assumptions and to understand which assumptions determine which results. Therefore, we go to some lengths in this section to explain how we have approached the economic analysis of energy and nuclear power, and to point out how our general conclusions follow more from quite general considerations than from detailed relationships in a computer model.

"Energy" as an Economic Concept

When analyzing energy, one must first decide whether ordinary rules of economics can be applied. Since every process involves the degradation of usable energy, some analysts conclude that energy is too valuable to use even though it is cheap. Other analysts hold that energy is the ultimate substitute for all other resources, that with enough energy we need never run out of anything, and that there is more than "enough" of everything if only we do not worry about what it costs to produce energy. Should anything so fundamental and irreplaceable be regarded as an ordinary economic good, to be produced and used on the basis of prices, interest rates, profits, and the like?

In our view, it is neither necessary nor wise to regard energy as something special, requiring novel intellectual approaches and analytical devices. In human affairs, we are not concerned with energy as a physicist's concept. We are not even concerned with an aggregate natural resource called "energy." What we are concerned with are many natural and man-made resources which can be combined in various ways to produce some intermediate goods in which usable energy is stored, and which in turn can be used in combination with other scarce factors to help satisfy human needs and desires. Some of the natural resources consist of material that flows out of the ground, burns easily, and can be put to use with a modest addition of transportation, refining, and utilization expense. Understandably, society is using these resources quickly, even though they are limited in amount. Other resources, such as coal and uranium, are less limited in quantity but are less valuable because they can only be used in combination with large amounts of land, labor, and capital in various forms. As society is forced to turn to these sources for its energy, it will discover that it is more economical to devote some of that land, labor, and capital to activities other than delivering energy. And finally, there are some natural resources, such as sunlight and the heat of the earth, which will probably some day play a major role in energy production but which will probably never (except in special, localized situations) have any scarcity value; the cost of the complementary man-made resources, not natural scarcity, will be the factor limiting their use.

The concepts and methods of economics are sufficiently flexible to accommodate most of the unusual features of energy. Environmental and foreign policy constraints, the fact that it takes energy to produce energy, the market distortions caused by government policy or monopoly power, the "irreplaceable" nature of certain high-quality natural resources, and uncertainty about technology and geology are all real complications and may require ingenuity in the application of general concepts in specific situations. But the basic ideas and models of economics provide a good starting point for thinking about energy problems, with all their complications.

Energy and the Market

One important economic principle is that prices established in more-or-less competitive markets provide reasonable guides to the relative values of goods, services, and natural resources. The owners of scarce resources are continually looking for higher prices for their assets; entrepreneurs are continually shopping for cheaper combinations of inputs and for more valuable outputs; and individual and institutional consumers are continually looking for less costly ways of satisfying their needs and desires. Subject to some important qualifications, the interaction of these competitive and self-serving actions produces two important results: market demands are met about as well as they can be, given the available resources; and market prices provide estimates of the terms on which the economic system can efficiently substitute one thing for another.

This competitive market model is a powerful analytical and predictive tool, which can be used for many purposes. If one is interested in recommending specific policies to improve the functioning of the economy, the competitive model can be used to identify divergences between the real world and the competitive ideal, and to suggest policies for closing the gap. If this were our purpose here, we would point out that regulated public utilities do not set prices on the basis of incremental costs, that natural gas is priced below its scarcity value, that institutional rigidities and lack of information may be causing uneconomical delays in implementing energy conservation measures, that basic research and development is handled poorly by the market, that certain sectors of the energy economy are insufficiently competitive, and that political decisions can introduce biases into the costs of alternative forms of energy. Our emphasis would be on the failure of the real market to match the ideal in some important respects, and what to do about it.

Our purpose here is different, however. Instead of recommending policies to improve the functioning of the economy, we want to predict in general terms how the economy—with all its imperfections—is likely to behave in the long run when confronted with a set of objective conditions, such as the relative costs of various technologies and the scarcity of certain resources. We also want to consider how society should behave in the long run, when faced with these conditions. For these purposes, using the market model is equivalent to assuming that society should try to maximize its economic output over the long run and will approximately do so through some combination of free markets, regulation, government-sponsored R&D, public investment, etc.

It is in this sense that the competitive market model is used here. It may take many years for public utility regulation to recognize high incremental energy costs, and the economic losses may be large in the meantime; but the competitive model is a reasonable guide to what regulation should and probably will approximate in the long run. Natural gas prices may be held below market levels indefinitely, but if so, nonmarket allocation devices will have to be developed to deal with the fact of scarcity, and the competitive model provides an estimate of what these nonmarket processes should and will approximately accomplish. Large firms may be somewhat insulated from market forces, and antitrust action may be necessary to remedy these situations; however, the assumption of cost-minimizing responses to price changes has been shown to be as good a way as any to predict the behavior of even large firms. Regulated utilities may be able to "pass on" costs to consumers, and may have some incentive to use uneconomically capital-intensive methods of production; but the evidence is that cost-saving innovations are adopted as readily[a] by regulated utilities as by other industries, and that a cost-minimizing model predicts as well as any other. For the purposes of estimating how resources should be allocated and how the economic and political processes of society will allocate them in the long run, the competitive market model is of great value—even in the face of market defects.

For energy policy, there are three areas in which the competitive model is particularly weak, and where we assume that govenment will play an active role. One is research and development, especially where basic science is concerned. Competitive markets do not efficiently provide for basic research, and strong government support is clearly called for. Similarly, the environmental problems associated with energy production and use require government action to see that external costs are, if only approximately, incorporated in energy prices. The costs in our analysis are intended to include at least the most obvious of these environmental costs. Finally, regulatory and political processes do not allow prices to be raised fast enough to keep up with rising incremental costs. In our analysis, this problem has been taken into account by using estimates of real scarcity or of incremental social costs in comparing alternatives, in effect assuming that, in the long run, regulatory processes will produce something like the efficient allocation of resources even where market forces are subject to controls.

One area in which the price mechanism probably does a better job than it is generally given credit for is in allocating resources over time, even where "irreplaceable" natural resources are involved. Here, market forces can play a constructive role, by raising prices in anticipation of scarcities. Such anticipatory price increases are not necessarily a sign of monopoly or antisocial behavior. When there is not enough of a resource to satisfy demand at a price equal to its production costs, then its true economic value and its efficient price are indeed higher than its production cost. Since some of those who would like to use it will be forced to do without it, those who do use it should pay enough to bid it away from alternative uses, now and in the future. By the same logic, setting aside some of an "irreplaceable" resource today for use in the future is economically justifiable only if its net price (market price minus costs, ad-

---

[a]And cost-raising changes—even capital-intensive ones such as flue gas desulfurization— are resisted as strenuously. Regulated Commonwealth Edison and publicly owned TVA are no more willing than unregulated, private enterprises to raise their costs by installing desulfurization equipment.

justed for changes in the general price level) in the future is expected to be higher than the current net price plus accumulated interest. If the future net price is not expected to be this high, the current generation can do more for the future by using the resource now to produce capital which will add more to output over the years than the conserved resource would have. Oil or coal left in the ground is like any investment, and decisions about whether to leave it there or to draw it out and apply the proceeds to other uses should be made on the same economic criteria that are applied to other investments.

In summary, markets work well enough to allow us to predict price trends on the basis of real costs and scarcities, and then to use long-run, cost-minimizing models to predict responses to these prices. Where specific market failures with significant distortions can be identified, we impose constraints or corrections in the analysis to take these into account. This procedure does not mean that markets are perfect, or that we invariably favor "free market" approaches to energy policy. We have made few explicit assumptions about just what combinations of free markets, regulation, government investment, etc., will be used to bring about the "market" solution. Nevertheless, the competitive market model, with some corrections, is a reasonable basis for estimating how society will allocate its resources.

Energy and Employment

In the short run, any unanticipated shock to the economy can have adverse effects because of market inertia and because the ability to predict and manage the economy and to shift resources in response to problems or opportunities is less than perfect. The oil embargo and the fourfold increase in world oil prices in the winter of 1973–74 was an unusually severe shock, and contributed to the world's subsequent economic difficulties in three ways: the higher cost of oil contributed to the already high inflation rate; suddenly larger payments for imported oil reduced aggregate demand for goods and services; and certain sectors of the economy suffered directly from energy shortages and from the sudden shift in consumer demand away from energy-using activities. This combination of effects was particularly difficult to deal with, since the expansionist monetary and fiscal policy which would ordinarily be used to offset the higher foreign payments would not provide any special stimulus for those sectors affected directly by energy price and supply problems. Since controlling inflation was given priority by the U.S. government, monetary and fiscal policy was kept restrictive. In addition, energy price controls and allocation mechanisms were used to try to reduce inflationary effects, contributing to shortages in specific situations. The result was unemployment and loss in output—in short, a recession. The worst effects were temporary, however. As industry and consumers began to adjust their production and consumption to the higher energy prices, as the payments for imported oil began to come back to the United States in capital transfers and demand for exports, and as energy controls were loosened, the specific effects of the oil embargo on the recession began to recede in importance.

This kind of temporary disruption can be caused by any sudden economic change, good or bad. If the United States were suddenly to discover huge, readily exploitable, cheap reserves of natural gas, the long-run impact on income would clearly be positive; but the short-term employment effect might well be negative, as investment plans were redone, coal mines were shut down, and consumers and industry shifted away from expensive energy-conserving goods and services. The frequency and severity of such short-term disruptions have no simple relationship to the price or amount of energy, so one cannot argue for expansion of supply or curtailment of demand in general as a means of reducing short-term vulnerability. Rather, one must examine supply and conservation alternatives to develop a balanced and flexible energy position, recognizing that there are costs as well as benefits to insuring against contingencies; any system that can generate and take advantage of opportunities is in danger of being "disrupted" by the emergence or disappearance of a particular advantageous opportunity.

We have not analyzed all the supply, conservation, and contingency planning possibilities and are not able to outline an energy policy that balances short-term stability against longer-term well-being. However, establishing a "permanent embargo" by eliminating imports is not the best way to respond to the threat of a temporary embargo; a diversified energy supply mix, stockpiles, and contingency planning are preferable. Nuclear power has a role to play in such a policy, but not a central role in the near future; it cannot be expanded quickly, produces only electricity, and is subject to its own uncertainties.

For the long run, we can say with confidence that there is no direct relationship between energy cost and the number of jobs. It may be that unemployment will remain a serious problem in the future, as the composition of the labor force, personal attitudes toward work, and the availability of socially provided goods and services change. But gradual increases in real energy costs need not make the employment problem more difficult. If energy should become free and unlimited in quantity, there could be plenty of jobs in finding ways to apply that energy to the satisfaction of human desires, while continuing to deal with the fact that copper and land and water and capital and smart people and time and some other important things are available only in limited quantities. Conversely, if it becomes increasingly difficult to get safe, usable energy there will be work, producing what energy we do produce, substituting human energy and other factors for energy, and continuing to deal with the other scarcities of life. Given time, jobs can be redefined, equipment can be redesigned, and habits can adjust to provide employment whether energy is cheap or expensive. The fear that it might be otherwise is the ironic counterpart of the long-standing popular view that increases in productivity will lead to unemployment because fewer workers are needed to produce a given volume of goods. The same processes which have adjusted the economy to technological change should respond to energy scarcities without significant increases in unemployment.

Energy, Incomes, and Growth

Even though the economy can, in the long run, adjust to higher energy costs without unemployment, it is conceivable that the loss in productivity could be so great that current incomes and economic growth would be severely reduced. Thus, it is not enough to conclude that everyone can be employed, without considering the effects of higher energy costs on incomes and growth. Will scarce and expensive energy mean that the full-employment economy of the twenty-first century provides real incomes or required styles of living which are significantly less desirable than if energy were cheap and plentiful?

The answer to this question rests on the fact that energy is simply not a very large cost item in our economy. In 1975, the United States used about 70 quads (a quad is 10^{15} British thermal units) of primary energy and had a GNP of around $1,500 billion. Although this energy came from a variety of low- and high-cost sources, we have assumed in our analysis that the average cost[b] of energy was as high as $1 per million British thermal units (MBTU). This would indicate that the total economic cost of energy in 1975 was as much as $70 billion or almost 5 percent of the GNP.

Suppose now that energy in 1975 had cost twice as much, $2 per MBTU, and the economy had used the same 70×10^9 MBTU even at the higher cost. The level of "productive effort" in the economy would have been about the same, the difference being that more resources would be devoted to producing energy domestically or paying for imported energy. The resources available for private consumption, investment, and government services would have been lower by the amount of the increased energy cost, $70 billion or almost 5 percent of the original GNP. While a 5 percent difference in national income is not something to be dismissed lightly, it is also not something that is likely to require profound social or economic changes—once the economy has adjusted to the changed pattern of demand.

Higher energy costs, by increasing the amount of productive effort which is required to produce a unit of income, will also increase the amount of investment per unit of income which is required to maintain the rate of economic growth, especially if energy-related activities are more capital-intensive than the average for the economy. In the short run this effect could be quite large, if the economy tries to build up its energy-producing capital stock quickly; rising

[b]For estimating impacts on national income, it is average cost and not the marginal cost or market price which is important. Thus, while world oil prices were about $2 per MBTU in 1975, most U.S. energy came from sources with costs below $1 per MBTU. On this basis, the actual weighted cost of energy in the United States in 1975 may have been as low as $0.70 per MBTU, but we have taken a convenient conservative figure of $1 per MBTU for illustrative purposes in our analysis.

interest rates, high rates of corporate retained earnings, foreign capital inflows, and some diversion of investment funds from other sectors of the economy could be expected as part of this "capital shortage." In the long run, however, higher capital costs become simply part of the higher energy costs the economy is paying, and a permanently higher savings rate is one of the ways in which these higher costs would show up.

Of course, if the savings rate does not increase, then economic growth could be affected. In the United States, energy-related investment is now less than 2 percent of GNP, and the savings rate is about 10 percent of GNP. Thus, a doubling of energy costs unaccompanied by any reduction in energy use would require about 4 percent of the GNP to go to energy-related investment activities. If savings did not increase at all, this would reduce by 25 percent—from 8 percent to 6 percent—the fraction of GNP available for investments in growth-generating activities. About half of economic growth is due to capital formation (the other half being attributed to education, technical progress, general improvement in information and procedures, etc.), so this decline might lower a 3.5 percent GNP growth rate to 3.2 percent, which would have a large effect on GNP over a period of years. Since most other developed countries already have savings rates significantly higher than the United States, it is very unlikely that savings would not increase to provide most of the larger investment required for energy needs. Therefore, suddenly higher energy costs may cause a temporary slowdown in economic growth, as the economy moves to a slightly lower growth path and adjusts its savings and investment patterns; but then growth should be able to continue as before, from a somewhat lower income base.

If, at energy costs of $1 per MBTU, income and energy use would grow together at 3.5 percent per year, then at $2 per MBTU they could still do so, but with income lower by the increased cost of energy. Using the 1975 U.S. figures to illustrate, GNP in the year 2000 would be reduced by 5 percent from $3,550 billion to $3,370 billion; this is equivalent to a decrease in the average compound growth rate for the twenty-five-year span from 3.5 percent to 3.3 percent or a postponement of the $3,550 billion economy from the year 2000 to the year 2002. If energy costs climbed to $5 per MBTU by the year 2000, and the country did nothing more economical than just to pay these costs (including the higher investment) for the same amount of energy, U.S. income in 2000 would be reduced by 20 percent to $2,840 billion, equivalent to a reduction in average growth rate over the period from 3.5 percent to 2.6 percent or a postponement of the $3,545 billion economy from 2000 to 2007. Of course, society may evolve in rather different patterns if energy is expensive, adjusting social institutions and personal lifestyles to conserve on energy instead of simply expanding its productive machine and energy use as though energy were cheap; but that is a choice which is open to the society and its members, not something which high energy costs will force on them.

In assuming that energy problems take the form of higher costs which can simply be paid, we are rejecting the assertion that there are absolute constraints on energy production or use which cannot be overcome at some cost. The constraints sometimes referred to as "absolute" are more correctly thought of as problems which can be overcome only at costs higher than anyone is currently willing to contemplate, but still within a manageable economic range. At energy costs of, say, $5 per MBTU, equivalent to $125 per ton of eastern coal or $30 per barrel of oil, almost any conceivable energy demand can be met, at least for the next fifty years, from any of several sources: surface mining can be expanded and the land reclaimed, even if water must be imported to arid regions; nuclear plants and their associated fuel cycle facilities can be put underground at remote sites; oil can be produced domestically from offshore sources or shale or coal; stockpiles and diversified supplies can be developed to reduce the risks involved in imports. Also, resources diverted from income to digging mines or paying for imported oil could just as well be used to insulate homes or modernize boilers, with the same effect on usable energy, so that conservation will be regarded as a "supply" option in this analysis. It will probably not be necessary to turn to primary energy sources as expensive as $5 per MBTU;[c] but at costs of this magnitude, which are still manageable economically, we see no absolute constraints which cannot be overcome by some mix of supply and conservation measures. Thus, energy problems, for the next fifty years at least, can be analyzed as cost problems rather than as problems of absolute constraints on energy production and use. And we see little likelihood that costs will get so high that they become important influences on the shape of the political, social, or economic future.

Higher energy costs and hence a slightly lower increase in income levels than would otherwise be the case appear to be inevitable. It is useful to know that the broad outlines of the future need not be markedly different as a result. Furthermore, since nuclear fission is only one among several energy sources and, for the next few decades at least, has no more than a small advantage over other high-cost energy sources, near-term actions on nuclear power can hardly be critical to society's economic future. It is as incorrect to argue that nuclear power is "needed" now to allow society to continue its development as it is misguided to hope (or fear) that society could be forced into some no-growth utopia if only the current nuclear program were abandoned. Energy costs are just not that important, and nuclear power will not do much to reduce them in any case.

In the long run of fifty years or more, the economic role of nuclear power could be more crucial. Nuclear power with the breeder, and coal, are the only almost-sure energy sources which could provide large amounts of energy at about current costs. If fusion turns out to be infeasible, solar energy proves too

---

[c]In the model discussed below, a weighted average of energy prices never increases more than a factor of three above 1975 levels of $1 per MBTU in 1975 dollars.

costly to exploit for electric power, and large-scale use of coal is unacceptable for environmental reasons, the economic and social importance of nuclear power and the breeder could become substantial. We believe that some mix of coal, solar, and fusion energy, assisted by conservation, would be capable of supplying society's long-term energy needs. But we also believe that the nuclear breeder will be competitive and is likely to prove eventually to be socially acceptable. Society does not have to act on these guesses today. Our children will be in a better position to worry about our grandchildren than we are. Some combination of conventional and new energy sources, conservation, and social evolution will take care of future energy problems—if today's decisions are wise enough to provide the options and the opportunities the future will require.

PREDICTING ENERGY DEMAND

This study has adopted the common practice of treating demand as an aggregate determined by a few simple relations while analyzing supply in more detail. But demand should not be treated as something which is certain or which "must" be met, especially when uncertain and questionable supply actions would be necessary to meet the unquestioned demand. Both supply and demand are uncertain variables influenced by market forces, social developments, and public policies. Therefore, we discuss in this section the uncertainties in demand and the reason for our choice of demand parameters; and in the next section we include a range of demand assumptions, to emphasize that energy policy is a matter of supply and demand.

Income, prices, demand, and supply are mutually interacting in an economy, making it necessary to consider all of them simultaneously. But when energy costs are only 5 to 10 percent of income, they are not a strong influence on income levels or growth rates, and hence income can be predicted independently of supply conditions. If it is known how energy demand would vary with income *at unchanged energy costs,* and then how higher energy costs would affect demand, an energy demand function can be determined. This demand function then can be combined with estimates of costs and constraints on the supply side to determine energy use and prices simultaneously. If this procedure yields energy costs higher than assumed in the original income projections, these can be subtracted to obtain a corrected net income; as long as this correction is not large enough to invalidate the original assumptions about income growth or demand, the result will be a consistent projection of the energy and economic futures.

This procedure is straightforward analytically and, in one form or another, is used in all economic modeling of energy and the economy. The difficulties on the demand side arise in predicting income, determining how energy use would vary with income and over time even if there were no energy problems, and then estimating how higher energy costs will influence society's use of energy.

Income, Lifestyles, and Energy Demand

The initial projection of income is usually made by assuming that economic growth rates of the past twenty-five years were normal and will continue for the next twenty-five or fifty years. Typically, 3.5 percent growth per year is taken as this "historical" rate for the United States—somewhat below the 1960–69 average, somewhat above the average for 1946 to 1975, and a reasonable mean of the good years for the postwar period. There are reasons to doubt that this high growth rate would continue indefinitely, even if energy costs did not increase. The postwar period was historically unprecedented, and it will become increasingly difficult to generate economic growth by the traditional means of expanding and upgrading the work force, shifting workers from low- to high-productivity jobs, and adding capital. Thus, to assume that historical rates of growth will continue into the next century is probably an overestimate, at least for the United States and other rich countries.

Predicting how energy use would change in the future if there were no increase in the costs and risks of energy supply is more difficult and speculative. Since 1945, energy use in the United States has expanded somewhat more slowly than GNP, in a time of steadily decreasing real energy costs; and foreign experience suggests this pattern is typical in industrialized countries. Cross-section analysis of personal consumption at various income levels indicates that energy consumed directly plus the energy used to produce the other goods and services consumed increases almost in proportion to income. Therefore, if the pattern of economic development remains unchanged, and if tomorrow's "poor" live about the way today's "rich" do, then energy demand at constant costs would increase slightly less rapidly than income— i.e., the "income elasticity of demand" would be less than 1.0, perhaps 0.8–0.9.

When predicting how the high incomes of the future will be spent, however, past behavior is not a reliable guide. With national income growing at 3.5 percent per year and population growing at less than 1 percent per year, per capita income doubles every twenty-five or thirty years. The fact that energy demand almost doubled while income doubled between 1950 and 1975 does not mean that another income doubling by 2000 will do the same, much less than another doubling after 2000 would have the same effect again on energy demand. A quadrupling of income per capita over half a century will be accompanied by many changes in social arrangements and personal lifestyles, and these would affect energy demand even if costs and environmental constraints did not increase. Without knowing what these changes will be, it is not possible to project energy demand with any certainty.

We have not attempted in this study to predict how our grandchildren will choose to live their lives and organize their society, nor do we regard such an exercise as useful for our purposes. It may be that rising incomes will be spent more on leisure and service activities and that future generations will opt for simpler, less energy-using lifestyles than a straightforward extrapolation of trends would indicate. But if social and economic forces unrelated to energy are spontaneously moving society to a future in which centralized energy production has little value, then there is certainly little to be lost by foregoing or delaying nuclear power. To be valid, however, our basic conclusion—that energy costs in general, and nuclear power in particular, are not critical to economic well-being—must hold true even if society is not spontaneously evolving toward a less energy-dependent future. Therefore, in the analysis of the next section, we assume that income will continue growing at high rates as far into the future as we need to look, and that energy use would continue growing in proportion to income if energy costs did not increase.

In our analysis, we have assumed a GNP growth of 3.5 percent per year until 2000 and slightly less thereafter, and have made assumptions equivalent to an income elasticity of demand for energy of about 0.9 until the year 2000 and 1.0 thereafter. With these assumptions, U.S. energy demand at constant costs would be expected to increase from 71 quads in 1975 to 160 quads by 2000 and 350 quads by 2025. However, because costs will increase, these demand levels will not be reached.

Energy Prices and Conservation

Higher energy costs in the future will prevent energy consumption from continuing to grow at the high rates of the past several decades. There are problems associated with energy supply, discussed in Chapter 2, which will result in higher energy costs to society. These costs will be reflected in market prices, in governmental policies, or in shortages, which will cause changes in methods of production, in the mix of goods produced, in personal habits and lifestyles, in the design of buildings and transportation systems, eventually reducing the energy needed per unit of "benefit" or income. We do not specify the mix of free markets, regulations, government incentives, and research that will reduce energy demand. We do assume, however, that the results will be similar to those which well-functioning markets would produce. For example, federal regulations mandating improvements in the efficiency of new automobiles by 1985 will result in automobiles similar to those that well-informed buyers would demand if they paid the full incremental cost of gasoline. Therefore, we can estimate demand on the assumption that energy prices are at economical levels and that energy buyers seek to minimize their costs.

The percentage by which demand is reduced when the price of something increases 1 percent is called the "price elasticity of demand," and is a measure of the ease with which people can get by with less of the good. Demand for a good with many near-substitutes, or for one used for something which is easily given up or done another way, will have a large price elasticity. Price elasticity also depends on whether the good is broadly or narrowly defined (e.g., "energy" or "eastern coal"), whether the time allowed for adjustment is short or long, and whether one assumes that the social setting will change or not in response to the price changes. In our analysis, price elasticity represents the full range of energy-saving actions which will be stimulated in the long run by energy cost increases, and hence must be large enough to include the expected effects of all such actions; in fact, this price elasticity must include the effect of nonconventional energy sources in reducing the demand for central-station electricity and "industrial" fuels. Therefore, "price" elasticity in our analysis includes more than the response of individual buyers to higher market prices, and must be given values accordingly.

The standard way to estimate price elasticities is to use prices and quantities from different areas or times and, with econometric methods, separate the effects of price from the effects of other influences on demand. Estimates based on U.S. experience are of little value for our purposes, because they are derived from a period of low and falling real energy prices and include little variation in expectations, institutions, and policy of the sort which high and rising energy costs could be expected to produce in the long run. Estimates based on international comparisons are more helpful and show larger variations in energy use, roughly correlated with price differences; but even here the price differentials are small compared with expected future increases, and it is unclear which demand differences can be attributed to energy costs and which are due to more general historical and cultural differences. Thus, econometric estimates of past behavior or price elasticity do not tell us with any accuracy what values to use in our analysis.

Another way to estimate price elasticity is to use technical and economic data on energy-conserving methods and nonconventional sources to judge how much demand could be reduced economically at various levels of higher prices. Although the evidence from studies of this sort is hardly decisive, it suggests that significant reductions in energy demand can be accomplished economically, given time, energy prices which reflect economic scarcity, and supportive government policies. In the commercial and residential sectors, which consumed 22 percent of U.S. energy in 1970, improved insulation, changes in lighting, cooling, and heating systems, and—in new buildings—energy-conscious design can reduce energy use per unit of floorspace by as much as half. In manufacturing, which consumes some 28 percent of U.S. energy, the known energy-efficient methods used in other countries could save 20–30 percent, and even better techniques can be developed. Cogeneration of electricity with industrial process heat could increase by half the energy-efficiency with which a substantial portion of new electricity is generated—although changes in utility regulation would be required. The automobile, which uses 12 percent of U.S. energy, is being improved in gas mileage by a factor of two, in response to market forces and government regulation, and other transportation could yield substantial savings. Faced with higher energy costs, the economy will rearrange itself over time to reduce reliance on energy-intensive techniques and habits.

In addition to these energy-saving possibilities, elasticity must include the effects of "nonconventional" energy sources. Only central-station electricity generation and industrial fuels such as coal, oil, and (eventually) hydrogen are reflected in our supply model. If such new sources as solar space heating or windmills begin to displace electricity and industrial fuels as costs rise, demand for the energy sources included in our analysis is reduced. These nonconventional sources are not likely large-scale substitutes for the central-station electricity and industrial fuels we analyze on the supply side, but they may become more important as energy costs increase.

Because our conclusions must not depend critically on the assumed values of uncertain variables, we use a range of price elasticities in our analysis, chosen so that projected demand spans a wide range. The lower end of this demand range is about 100 quads of primary energy in the year 2000, approximating both the zero energy growth scenario of the Ford Foundation's Energy Policy project and the "most probable" projection of the more recent study by the Institute for Energy Analysis. Such low levels of energy use are projected by our analysis when energy costs are high and price elasticity is 0.5; since our important conclusions are merely strengthened if elasticity is higher than 0.5, we need not consider anything higher. At the other end of the range, we use a price elasticity of 0.25, which results in a doubling of energy demand between 1975 and 2000 if energy costs increase only 30 percent or so. As a "base case" we use 0.35.[d]

In choosing our demand assumptions, we have consciously tried to err on the side of *high* demand, so that our conclusions about the economic significance of nuclear power would not appear to be based on the view that energy will be unimportant in the future. Under all of our assumptions, incomes and energy use will increase over time. Social arrangements and personal lifestyles will change in ways we do not attempt to predict, but which probably depend more on changing social values and the growth of income than on the costs and sources of energy.

We recognize the possibility of significant lifestyle changes in response to energy supply problems. We believe that it is unwise, however, to base policy on the assumption that such changes will occur, and we see no reason to argue that they should occur. Every energy supply system and conservation measure has its difficulties: implementation of solar home heating systems on a large scale might require new laws and more government control of building design and locations; cogeneration of electricity with industrial process heat or district heating systems would impose constraints on location and timing decisions of industry and individuals; widespread use of mass transit would put individuals more at the mercy of large organizations; and nuclear energy requires special security measures. Social institutions and individuals will continue adjusting to social and economic change in the future as they have in the past, and these

---

[d]It is also assumed that only 3 percent of the capital stock can be adjusted each year to take advantage of energy-saving opportunities, a conservative assumption.

adjustments are not obviously easier if energy demand grows rapidly or slowly, or if energy supply is nuclear or nonnuclear. Therefore, while we are aware of and have considered the arguments of those who feel nuclear power should be either pursued or abandoned because of its effects on lifestyles and social arrangements, we conclude that nuclear power will be a secondary factor influencing the shape of the future.

A MODEL OF ENERGY AND THE ECONOMY

The general conclusion of our economic analysis—that nuclear power is not crucial, at least not until well into the next century—follows from a few basic facts and assumptions, and is not dependent on detailed models. But to be specific about the importance of certain cost and demand parameters, and to be certain that the detailed interrelationships of energy supply do not hold any surprises, it is necessary to deal with quantitative relationships carefully. The Energy Technology Assessment (ETA) model, developed by Alan Manne, was modified to address the questions of particular interest to us and has served as our basic tool.

The ETA model has a number of useful characteristics. It is simple, as energy models go, including just enough detail to deal with the questions we are interested in. It is designed for exploring broad alternatives, not for detailed planning. The critical quantitative parameters can be varied easily, to explore a wide range of assumptions about demand, costs, timing, and the like. On the other hand, the ETA model does not tell us about regional differences, does not distinguish among energy types in detail, and does not analyze demand in a disaggregated way. But these disadvantages of simplicity are outweighed by the advantages. As with any tool, one must keep its limitations in mind and use it with care. A detailed technical description of the ETA model appears in the *Bell Journal of Economics and Management,* Autumn 1976; here, we give an overview of the structure and key assumptions, and highlight the results we obtained.

The Basic Features of the ETA Model

The model makes the assumptions about energy and income we used above: at constant energy costs, income and energy use would grow at 3.5 percent per year for the rest of this century, and at slightly lower rates thereafter. At higher energy costs, the level of employment and resource use continues to grow at the same rate, energy demand is less, and income is lowered by the higher unit cost of the smaller amount of energy and by the cost of getting by with less energy. Thus, the aggregate economic effect of higher energy costs is a lowering of income, without much effect on growth rates.

There are two forms of energy in the model, each capable of being produced in several ways. Electric energy can be produced by coal-burning power plants, light-water reactors, fast breeder reactors (after 2000), and an "advanced" technology which is available after 2020. This advanced technology is assumed to have the same cost characteristics as the fast breeder, but no fuel limitations, and could be an advanced breeder, solar, fusion, or some combination—it will be called the "Delayed Breeder or Alternative," or "DBoA." Non-electric energy, liquid or gas, can come from oil or natural gas, synthetic fuels based on coal or shale, or hydrogen produced by electrolysis. Electric and non-electric energy are used in combination to produce output or "benefit" and can be substituted for each other to some extent. The model chooses that combination of electric energy, nonelectric energy, and conservation which maximizes the discounted difference between output and energy costs, and hence the demand for energy in each of the two forms will vary negatively with its own price and positively with the price of the other. The overall price elasticity of demand for energy is the percentage decrease in total energy demand caused by a 1 percent increase in *both* the energy prices; this is the demand elasticity discussed in the preceding section.

This model allows a number of quantitative constraints to be imposed, approximating effects not fully modeled. For example: limits are put on the ability of new sources to expand rapidly; a gradually increasing constraint can be put on annual coal production to account for environmental and expansion problems; a limit is placed on the total amount of oil and gas over the period to approximate natural scarcity and import limits; the uranium available at two cost levels is specified; a small allowance for nonprice-induced conservation is included; and a limit is placed on the rate at which energy-conserving capital stock can be put into place. Other constraints approximate such technical details as the need for some oil- or gas-fired electricity generating capacity for peak-load plants and the need for coal for metallurgical and industrial processes.

When specific values are chosen for the parameters of cost, demand, and constraints, the model calculates an "optimal" future path of energy use, divided into electric and nonelectric, and the supply mix which produces this energy at least cost. This future can be interpreted either as the one a planner would choose, or as the one a competitive economy, acting within the same constraints, would produce. However, there are some features of the model which can cause the details of the projected future to be misleading. For example, the model assumes that costs are the same everywhere, so that small differences in costs can produce large differences in supply mix which have little economic significance: e.g., if nuclear power has a 1 percent cost advantage over coal it will expand everywhere as fast as the constraints of the model allow, suggesting that nuclear power is very "important" even though its costs advantage is trivial on the average and may be negative in some regions. Similarly, the constraints on fossil fuel, especially coal, can produce results which must be interpreted with care. For these reasons, we omit the details of the model's quantity and price projections in this presentation, although we have taken them into account in the discussion in the text.

A "Base Case"

Our illustrative or "Base Case" is based on illustrative assumptions that are within the ranges of uncertainty about costs and supplies discussed elsewhere in this report. These assumptions are not always consistent with cost figures developed in other chapters, however; to make sure that the economic value of nuclear power has not been understated, the assumptions are somewhat biased toward nuclear power and the breeder—nuclear costs are perhaps low, non-nuclear costs may be high, and constraints on fossil fuels are more restrictive than may prove necessary. They are intended only to illustrate the magnitudes involved in nuclear policy choices.[e]

The resulting energy demand in the Base Case is presented in Figure 1-1. This projected future has four distinct phases: from 1975 to 1995, oil and gas use continue to grow while light-water reactor (LWR) capacity expands to meet a growing demand for electricity; beginning in 1995, the increasing scarcity of oil and gas drives up their prices, makes synthetic fuels economic, and increases the demands for and prices of electricity and uranium; after 2005, fast breeder reactors (FBRs) take over the market for new power plants from LWRs as uranium prices begin to climb, while the high demand for coal increases coal prices

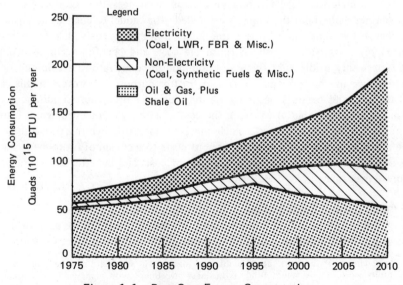

Figure 1-1. Base Case Energy Consumption

[e]The detailed technical assumptions used in the Base Case may be found in the Annex to this chapter.

rapidly; later still, the FBR lowers electricity prices, it becomes economical to produce hydrogen, and total energy demand increases rapidly.

This is not an implausible future, although no more plausible than others. Its most unlikely feature is the high price of coal it projects after 2000; at prices well below the $4 per MBTU projected and with lead times of forty years we see no constraints which could not be overcome, allowing expansion of coal prdouction beyond the constrained levels of the Base Case. The Base Case also indicates that oil imports would continue to grow for the rest of this century, to 75 percent of total oil use, unless domestic production increases unexpectedly. While current trends do point in this direction, the Base Case may be somewhat optimistic about oil and gas availability.

In the Base Case, energy-related investment increases from about 2 percent of GNP in 1975 to 4.5 percent in 2000 and 5.5 percent in 2025. The U.S. savings ratio, which has stayed around 10 percent in the postwar years, would have to increase to 13 or 14 percent in order to leave nonenergy investment and economic growth unchanged. If this increase does not occur, then interest rates will rise, the cost of capital-intensive energy sources such as nuclear and solar will increase, and growth in income and energy use will slow. Savings rates of 15 to 20 percent are common in other wealthy countries, so the United States should be able to generate the savings it needs. If not, the value of nuclear power is further reduced, both because nuclear costs increase and because energy demand growth is reduced.

To estimate the effects of policy or technical changes, the assumptions in the model are varied and the impacts estimated. The model calculates the costs of a change by computing the resulting increase in energy costs plus the increase in energy conservation costs. Because the costs and benefits occur over time, it is necessary to discount them at a common discount rate to a common time period; we use both 5 and 10 percent for this purpose.[f] However, in this case both costs and benefits occur over a long future, more or less simultaneously. This makes it possible to calculate the cost of a policy as a percentage of future annual income rather than as a single present value. Although this procedure can be misleading if used to compare present costs to a stream of future benefits, it is an appropriate way to compare streams of costs and benefits which are both spread over time.

The "Delayed Nuclear" Case

To estimate the economic importance of nuclear power in the near term, we have projected a future with less dependence on nuclear power, to determine

[f]The choice of discount rate is a difficult matter. The issues are whether the true opportunity cost of investment funds is business investment with high yields or consumer savings with low yields; and whether society has greater interest in future incomes than is reflected in market transactions. In real terms, i.e., net of inflation, we favor a rate nearer 5 percent than 10 percent for comparisons among value streams. Ten percent is probably closer to the rate used in making investment decisions and is used in our Base Case. When annualized effects are compared the choice of discount rate is less critical.

how it would differ from our Base Case. The "Delayed Nuclear" Case assumes: (1) all LWR construction beyond the 100,000 megawatts firmly committed for 1985 is suspended; (2) no plutonium recycling is undertaken; (3) no FBR is available early in the next century; but (4) research and development on advanced technologies is continued or even accelerated so that the Delayed Breeder or Alternative (DBoA) is available in 2020 as in the Base Case; and (5) LWR production can begin expanding again in the year 2000, if economical at that time.

The Delayed Nuclear Case is illustrated in Figure 1-2. In this case, demand reduction, rather than expansion of other supplies, is the principal economic response to the higher energy prices. Delaying nuclear power results in less LWR capacity in the year 2000, and the elimination of about 50 quads of FBR-produced energy in 2010. This drives up electricity prices around the year 2000, increasing demand for nonelectric energy and causing fluid fuel prices to increase. The demand for coal at $1 per MBTU reaches the constrained level about 2005, and this causes coal prices to soar. After 2000, LWRs are again produced and the DBoA appears after 2020; electricity prices and then fuel prices decline until eventually (after 2050) the effects of the nuclear delay fade away.

As a result of the delay in nuclear power, energy production and conservation costs over the period 1975 to 2045 are increased by an amount which has a 1975 value (discounted at 10 percent per/year) of $76 billion. This is equivalent in present value to an average loss of 0.3 percent of income in every year. Changing only the interest rate from 10 to 5 percent, the discounted cost of delaying nuclear power becomes $522 billion from 1975 to 2045, equivalent to 0.8 percent of income in every year over the period. While 0.8 percent or even 0.3

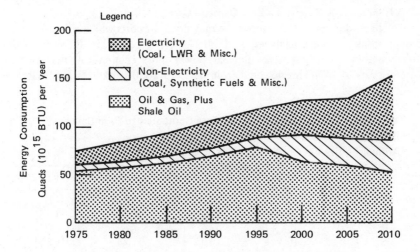

Figure 1-2. Delayed Nuclear Case Energy Consumption

percent of a growing income over three-quarters of a century is a lot of income, it is only a small fraction of the effects of higher energy costs in general, and surely would not be enough to cause fundamental social or economic changes.

Table 1-1 displays the 1975 present value of costs and the annual equivalent cost fraction in each of three time periods between 1975 and 2045 and over the entire period, for 10 percent and 5 percent discount rates. The effects of delaying nuclear power are essentially insignificant in this century but increase to more than 1 percent of GNP in the next. Since GNP is so large in the future, these costs are very large in absolute value in the years in which they occur. Changing the discount rate has a dramatic effect on present value but less effect on future costs; a lower interest rate does not change future decisions or costs as much as it changes their discounted value today. Since most of the noneconomic benefits which may result from delaying a commitment to nuclear power also accrue in the future, it may be beside the point to argue about which discount rate and hence which present value is more nearly "right." Instead, today's decisionmakers should ask themselves whether the citizens of 2025 would feel that a delay in deploying nuclear power in the late twentieth century had reduced nuclear risks sufficiently that the citizens of 2025, on reflection, will think it to have been worth the sacrifice of 1 to 2 percent of current income every year. Of course, the opposite course cannot easily be judged *ex post facto*. Risks are not certainties; if nuclear power has no serious negative consequences as a result of either accidents or proliferation, the generation of 2025 to 2050 will be glad to be richer. If, on the contrary, the passage to that period is marked by costly events on the proliferation or safety fronts, that generation will deplore our narrow-minded preoccupation with economics and our blindness to other and more important concerns.

A High Nuclear Case and a Low Demand Case

To check our conclusions about the magnitude of costs which a nuclear delay would entail, we have analyzed a "High Nuclear Case." This differs from the Base Case in ways that make nuclear power more economical: the assumed capital costs of LWRs are lowered from $700 to $600 per kw, giving LWRs the

Table 1-1. Costs of Nuclear Delay Case Compared to Base Case
1975 Present Values in Billions of 1975 Dollars
[Annual Equivalent Costs as Percent of GNP]

Discount Rate	1975-2045	1975-2000	2000-2025	2025-2045
5%	522	68	273	181
	[0.8%]	[0.21%]	[1.2%]	[1.5%]
10%	76	14	53	9
	[0.3%]	[0.07%]	[1.2%]	[1.2%]

same capital cost as coal-fired plants with scrubbers; FBR and DBoA capital costs are lowered from $800 to $700 per kw; all generating plants achieve a 70 percent capacity factor, instead of 65 percent; and demand elasticity is reduced from 0.35 to 0.25, making energy conservation more costly. For a check in the other direction, we analyze a "Low Demand Case," which differs from the Base Case only in the demand elasticity, which is increased from 0.35 to 0.50.

The High Nuclear and Low Demand cases are projected, and then the Delayed Nuclear constraints are imposed: no LWRs allowed beyond already planned 1985 capacity until 2000; no plutonium recycle; and no FBR. The cost of nuclear delay is, as before, the increased costs of energy production and conservation. The results are given in Table 1-2, with the Base Case. As expected, the High Nuclear assumptions result in higher demand and higher cost of delaying nuclear; but the effect is still no more than 2.5 percent of GNP in any period. The Low Demand Case results in lower energy demand and lower costs of a delayed nuclear commitment, which are not higher than 1 percent of GNP in any period. Thus, our qualitative conclusions concerning the relatively small economic importance of nuclear power are valid over a wide range of assumptions.

Table 1-2. Costs of Delaying Nuclear Power
1975 Present Values in Billions of 1975 Dollars
[Annual Equivalent Cost as Percent of GNP]

	Discount Rate	Cost of Delaying Nuclear Power			
		1975-2045	*1975-2000*	*2000-2025*	*2025-2045*
High Nuclear Case	5%	876 [1.3%]	104 [0.3%]	567 [2.5%]	205 [1.7%]
	10%	148 [0.6%]	43 [0.2%]	94 [2.1%]	11 [1.5%]
Base Case	5%	522 [0.8%]	68 [0.2%]	273 [1.2%]	181 [1.5%]
	10%	76 [0.3%]	14 [0.1%]	53 [1.2%]	9 [1.2%]
Low Demand Case	5%	NA	NA	NA	NA
	10%	44 [0.2%]	9 [0.05%]	28 [0.7%]	8 [1.0%]

Notes: High Nuclear Case differs from Base Case as follows:

    Price elasticity decreased to 0.25 from 0.35;
    Nuclear capital costs reduced $100/kw;
    All capacity factors increased to 70 percent from 65 percent.

    Low Demand Case differs from Base Case only in that price elasticity is 0.50 instead of 0.35.

    NA indicates figures are not available.

The Importance of Fossil Fuels

Fossil fuels remain the principal alternative to nuclear power over the period of our analysis. The ETA model is not designed to treat fossil fuels in detail; but by varying the supply constraints one can estimate the relation between fossil fuel supplies and the value of nuclear power.

For oil and gas, we assume a cost of $2 per MBTU ($12 per barrel) but a limit on the total amount available from domestic production and imports of 3,000 quads over the entire period 1975-2045. At this level, imported oil and gas might reach about three-quarters of domestic oil consumption, suggesting that the limit on oil and gas might be unrealistically high. The model also calculates that prices would go to $6-8 per MBTU ($30-40 per barrel) by 2010; and at prices this high the long-run potential for offshore discoveries and high-cost extraction from known sources and shale oil is large. The quantitative constraint may therefore be too tight. We cannot resolve these uncertainties, but can point out that the amount of oil and gas available has a strong influence on the value of nuclear power: varying the stock of oil and gas from 3,000 quads to 4,000 quads reduces the cost of delaying nuclear power by half, while reducing oil and gas to 2,500 quads doubles the cost of delay.

Similarly, the Base Case assumes constant coal production costs of $1 per MBTU ($25 per ton for eastern coal), but limits coal production. The production allowed under the constraint gradually increases from the 13 quads in 1970 to an eventual limit of 75 quads per year (or six times current production) while projected coal prices after 2010 greatly exceed production costs. With high prices and fifty years' time, it should be possible to overcome the technological, logistical, and institutional problems which limit coal, but the possible irreversible climatic effect resulting from the buildup of carbon dioxide in the atmosphere may eventually prove to be a limiting factor. This problem, about which little can be said with confidence, is discussed in some detail in Chapter 6.

The importance of coal can be illustrated by relaxing the coal constraint in the model. If coal were available in unlimited amounts at $1 per MBTU in the Base Case, its use would expand dramatically, primarily in the production of synthetic fuels rather than as a replacement for nuclear power in electricity generation. Coal use would expand to 250 quads in 2025 (10 billion tons, up from just over 0.6 now) and the substitution of electric for nonelectric energy would be slowed. If nuclear power were then delayed as before, coal production would expand more quickly and reach 270 quads in 2025. Because there would be little economic cost in using coal instead of nuclear for electricity generation, the cost of delaying nuclear would be insignificant, only 0.1 percent of GNP over the period.

The levels of coal production projected when unlimited amounts are available at $1 per MBTU cannot be taken literally. The local and global environmental effects of coal use in the United States on the scale of 250 quads per year might be unacceptable long before such a production is reached, even though coal resources and industrial capacity might be adequate. But this case does illustrate that coal is an economical substitute for nuclear and, to the extent that the environmental effects of coal are made less onerous than the risks of nuclear, coal can be substituted for nuclear to advantage.

The Breeder, Plutonium Recycle, and Laser Enrichment

The breeder reactor and plutonium recycle are methods of extracting more energy from natural uranium; improved uranium enrichment technologies play the same role. Because of current policy interest in the breeder and recycle, and because of possible breakthroughs in enrichment technology, we analyze the economic importance of these technologies separately.

Plutonium recycle in light-water reactors can extend energy obtainable from natural uranium. As discussed in detail in Chapter 11, the costs of recycle are uncertain, as are the costs of uranium and enrichment for which recycle is a substitute. At best, plutonium recycle in light-water reactors has a small economic benefit, perhaps lowering LWR fuel cycle costs by 10 percent and electricity costs by 2 percent. The economywide impact of recycle is insignificant, and if foregoing recycle reduced environmental and safeguards problems, there is no economic reason for proceeding.

The breeder reactor is, like plutonium recycle, also a substitute for mining and enriching uranium, but on a scale that is qualitatively different. Breeders could in principle increase fiftyfold to one hundredfold the energy obtained from uranium as compared with LWRs, virtually eliminating scarcity as a constraint on energy supply for the twenty-first century. Because of the safeguards problems involved in the breeder fuel cycle, there is potential advantage to postponing the breeder to allow time to resolve them. To estimate the costs of such a postponement, we have used the ETA model with both the Base Case and the High Nuclear assumptions, but with the breeder (FBR) removed. In effect, the FBR is bypassed, with reliance on LWRs and coal until the DBoA becomes available in 2020.

Yet another way to use uranium more efficiently is to increase the percentage of uranium-235 that is separated in the enrichment process. As discussed in Chapter 13, a new technique, laser isotope separation, may make it possible to achieve virtually complete separation of uranium-235 at a fraction of the cost of today's enrichment processes. If the laser process works it could expand the effective uranium resources by as much as half and reduce importance of the breeder. To examine the potential significance of this new technology, we have used the ETA model with the assumption that this technology becomes available in 1990, with and without the FBR available, for the Base Case and High Nuclear Case.

Because of the economic interactions between the FBR and the laser, it is convenient to discuss them together. Table 1-3 displays the economic benefit associated with the FBR and laser enrichment in various combinations, using the "No FBR, No Laser" combination as the base; the annual equivalent cost is the constant fraction of GNP over the period 2000-2045, which has the same present value, reflecting the fact that the laser and (especially) the breeder have their economic impact after 2000. The most interesting implication is that the cost of delaying nuclear power would be reduced by 75 percent if only the breeder is delayed and LWRs are allowed to expand to the extent they are competitive with coal; and the effect of delaying the FBR only is even smaller if laser enrichment is successful in stretching the supply of lower cost uranium. For example, without laser enrichment, the High Nuclear Case with a 10 percent discount rate indicates that the cost of delaying both the LWR and the FBR after 2000 would be $105 billion or 2.0 percent of GNP, while the cost of delaying the FBR only would be $28 billion or 0.5 percent of GNP. With successful laser enrichment, the cost of delaying both the LWR and FBR would be $119 billion or 2.2 percent of GNP (since lower cost uranium would improve the competitive position of nuclear power), but the cost of delaying only the FBR would drop to only $20 billion or 0.3 percent of GNP, since the competitive position of the LWR would improve compared to the FBR. Thus, most of the economic value of nuclear power can be obtained from LWRs alone, with the breeder postponed until 2020 or so, by which time an alternative to plutonium-based systems may be available. And if advanced enrichment processes are developed—as we think they will be—then the value of nuclear power in general is increased, while the relative value of breeders is decreased.

Table 1-3
Benefits of the Breeder and Laser Enrichment
1975 Present Values in Billions of 1975 Dollars
[Annual Equivalent Cost as Percent of GNP]

	Base Case		High Nuclear Case	
	5%	10%	5%	10%
FBR and Laser	NA	NA	434 [1.3%]	42 [0.7%]
FBR, No Laser	238 [0.7%]	17 [0.3%]	346 [1.0%]	28 [0.5%]
No FBR, Laser	NA	NA	189 [0.5%]	22 [0.4%]
No FBR, No Laser	0 [0%]	0 [0%]	0 [0%]	0 [0%]
Delayed Nuclear Case	−216 [−0.6%]	−45 [−0.9%]	−426 [−1.2%]	−77 [−1.5%]

Note: In Delayed Nuclear Case, pre-2000 costs are ignored.

Nuclear Power Economics Outside the United States

Our economic analysis has been focused on the United States, and hence is not directly applicable to other countries. However, for other industrialized societies, our basic conclusions about energy costs, economic well-being, and growth are generally applicable: higher energy costs will cause problems if they arrive unexpectedly and in the long run may reduce income levels by something like the cost of paying them; but an otherwise well-functioning economy should be able to adjust to cost increases of the magnitude we foresee and then resume its growth. The principal difference between the United States on the one hand and Western Europe and Japan on the other is that the United States has the option of relying on domestic fossil fuels. However, transportation costs of fossil fuels are not large in relation to the future energy cost levels we foresee, so that an open world market in these fuels would put all industrialized countries on about the same basis economically. But major countries may not be willing to depend too heavily on world markets for their energy, especially if the nuclear alternative makes it relatively inexpensive to reduce the extent of their dependence. Thus, individual national decisions concerning nuclear power will depend on the particular economic and political situation in each case.

For the developing world the situation is more difficult. Although the less developed countries account for less than 10 percent of current world consumption of "market" energy, the rate of growth of their energy use is much higher, so that their share may reach as much as 20 percent by the year 2000. The development plans in most poor countries are also more dependent on increased use of energy, particularly in the industrial and transportation sectors. Hence, the possibility of reducing energy growth in response to higher prices without also lowering GNP growth is probably considerably less than in developed countries. Thus, annual growth rates of energy demand in developing countries may be 7 percent or more for the rest of this century.

The share of this growing demand which can be supplied economically by nuclear power is a matter of some dispute. A 1974 estimate by the International Atomic Energy Agency concluded that more than half the total energy supply of the countries with larger markets might come from nuclear power by the year 2000. More recent estimates by the World Bank, by Richard Barber and Associates, and by Alan Strout suggest a number of reasons to doubt such a high estimate, including the high prices and hence slower demand growth for energy in all forms, the potential for increased production of fossil fuels in the developing countries, and the fact that nuclear reactors of less than 600 MWe capacity are not likely to be available. The fact that smaller reactors are not being manufactured is particularly significant since large 1,000 MWe power plants are not at all suited for countries with small aggregate demand and poorly developed electrical grids. If small reactors are manufactured especially for this market the capital cost per kilowatt will increase significantly, probably eliminating any economic advantage that nuclear power might have had. Taking these factors into account suggests that nuclear power will be economical in perhaps fifteen to twenty of the larger developing countries by the end of the century, although others may install it for noneconomic reasons without regard for whatever economic penalty results. In none of these countries is there a persuasive cost argument for plutonium recycle or breeder reactors—although considerations of international prestige, energy independence, or a potential weapons capability may argue otherwise.

INTERPRETATIONS AND CONCLUSIONS

Our quantitative analysis has confirmed the conclusion we arrived at by more general reasoning: energy costs within the range we foresee are not critical to determining the economic or social future. The fears that energy scarcity will force fundamental changes in economic and social structures or the lifestyle of the industrialized world are not well founded.

Specifically, our analysis indicates that the costs of delaying nuclear power would not be significant in this century, but in the next century could reach as much as 2.5 percent of annual GNP—although costs on the order of 1 percent of GNP are more likely. Even under the "High Nuclear" assumptions, there are policies that would reduce the social costs and risks of nuclear power while maintaining most of the economic benefits. Plutonium recycle can be delayed indefinitely, at essentially no economic cost. Breeders can be postponed several decades into the next century at costs that are small (less than 1 percent of GNP) under the worst conditions and very small (on the order of 0.1 percent of GNP) under more likely assumptions about costs, elasticities, fossil fuel supplies, and enrichment technologies.

Our analysis also shows that the benefits of nuclear power are highly dependent on the ease of reducing demand for conventional energy forms and expanding our supplies of fossil fuels. The uncertainties inherent in assumptions about price elasticities, costs, and coal production mean the economic costs of delaying nuclear power could approach the values calculated in this study or could be nearly zero.

One feature of our analysis, which is particularly important from the standpoint of policy, is the assumption that an "advanced technology" will be available around 2020. This advanced technology could be an advanced breeder or it could be solar or fusion, advanced coal production and use, or some combination. The prospects of these advanced technologies and the nature and timing of the U.S. breeder program are discussed in some detail in Chapters 4 and 12. For the perspective of this broad economic analysis, the precise costs of these advanced technologies are not important. It is important, however, that one or more of these technologies be available within the next fifty to seventy years to provide assured energy supplies and keep energy prices from increasing rapidly.

Annex — Technical Assumptions for Base Case*

General Assumptions

GNP Growth Rate: 3.5%/yr from 1970 to 2000; 3.2%/yr from 2000 to 2025; 2.9%/yr from 2025 to 2040

Price Elasticity of Demand: 0.35

Income Elasticity of Demand: 1.0 but with a costless conservation effect which grows linearly from 0% in 1970 to 10% in 2000 reducing demand 10% below projected levels after 2000.

Discount Rate: 10%/yr

Resource Assumptions

Resource	Production Cost	Constraints
Coal	$1 per MBTU	"Environmental" constraint limits annual use to amount given by logistic curve starting at (actual) 13 quads in 1970, increasing to 50 quads in 2000, approaching 75 quads asymptotically.
Oil & Gas	$2 per MBTU	Resource and political constraints limit total stock available to 3,000 quads.
Shale Oil	$2.5 per MBTU	Production begins in 1995, limited to 2 quads in 1995, 4 quads in 2000, 6 quads in 2005, etc.
Synthetic Fuels	$3 per MBTU	Produced from coal with efficiency of 0.67; coal subject to flow constraint above.
Uranium	$30 per pound $100 per pound	2.5 million tons supply 5.5 million tons supply
Hydrogen	$5–$8 per MBTU	Cost is calculated in the model by assuming 100% efficiency of conversion of electric energy to hydrogen fuel, at additional capital cost at the power plant of $150 per kw.
Hydro, Geothermal, Wind, Garbage, etc.		These "other" sources produced 16% of U.S. electricity in 1970; it is assumed these sources will expand at 2%/yr, independent of costs and prices.

Capital Cost Assumptions

Facility	Capital Cost	Comments
Coal-fired Power Plant	$600 per kw, 65% capacity factor	36% conversion efficiency. Stackgas scrubbers required. O&M cost of 3.0 mills/kwh.
Light-water Reactors	$700 per kw, 65% capacity factor	Required uranium feed and enrichment capacity act as constraints. O&M cost of 2.0 mills/kwh.
Fast Breeder Reactors	$800 per kw, 65% capacity factor	Introduced in 2000, expansion limited by plutonium availability. O&M cost of 2.0 mills/kwh.
Delayed Breeder or Alternative	$800 per kw, 65% capacity factor	Introduced in 2020, costs identical to fast breeder, but unconstrained by fuel availability.

(continued)

Annex — Technical Assumptions for Base Case* (cont'd.)

Nuclear Fuel Cycle Cost Assumptions

Light-water Reactors:

The model calculates fuel costs (in mills/kwh) from uranium prices, enrichment price, fabrication and handling costs, tails assay, etc. With uranium price at $30/lb U_3O_8, enrichment price at $75/SWU, no plutonium credit, and fabrication, etc. at 0.75 mills/kwh, fuel cost is about 3.4–3.5 mills/kwh, or about 4.2 mills/kwh counting the interest charges on fuel in the reactor.

Fast Breeder Reactors:

Total fuel cycle cost, excluding plutonium credit, is assumed to be 1.5 mills/kwh, and is independent of any resources prices. Plutonium availability limits expansion.

Delayed Breeder or Alternative (DBoA):

Costs identical to fast breeder, but no fuel constraints; could be fission breeder started with stockpiled plutonium or with uranium-235.

---

*As discussed in Chapter 1, p. 21, these illustrative assumptions are not always the same as those developed elsewhere in the report.

Chapter 2

Uranium and Fossil Fuel Supplies

The cost of energy and its broad impact on the overall economy, which we considered in Chapter 1, depend on the availability and cost of fuel supplies. This chapter assesses uranium and fossil fuel supplies and their potential to meet future energy demand. Considerable disagreement exists over the extent of uranium resources, the future costs and availability of coal, the time scale for the exhaustion of oil and gas, and the prospects for energy from alternative sources (solar, geothermal, and fission) discussed in Chapter 4. All these questions are central to evaluation of the role of nuclear power in the energy economy and to discussion of such key issues as plutonium recycle, the breeder, and uranium enrichment.

Estimates of supplies can be categorized as "reserves" and "resources." The definitions differ from fuel to fuel. For oil and gas, for example, "reserves" are understood to refer to material in the ground that can be produced and marketed with current technology at current or near-current prices. Reserves of coal are defined principally by physical characteristics such as depth and thickness of seam. Economists have long asked resource geologists for a concept that would explicitly include price as a parameter and would permit the construction of a supply function. Attempts in this direction have remained episodic.

As for "resources," the picture is even fuzzier. Some practitioners include all of the material that exists as "resources." Others have favored the term "resource base" for this concept. Still others have been more restrictive, pointing out that in the case of oil, for example, not every drop can be considered a "resource." The U.S. Geological Survey, the fountainhead of information in this area, has had recourse to the liberal use of qualifying adjectives, ranging from "probable" to "hypothetical" and "speculative," and stopping short only of "impossible."

Even when estimates are characterized as "geologic" (as in the recent U.S. Geological Survey estimate of undiscovered, recoverable oil and gas resources), an economic dimension creeps in. It may not be explicit but some concept of "commercial prospect" is clearly implied. Since the dimensions of that prospect will vary over time and among observers, even these noneconomic estimates are subject to variation, apart from new knowledge or new discoveries.

The most satisfactory attempt to create some order in the field was made by Vincent McKelvey, now director of the U.S. Geological Survey, when he developed the "McKelvey Box," which is schematically reproduced in Figure 2-1. It incorporates the two dimensions that lead to useful distinctions: physical identification and economic feasibility.

Reserve and resource estimates for minerals have a long record of being understatements. Estimates of reserves typically originate with industry and reflect its view of what is marketable as well as of what it is prudent to characterize as reserves. Estimates of resources beyond reserves are typically government estimates. Both kinds of estimates are conditioned by the technology of exploration and recovery and by the extent of exploration. Behind it all lies

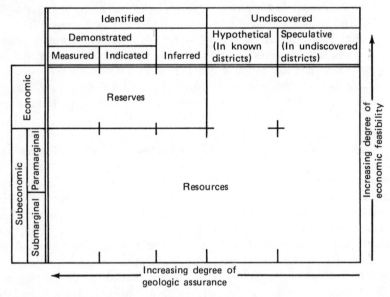

Figure 2-1. United States Department of the Interior Classification System

Source: John J. Schanz, *Resource Terminology: An Examination of Concepts and Terms and Recommendations for Improvement,* Report prepared for the Electric Power Research Institute (Springfield, Virginia: National Technical Information Service, 1975), PB 224–433.

some concept of the mineral's future significance and the future economic picture. Generally, as markets expand or as prices rise, an industry is motivated to look for, and tends to find, new reserves. This explains why reserve and resource estimates rise along with rising production. One should, therefore, keep an open mind regarding the potential for incremental discoveries and large surprises.

The government's resource estimates have generally been the responsibility of the U.S. Geological Survey and the Bureau of Mines within the Department of the Interior. In the case of uranium, the estimating task was originally assigned to the Atomic Energy Commission for reasons of security. The government relies on the mining industry, however, for much of the basic data. To the extent that industry's judgment is conditioned by market prospects, so are the data it furnishes.

The search for resources can be facilitated or inhibited by the geologic characteristics of the material being sought. Coal deposits can be inferred relatively accurately; they extend many miles under geologic conditions similar to those of previously identified reserves. In contrast, oil, gas, and uranium are generally found in relatively discrete deposits; extrapolation to undiscovered deposits involves much greater uncertainty. Improved geologic understanding has in the past opened up wholly new environments to successful search in oil and gas.

In assessing uranium, emphasis until recently was almost exclusively on sandstone. In other countries, uranium is produced from a variety of other geologic settings including granite, conglomerate rock, calcrete, shale, and mineral veins. This experience suggests that new discoveries in the United States might be expected in previously neglected environments. Despite their importance, however, the new exploration efforts may not be attractive to private investment, in view of their high capital requirements and uncertain payoff relative to other investment opportunities. By supporting exploration to improve the reliability and scope of resource projections, government can play an important role in influencing industry actions.

Estimates of domestic and foreign energy resources provide the basis for important decisions not only in energy policy but also in foreign policy, economic planning, R&D strategy, transportation policy, and the many interrelated domestic issues influenced by energy. The importance of these decisions argues for a sound program to improve the reliability of these estimates. In mid-1975 the U.S. Geological Survey prepared a detailed estimate of domestic oil and gas resources, and a new look at coal has begun, taking into account criteria beyond those conventionally employed in judging current profitability. The National Uranium Resource Evaluation (NURE) program was initiated in 1974 to clarify questions about the uranium supply; ways in which this program might be improved are discussed below.

URANIUM RESOURCES

A key question in the nuclear power debate is whether enough uranium will be available to meet future nuclear power needs. The answer to this question is central to decisions on recycle and the breeder reactor, since the urgency of development of these technologies is premised on an anticipated shortage of world uranium supplies.

Unlike conventional power plants that can be modified to run on coal, oil, or gas, nuclear reactors can run only on uranium fuel. Nuclear reactors may take ten years to build and have a thirty-year operating life; thus, concern over adequate uranium supplies extends far into the future. Moreover, potential customers presently cannot count on securing their needs "in the market" as occasion arises, as they would do for other fuels, since a broad uranium market has not yet come into being.

While we recognize the concern and the uncertainties, our review suggests that the official statistics err substantially on the low side. A history of low prices for uranium supplied to a single customer—the U.S. government—has discouraged exploration, especially for lower-grade ores that are more expensive to mine. Assuming the incentive of higher prices, more intensive and extensive exploration should produce enough uranium, both in the United States and abroad, to supply the domestic and foreign requirements of light-water reactors (LWRs) well into the next century.

Examination of several possible U.S. supply and demand situations, based on typical ERDA planning scenarios, indicates that all predicted demand levels for LWRs could be accommodated with a reasonable expansion of present reserves at prices not exceeding $40/lb (1975 dollars) for the balance of this century and that uranium resources will probably be adequate well into the next century. A comparison of uranium supply and demand projections outside the United States leads to similar conclusions.

We do not therefore foresee a pending uranium shortage, which has been cited as a reason for early commercialization of plutonium recycle or of the breeder. Nevertheless, we recommend intensified exploration efforts, since better understanding of the extent of uranium deposits will provide a sounder basis for future decisions on recycle, breeders, and alternative energy sources.

Reserves

Within the general climate of uncertainty that surrounds all mineral estimates, uranium reserves and resources are afflicted by a special set of circumstances. Uranium was originally a material with a single buyer, the United States government, for a single noncommercial use, weapons-making. Prior to the emergence of commercial customers in the late 1960s, the U.S. government set what it judged to be a "fair" price, and entrepreneurs found material to meet the buyer's terms. This meant finding uranium that could be profitably sold at a price of around $8 per pound of uranium oxide (U_3O_8).[a]

In this situation, "economic feasibility" as a criterion for reserves was equivalent to purchasability by the government, and until a few years ago "reserves" were defined strictly in that way. Higher-cost U_3O_8 was not introduced into the government statistical scheme until 1968, and estimates of resources were not even published until 1969 and were not introduced in their current full-fledged form until 1973.

It is necessary to recall these circumstances since some observers have been led to believe that more is known about uranium than about most other minerals. This misunderstanding results from presenting uranium statistics in cost terms that seem to answer the demand for data from which one can fashion a supply function. Current surveys estimate the reserves that would be available at $8, $10, $15, and $30 per pound of U_3O_8.

Some qualifications emerge, however, that drastically impair the validity of these data. The ranking by cost of recovery is based upon a restrictive definition of cost, called "forward cost." This concept, which was evolved by the government as purchaser early in the history of uranium mining, refers to operating and capital costs not yet incurred at the time an estimate is made. Other factors are specifically excluded, such as past expenditures for property acquisition, exploration, and mine development. For these reasons, uranium costs used in estimates are not equivalent to the actual cost of production or, obviously, to selling price. In addition, the costs used in estimates represent the maximum or cutoff cost of the specified category, in which average costs would be substantially lower.

These two qualifications have led to much confusion. Forward costs tend to be confused with full costs and with price; and the upper boundary tends to be taken as applicable to the entire category, which it is not. Thus, one might wonder how sellers made a profit at a price of $8 per pound, when the lowest-cost reserves had an $8 cost label. The uranium in that category, of course, actually included quantities costing much less to produce. Sales at prices below $8 point to the existence of lower costs. Analogously, the "$30 group" includes all the lower-cost groups whose costs probably average below $15 since, of the total reserves (see Table 2-1), only a little less than one-third lies in the $15 to $30 category.

Comparisons over time are complicated by changes in the cost of labor and material, and by inflation. Thus, a cutoff of $10 one year will bracket a different segment of reserves than the $10 category another year. Recent changes in reserve estimates have primarily reflected cost changes, not discoveries and

---

[a]Unless otherwise specified, all quantitative data in this chapter refer to U_3O_8, the oxide of uranium delivered by the mills for further processing.

Table 2-1.   U.S. Uranium Resources (Thousand Tons U$_3$O$_8$)

$/LB U$_3O_8$ Cutoff Cost*	Reserves	Potential Probable	Possible	Speculative	Total
10	270	440	420	145	1,275
15	430	655	675	290	2,050
30	640	1,060	1,270	590	3,560
By-product from copper and phosphate production 1975–2000	140	–	–		3,700

Source: *Statistical Data of the Uranium Industry–1976,* ERDA, Grand Junction, Colorado, January 1, 1976.
*Each class includes material in the lower class or classes.

depletion. For example, between ERDA's January 1, 1974 and January 1, 1975 estimate, the so-called $8 reserves shrank from 277,000 tons to 200,000 tons, because of "reevaluation" due to inflation. Similarly, out of a reduction of 30,000 tons in the $10 category, 28,000 were due to inflation. In contrast, depletion by production was only 13,000 tons in the $8 category and 15,000 in the $10 category. (The reason that depletion by inflation is lower at the $10 than the $8 cost cutoff is that some material in the lower group gets "inflated" into the higher cost group.)

Until 1973, ore above the $8 cutoff was of little interest to exploration and mining enterprises since there were no buyers. Actually, most of the time until 1973 there were no buyers even at prices near $8. Since then, the $8 group has been abandoned altogether in the statistics and the $15 cutoff is now frequently referred to as "low-cost." Moreover, since inflation between 1965 and the end of 1975 has been approximately 75 percent, a $15 cost in 1975 corresponds to $8.60 in 1965. ERDA now corrects its estimates for inflation.

The cost group into which a uranium property falls depends on such classifications as grade, depth, and geologic setting. The bulk of reserves in the "up-to-$15" group ranges between 0.10 and 0.15 percent in grade, lies less than 700 feet below the surface, and is found almost exclusively in sandstone. Raising the cost cutoff lowers the average grade of the group. Based on ERDA estimates of January 1, 1976, up-to-$10 reserve class ore averaged 0.17 percent U$_3$O$_8$; up-to-$15 ore averaged 0.13 percent; up-to-$30 ore averaged 0.08 percent.

While a close correlation obviously exists between grade of material and resource classification, some very low-grade material forms part of even the low-cost reserves, because it is collocated with adjacent higher grades and it is advantageous to mine them together. Raising the cost cutoff from $15 to $30 raises reserves from 430,000 to 640,000 tons (as of January 1, 1976). Of the 210,000 ton increment, all but 55,000 tons are in properties that also have higher-grade reserves.

Potential

Beyond reserves lie the potential resources. Estimates of potential resources were first published in 1969. Those estimates referred only to conventional geological settings in the western United States—the Colorado Plateau, the Wyoming Basin, and the Texas Coastal Plains. The late date and geographic confinement of that effort indicate the relaxed attitude regarding uranium availability that prevailed until a few years ago. Establishment in 1974 of the government's NURE program, located first in the Atomic Energy Commission and now in ERDA, reflected a rapidly developing interest in resources nationwide. Between January 1974 and January 1975 estimates of total U_3O_8 resources were raised by 1.3 million tons, owing to the initial effort of NURE in evaluating and organizing evidence already in the files.

In developing estimates, NURE adopted a threefold classification of resources with declining levels of confidence: probable, possible, and speculative.[b] Table 2-1 shows current ERDA estimated reserves and resources by cutoff group as of January 1, 1976; a Nuclear Regulatory Commission report, *Final Generic Environmental Statement on the Use of Recycle Plutonium in Mixed Oxide Fuel in Light Water Cooled Reactors*, (October 1976, referred to hereafter as GESMO), raises these estimates by almost 50 percent. A range of ±20 percent is assigned to the reserves, the conventional range in mineral reserve estimates. None of the three classes of resources, however, has a confidence level attached. Users of the data must therefore make their own assumptions as to future "availability." The failure to specify confidence limits is somewhat mitigated by the growing inclusion of higher-cost material in the reserve category. This increases confidence in total uranium supplies since reserves are by definition high confidence supplies. Moreover, the "probable" category resembles closely what are "inferred" reserves for other minerals, which are often treated as reserves.

To the extent that material is high-grade and low-cost, it is drilled up and finds its way into the reserve category; if it is low-grade and high-cost, it remains in one of the potential classifications. It is, however, wrong to conclude from the data that low-cost material is nearing exhaustion. There is too little knowledge of uranium resources over the whole country. The effort reflected in available statistics is too early to be definitive.

Low-Grade Resources

Prior to the boost in oil prices in late 1973, U_3O_8 with an estimated production cost much above $10 was not seriously considered a competitive fuel. Since then, U_3O_8 at several times that price has come to be acceptable for

[b]To qualify as "probable," deposits must lie in areas in which past production plus reserves add to no less than ten tons of U_3O_8, and in extensions of known deposits, or in new deposits that have been identified by exploration. "Possible" deposits are defined as lying within the same geologic province or subprovince. "Speculative" deposits are those that are found in formations or geologic settings not previously productive.

power generation, given the new prices for oil, gas, and coal. It should be noted that the lower environmental impact of uranium mining compared with coal mining will be eroded as low-quality uranium ore is mined. There is a point below which a greater volume of uranium ore must be mined and processed than coal for an equivalent amount of energy.

A useful way of visualizing the situation is to compare the tonnage of fuel required to operate a 1,000 MWe coal-fired plant for a year with the one needed by a nuclear power plant of equal capacity. Limiting the comparison to surface mining, it turns out that the coal-burning plant consumes about 3 million tons of coal per year, which translates into something like 3.75 million tons of coal in place. A nuclear plant would not require this tonnage of ore to be mined until the average U_3O_8 content of uranium-bearing ore declined from the present level of a little under 0.2 percent to a range of 0.006–0.007 percent, just about the quality of Chattanooga shale. This is more than twenty times as lean as the average ore now being mined. No such low-grade material now appears in any uranium reserves or potential. The volume of solid waste products from the combustion of coal is small compared with amounts of waste from uranium. Even with scrubbers, coal plant waste constitutes less than 30 percent of the weight of the coal burned, whereas typically all but 0.2 percent of the uranium ore mined is discarded.

The comparison raises the question of whether the estimated cost of mining low-grade ore properly reflects the environmental effects of this volume of mining and processing. If the 3 million tons per 1,000 MWe relationship were to be applied to the roughly 200,000 MWe of nuclear capacity now in operation, construction, and on order (including deferred plants), the resulting requirement would be 600 million tons of uranium ore, about the same as current annual coal production. This example suggests some caution in assuming use of ever-expanding quantities of increasingly lower-grade resources.

There is little knowledge of useful uranium in the range of richness between what is currently being mined (down to 0.1 percent or 1,000 parts per million) and the truly low-grade ores which run below 0.01 percent. Even theoretical knowledge is not as yet on firm ground. The geological origins of uranium remain controversial, but are crucial to exploration. Whether sandstone deposits may have halos of lower-grade ore remains a question. Whether uranium will be found in the United States in other geologic environments in which uranium has been found in other countries is not yet clear.

An alternative to the technique of aggregating information on specific ore bodies into a resource picture is to use a "from-the-top-down" technique, which at least until recently has been the dominant method for oil and gas. An attempt to extend this technique to uranium was made in 1974 by Milton Searl, who argued that the geographic confinement of exploration and shallow drilling (predominantly to 400 feet) have produced a distorted picture. By extrapolating both depth (to 4,000 feet to begin) and areal extent, he arrived at a much greater potential. In addition, he utilizes a statistical technique which designates what has been discovered so far as the richest block of a stipulated set of blocks, each of which is smaller, by an arbitrarily set percentage, than the preceding one. Combining these techniques Searl arrived at an estimated potential ranging from a low figure of 7.7 million tons for the entire United States, with a probability of 0.5 to a high of 29 million tons with a probability of only 0.05.

Criticism of Searl's approach has been directed at (1) his treatment of depth (which assumes that material is distributed in sandstone with equal frequency throughout the first 4,000 feet), (2) the concept that total discoveries to date can be treated as a single "district," and (3) the "block-by-block" idea, derived from mature resources like oil and gas but applied in this instance to a quite immature one. Nevertheless, his analysis is a reminder not to foreclose prematurely the chances of much larger discoveries, and his effort to assign probabilities should be emulated.

Potential Geologic Sources, United States and Worldwide

More than 95 percent of present domestic uranium reserves are in sandstone. It is only in the estimates of potential resources that other host rock comes into view. Table 2-2 illustrates how decreasing certainty—represented by the three "potential" categories—is associated with the increasing importance of ore from environments other than sandstone.

Early uranium discoveries in the United States were made in sandstone, and commercial exploration stuck to what had proven rewarding. As a consequence, little drilling has been done in different host-rock environments. This U.S. emphasis on sandstone has resulted in a pattern different from that in other areas. Canada's uranium is found in veins and conglomerates; Australia's in veins and calcrete; Africa's in a variety of rocks; Sweden's in shale (at five to six times the

Table 2-2. U.S. Estimated Distribution of $30/lb. U_3O_8 Potential Resources by Host Rock (Thousand Tons U_3O_8)

Host Rock	Probable	(%)	Possible	(%)	Speculative	(%)
Sandstone	847	(80)	820	(64)	358	(61)
Conglomerate	56	(5)	76	(6)	53	(9)
Veins	100	(9)	202	(16)	162	(27)
Limestone	16	(2)	5	(<1)	13	(2)
Lignite	15	(1)	2	(<1)	4	(<1)
Volcanic Rocks	26	(3)	165	(13)	0	
Total	1,060	100	1,270	100	590	100

Source: Energy Research and Development Administration, *Uranium Industry Seminar,* October 7–8, 1975 (Grand Junction, Colorado, 1975), p. 113.

concentration of Chattanooga shale); Brazil's in granite and perhaps conglomerates. The NURE program is designed especially to produce information on areas outside the conventional western uranium districts in Wyoming, Colorado, New Mexico, and Texas and on geologic environments not previously known to be productive in this country. Experience suggests that discoveries will follow more extensive surveying, exploration, and accumulation of knowledge. Conventional wisdom puts the elapsed time between exploration and production at eight years. One should not, therefore, expect quick results; nor is it possible to predict costs. In other countries, however, some of these different geologic settings are yielding competitively priced uranium.

Geologists have made judgments as to geologic environments in the United States in which uranium is likely to be encountered, but most of these settings have yet to be thoroughly investigated. Among them are calcrete, massive vein-like occurrences, alaskite, and possibly syenite. Conglomerates in the Great Lakes region, adjacent to Canadian deposits, are possible uranium sources, and so are veins. Uranium deposits too expensive to recover (i.e., substantially above $100/lb) are found in shale and granite, both abundant in this country.

A possible change in focus is found in the suggestion by Frank C. Armstrong of the U.S. Geological Survey in a contribution to the IAEA's 1974 *Formation of Uranium Deposits*. He suggests that "porphyry" uranium deposits, such as those of the huge Rössing mine in Southwest Africa, may become the chief source of the world's future uranium supply. Geology suggests that such deposits may exist in the United States.

Non-U.S. uranium data are even less detailed than those on U.S. resources. They are divided into only two classes: "reasonably assured," which corresponds roughly to reserves, and "estimated additional," which is comparable to the U.S. "probable"—i.e., the least uncertain portion of the U.S. potential category. Table 2-3 summarizes estimates of world uranium reserves and resources. While there are some differences between quantities published by OECD and by ERDA, these differences are small, and Table 2-3 uses the ERDA tabulation in which, as in U.S. statistics, data for the higher-cost category include the lower-cost ones.

Foreign reserves of 1.1 million tons of U_3O_8, up to the $15 cutoff level, plus an additional 680,000 tons up to the $30 cutoff level, compare to U.S. reserves of 430,000 and 210,000. Estimated foreign "additional" resources, which in the United States are about 150 percent of reserves at the $30 level, are very small abroad, reflecting undoubtedly a similar lack of exploratory effort.

Current U.S. production accounts for about 45 percent of the world's production of an estimated 26,000 tons (1975). Canada contributes about 25 percent, South Africa less than 15 percent, France less than 10 percent, and Nigeria 6 percent. Gabon, Spain, Portugal, and Argentina make up the balance. OECD's projection of production capacity for 1985 is 115,000 tons for the world compared with about 30,000 in 1974. According to this projection, the

Table 2-3. World Uranium Resources and Production
(Not Including Communist Countries)
(Thousand Tons of U_3O_8)

	Resources		Production (From all Material)	
	Reasonably Assured	*Estimated Additional*	*1974*	*1975*
Up to $15/lb U_3O_8				
Australia	430	104	–	–
S and SW Africa	242	8	3.5	3.4
Canada	189	394	4.4	6.1
Niger	52	26	1.6	1.6
France	48	33	2.1	2.2
Algeria	36	–	–	–
Gabon	26	6	0.6	1.0
Spain	13	11	0.1	0.2
Argentina	12	20	0.1	0.1
Other	56	26	0.2	0.2
Foreign Total	1,100	630	12.6	14.8
United States	430	655	11.5	11.7
World Total	1,530	1,285	24.1	26.5
Up to $30/lb U_3O_8				
Australia	430	104		
Sweden	390	–		
S and SW Africa	359	96		
Canada	225	887		
France	71	52		
Niger	65	39		
Algeria	36	–		
Spain	30	55		
Argentina	27	50		
Other	150	110		
Foreign Total	1,780	1,390		
United States	640	1,060		
World Total	3,310	2,450		

Sources: Energy Research and Development Administration, Mid-1976, communicated by John Patterson; and Organization for Economic Cooperation and Development, *Uranium: Resources, Production and Demand,* A Joint Report by the OECD Nuclear Energy Agency and the International Atomic Energy Agency (Paris, December 1975). The OECD/IAEA statistics employ a cost concept that is closer to full costs than that used by ERDA, except for Canadian resources, which are based on prices. The resulting lack of comparability, according to OECD/IAEA, is "not significant" within this general range of uncertainty in these estimates.

future composition, by countries, would change mainly by emergence of Australia and a relative decline of both Canada and France, to look as follows:

United States	46%	Australia	6%
South Africa	16%	France	4%
Canada	13%	All Other	8%
Niger	7%		

Recent information from Canada, however, indicates that, contrary to the OECD projection, Canadian uranium may become more rather than less important.

There are undoubtedly large amounts of uranium in the Soviet Bloc and China. The Soviet Union has developed large nuclear military and power programs using its own uranium and that of Eastern Europe, but there is no reliable information available on Soviet reserves and resources.

Prospects for Increasing Uranium Supplies

The history of most mineral industries reflects a period of pessimism after the cream has been skimmed and the obvious locations investigated, and before new ideas have turned explorers to new structures and minerals. The search for petroleum was extended to ever-deeper strata, different traps, and from onshore to offshore fields; for many years reserves rose faster than needed to replace production. Copper ore mined one hundred years ago contained 3.5 percent copper; today it contains 0.5 percent and is produced at real costs not much higher. The immature uranium industry, which has barely "gone commercial," is not likely to prove the exception.

It is also a general experience in the mining industry that in periods of rapidly rising prices the updating of resource estimates lags, especially when there is great uncertainty about demand, which dampens the incentives for exploration and development. Thus, the failure of estimates to show an immediate rise should not be interpreted as an exhausted resource base. Actually, uranium reserves have already risen impressively, if the progressively higher cost cutoffs are properly taken into account. In a recent paper, John Patterson of ERDA shows that U.S. reserves have risen from 195,000 tons in 1965 to 430,000 in 1976, despite production of 130,000 tons and without reserves above the $15 cutoff.

ERDA is dependent upon information from industry for reserve and much of its resource estimates, and government statistics may not reflect all the knowledge possessed by industry. ERDA/industry relations are close, but ERDA's dependence on industry for information is unfortunate, given the critical role uranium resource estimates play in the formation of energy policy. The government's "need to know" is greater than industry's sense of what is worth communicating.

Recent events have stimulated the uranium mining industry. Drilling is up sharply, in both physical and dollar terms. Expenditures for exploration in the United States rose from $12.2 million in 1975 to $32.4 million in 1976, and are planned to go higher. The NURE program, including airborne radiometric reconnaissance, hydrogeochemical surveys, stream sediment sampling, and geologic, geochemical, and remote sensing investigations, is expected to add significantly to exploration.

The national character of NURE enables it to widen attention from the Rocky Mountain area to include the nation as a whole. ERDA's uranium exploration budget is now one-third the size of the budget available to the U.S. Geological Survey for all its mineral resource surveys. With the exception of the National Laboratories, which are funded directly by ERDA, all NURE funds are allocated to a prime ERDA contractor, Bendix Field Engineering Corporation, which subcontracts with a variety of organizations. The findings, most of which are based on techniques other than drilling, are then made publicly available, in the hope that mining companies will proceed from ERDA's work to the kind of exploration that confirms reserves and potential resources.

While it is too early to evaluate the NURE program, it does appear that its failure to work closely with the U.S. Geological Survey represents poor use of government resources. The USGS has spent some $5 million annually for at least a decade to develop an improved methodology of resource estimation. The AEC, and now ERDA, have worked independently on their own estimates with the result that these estimates are now conceptually incompatible with those of other resources. Further, the USGS is staffed by regional specialists whose knowledge would be particularly valuable.

A major role for the USGS in preparing uranium estimates would also refute the suspicion that ERDA's narrow "sandstone focus" is not accidental but has been associated with the agency's drive toward the breeder. While understandable when the market for uranium was exclusively military, the buildup of a separate, large "Uranium Geological Survey" is a questionable procedure.

The results of the different government surveys are made available to industry with the expectation that mining companies will engage in further necessary field work, especially drilling, to delineate new reserves and resources. It is not clear, however, that there are incentives to invest in exploration in unconventional environments. The mining industry has traditionally been reluctant to tie up capital beyond the minimum required to support operations. Furthermore, industry undoubtedly knows of higher-cost sources that could be attractive if markets rise sufficiently. If accurate estimates of uranium resources are needed as a basis for future energy policy decisions, it may be advisable for government itself to carry out the search that would lead to the necessary information.

While estimates could be changed dramatically by new exploratory efforts, they could also be affected by new technology which could extend current supplies. For example, laser enrichment might permit more complete extraction of uranium-235 from uranium. This would extend uranium resources by up to 50 percent; even the existing "tails" (discarded residue from past enrichment) might be economically reprocessed.

Mining and Milling: Reaching Required Goals

Ascertaining uranium resources is one thing; producing uranium is another. At the likely rate of increase in demand, investment in drilling and other facilities is critical. Manpower recruitment, new mills to process the ore, and developmental drilling are required to meet the demand for uranium fuel. Increases will have to be 10 percent per year for a number of years. These are not rates easily sustained. Moreover, the uncertainties are likely to depress investment in processing facilities.

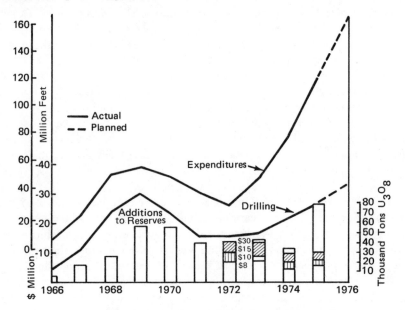

Figure 2-2. U.S. Exploration Activity and Plans

Source: Personal Communication, John A. Patterson, ERDA.

Some indication of increased activity can be gleaned from Figure 2-2, which shows the recent upswing in drilling together with associated additions to reserves, a development that runs counter to the concept of a fade-out of reserve additions, typically derived from a faulty interpretation of ERDA statistics.

Concern about a uranium shortage might be more productively directed toward possible processing limitations. Fortunately, society has better means of dealing with industrial constraints than with a lack of natural resources.

Evaluating the Estimates Against Requirements

Even though estimates of requirements are very uncertain and have recently been changing significantly, they are easier to predict than supplies. Requirements need to be described in terms of specific assumptions such as capacity factors, tails assays, fuel recycling, and reactor mix. Differences in tails assay can affect requirements by 25 to 50 percent. If the reactor mix included many breeders, this would overwhelm the other variables. To test the range of supplies that might be required, we have examined a number of scenarios without assigning any probability to them. In these examples, assumptions are specified for purposes of clarity, not as predictions of likely future developments.

To Meet Demand Projections for Currently Planned Capacity. As our first example, we assumed that all plants now in place, under construction, or on order (including deferred plants) operate under a reasonable set of conditions.[c] This program, which amounts to about 200,000 MWe when complete, would call for 35,000 tons of U_3O_8 per year. Some 1.1 million tons would be needed to support this full capacity for thirty years, plus about 200,000 tons during the buildup to 1985-90, for a total of 1.3 million tons required from now through the year 2015 or 2020.

On the basis of the estimates shown in Table 2-1, requirements could be met in several ways:

1. Reserves plus probable potential resources, both up to the $30 cutoff;
2. Reserves plus probable and possible resources, all up to the $15 level; and
3. Other combinations, including reserves above the $30 cutoff.

While the scenario implies large demands on U.S. resources, it requires no resort to speculative resources or to probable and possible resources above the $30 cutoff. Each additional 1,000 MWe facility would, over a thirty-year operational period, add 5,300 tons to total requirements; and the entire 200,000 MWe program would add 35,000 tons every year beyond 2015 or 2020.

To Meet the ERDA "MID" Scenario. As our second example, we compared the total supplies with the requirements that would develop under the mid-range growth scenario that ERDA has used since mid-1976 (the so-called MID scenario). This scenario, which results in an installed capacity of 510,000 MWe in the year 2000, would require the tonnages of U_3O_8 shown in Table 2-4 under conditions similar to those of the previous example.[d] The 1.2 million tons to the year 2000 could be met by various combinations. One would be reserves to the $30 cutoff plus half the probable resources. Alternatively, a combination

---

[c]0.2 tails assay, 65 percent average lifetime capacity factor, and no recycling.

[d]A variable capacity factor that works out at about 65 percent lifetime average, 0.2 tails assay, spent fuel recycled only for what ERDA calls the "nominal" quantities of uranium to be used in LWRs, and a breeder introduction so close to 2000 (i.e., 1995) as to make little difference for requirements up to the year 2000. Excluding what little uranium recycling is included in this ERDA scenario would raise the estimates only slightly, since that process would not begin until the mid-1980s at the earliest.

Table 2-4.   Annual and Cumulative U_3O_8 Requirements United States, 1975–2000 (ERDA MID Scenario With 0.20 Tails)

	1975	1980	1985	1990	2000
Installed Capacity	39,000 MW	67,000 MW	145,000 MW	250,000 MW	510,000 MW
U_3O_8 Requirements, Annual (Thousand Tons)	9	19	36	55	82
U_3O_8 Requirements, Cumulative (Thousand Tons)	9	73	220	455	1,195

Source: 1975:   Edward J. Hanrahan, Energy Research and Development Administration, *Demand for Uranium and Separative Work,* Atomic Industrial Forum Fuel Cycle Conference, 1976, Phoenix, Arizona.

1980–2000:   Edward J. Hanrahan, Richard H. Williamson, and Robert H. Bown, *World Requirements and Supply of Uranium,* September 14, 1976, AIF International Conference on Uranium, Geneva, Switzerland.

of only lower-cost reserves plus less certain lower-cost potential resources could meet the demand. In either event, conversion of potential resources to reserves would be required to meet demand beyond 1990. Reserves above the $30 cost cutoff are a further resource in the years beyond 1990. "Probable resources" are a close relative of "inferred reserves" and are often regarded as reasonably certain in analyses of other minerals.

Recycling of uranium and plutonium would reduce U_3O_8 requirements by about 300,000 tons to 900,000 tons at 0.2 percent tails. Recycling would ease demand, but reserves would still have to be augmented by the "probable potential" category.

To Meet the ERDA "MID" Scenario Projected to 2030. As our third example, we estimated what would be required for the capacity built up by the year 2000 under the MID scenario if it were to run for an additional thirty years, to 2030. Assuming similar conditions (0.2 tails and 65 percent average lifetime capacity) and no recycle or breeders, the 510,000 MWe existing in 2000 would call for 90,000 tons annually, or 2.7 million tons in thirty years. To this amount must be added the 1.2 million tons required between 1975 and 2000, the period when capacity builds up to 510,000 MWe, for a total requirement of 3.9 million tons from now to 2030. Current estimates of reserves and all classes of potential total 3.6 million tons (see Table 2-1), some 10 percent less than the calculated requirement.

Mineral history supports the expectation that increases will be forthcoming and potential quantities will be confirmed, especially if moderately higher-cost resources are included. It is true, nonetheless, that at this time these quantities are speculative. What adds unusual pressure to demand is that utilities are attempting to secure an assured lifetime supply of uranium for each plant. These attempts have been intensified under the impact of Westinghouse's announcement in September 1975 that it could not make good on its uranium supply commitments at contracted prices. In the pursuit of assured supply, some utilities have become partners in mining ventures. Abroad, Canada has decreed a thirty-year set-aside of domestic reserves for each domestic reactor. The vicious circle is obvious: a temporary tightness sends consumers into panic buying, only to aggravate the tightness. Statistics showing purchase commitments add further to a sense of impending shortage, as they reveal the extent to which future requirements are not covered by contracts in the more distant years. Greater reassurance with regard to uranium resources would take some pressure off the drive to be covered for decades ahead. There is no comparable desire to have assured contracts for coal, oil, and gas, although oil prices have quadrupled in the past few years and there is no indication what prices will be in thirty years.

Even for those who believe that the published resource statistics accurately describe the supply of uranium, judgments on adequacy depend on the confidence ascribed to potential resources. It seems to us that the "probable" category is not far from reserves, since the inferences involved are very likely correct. This is somewhat less true for the "possible" group, although even here the estimates rest on sound geologic reasoning. Fortunately, the "speculative" category represents only 15 percent of the total, and its level of uncertainty is not critical. In all, one would need to be excessively distrustful to draw conclusions of inadequacy from the present data.

Expressed public concerns about running out of uranium in the next decade appear to be based on only $8 or $10 reserves or on assumed milling bottlenecks or enrichment constraints. It is hard to see how they can rest on resource considerations. Even a continuing, steep ascent from the current annual production of 11,000-12,000 tons would not create supply problems within the next decade. What distorts the debate further, and is worth reemphasizing is, that the discussion of uranium adequacy takes place in the context of the unique cost definitions that the AEC introduced early in its existence and that ERDA has inherited, as explained earlier.

A few simple assumptions help to illustrate these costs. As shown in Table 2-1, 270,000 tons of U_3O_8 are available at cost up to $10; since the government's price as recently as 1971 was only $6, the average (partial) costs in this group are unlikely to exceed an average of $5, or else sellers would not have sold. There are available 160,000 tons of U_3O_8 with costs between $10 and $15 —let us say $12 on the average—and 210,000 tons with costs between $15 and $30, averaging, let us say, $23. The weighted average cost for the entire quantity of 640,000 tons is about $13. Applying a rule-of-thumb add-on of 70 percent to obtain full cost will yield $22 as the average full cost of recovery for all reserves.

This kind of calculation helps to explain why "$30 uranium" is compatible with a reasonable price of $40 for some time.

A Perspective on Prices

It is remarkable how little is known about the present and future costs and prices of uranium. Since every producer knows his present costs and how such costs relate to prices, a large body of relevant information obviously exists. Even a small fraction of the funds spent on technology could make this information available and improve the information base that critically influences major energy policies.

Fortunately, since a $10 difference in uranium prices translates into less than a 1 mill difference in the cost per kilowatt hour of nuclear-generated electricity, wide price ranges have only small impact on the economic comparisons. The indicators and inferences utilized here suggest that the price of uranium contracted for by the end of the century is likely to be no lower than $40 (in 1975 dollars) and will probably be higher, although the average price of uranium burned in reactors in 1999, which was contracted for many years earlier, could be lower. The distinction between these different types of prices is important and often not specified by estimators. The range in any year can be wide. Today, prices vary from spot prices (for various future delivery dates) of $40 and more to a range of annual average prices for all uranium now committed to buyers (in various years in the future) of $11 to $19. This range undoubtedly reflects commitments within the low-cost category as shown in Table 2-1.

The information which helps define a plausible price range includes:

1. The forward cost classification described above;
2. The rate at which a given nuclear capacity scenario might deplete the reserves and resources now identified;
3. A recent estimate of typical costs for recovery of U_3O_8 from 0.15 percent uranium ore; and
4. Estimates made in the 1976 GESMO report.

Large discoveries of low-cost uranium, heavy imports of low-cost uranium, improved technology for enrichment, and a slower buildup of capacity could all lower the estimated price range. Overestimation of reserves, inadequate investment in producing facilities, stricter environmental or other cost-increasing regulations could raise it.

First, let us conjecture what the full cost of U_3O_8 might be as demand taps the $15-$30 increment of reserves and resources. As production approaches the $30 boundary, full costs will approach $50; at the same time, the average full costs of all reserves up to $30 are unlikely to be lower than $22.

The manner in which the industry will meet future demand has much to do with the price at which demand will be met. Will the lowest-cost reserves and lowest-cost "probable" potential be used up first, the medium cost next, and so on? Or will all reserves be used first, regardless of cost, before potential is tapped? Since most higher-cost ore is collocated with low-cost ore, it may be coproduced. If not, it would have to be recovered later at higher cost. Thus, the most probable course is a movement in both directions, from the lowest-cost, best identified reserves up the cost ladder, and also "sideways" toward less well-delineated, low-cost deposits.

To produce the 1.2 million tons required for the ERDA year 2000 MID scenario, production must reach into the $15-$30 class. However, the more it has recourse to reserves only, the faster it will approach the $30 boundary (equal to $50 full cost). That is as far as we can conjecture, and a price approaching $40 for the year 2000 does not seem unlikely. The average for the entire period to 2000 would certainly lie below $40. At the same time, contracts made in 2000 for future delivery are likely to be higher.

Second, a recent ERDA study[e] suggests production *costs* of 0.15 percent

[e]*U.S. Uranium Production Outlook,* John A. Patterson, September 14, 1976, AIF International Conference on Uranium, Geneva, Switzerland.

grade ore of $10 to $18, depending on whether the material is produced by surface or by underground mining. Since the average grade of all material up to the $30 cutoff level is not 0.15 but 0.08 percent, costs are bound to rise. A portion of the cost should double, since nearly twice as much material needs to be moved and processed; other cost elements will remain stable or rise only moderately. The range for 0.08 percent ore might thus rise from $10-$18 to, say $17-$30.

Third, the GESMO suggests that the price will rise to between $33 and $37 by the end of the century. This projection rests strongly on the low prevailing average price, which has barely begun to reflect the current, high spot prices, and an assumed mining sequence in which low-cost blocks are mined out before higher ones are tapped. On the first point, experience in the coal market suggests that high spot prices tend to erode low contract prices. Therefore, the average uranium price will probably rise, even if spot prices go no higher or even decline. On the second point, one must consider the effects of collocation discussed earlier.

Finally, one should recall that between 1955 and 1973 the price of uranium was cut in half, in current dollars, and even more in constant dollars. The price was bound to bounce back at the first sign of tightness. Moreover, there is a natural tendency for one fuel to escalate in sympathy with its competitors. Coal at $20 a ton no more reflects a corresponding rise in the underlying cost of all coal than the U_3O_8 at $40 reflects the cost of all uranium. Marginal extraction plus scarcity cost are what matters.

Some information is bound to emerge from the current spate of investigations and litigation. Ever since Westinghouse, citing its inability to find sufficient low-cost material, announced that it would not deliver uranium at previously contracted prices, the market has been in turmoil. In these circumstances, it is wise to contemplate future sales transactions with considerable caution. The current suits and countersuits by utilities, by mining companies, and by Westinghouse against more than two dozen domestic and foreign producers, center on availability, costs, and prices. These cases should throw light on the real rather than the statistical world. The jump in spot prices from $6 to $40 has prompted the Justice Department to suspect worldwide collusion and to begin an antitrust investigation. (These developments by themselves could affect future prices.)

The data here developed suggest that in this century the full-cost level should move toward $25, and the price, beyond $25. However, these movements will depend on the urgency with which utilities insist on securing a committed fuel supply two or three decades ahead. The combination of a consumer wishing to be safe for decades ahead and a supplier traditionally extending its planning horizon by no more than a decade could push prices into the $40 range sooner than the above speculations suggest.

Non-U.S. Uranium Requirements and Foreign Trade

Until recently, imports of uranium into the United States were barred to protect domestic mining, and foreign supplies were therefore of only academic interest, even though foreign demand constituted a drain on domestic resources. This has changed with the 1974 decision to allow imports. Knowledge of the situation outside the United States is now relevant to U.S. supply.

On the basis of two recent projections of uranium requirements outside the United States, identified reserves plus potential in the rest of the world appear sufficient to support the growth of nuclear power in about the same manner as is true for the United States. These projections, by OECD and ERDA, include uranium up to the $30 cutoff level. Though based on similar scenarios,[f] the two estimates have recently diverged significantly.

Each of the elements is subject to question. The demand projections are reduced with each revision. The resource compilations reflect an exploration effort of relatively short duration, especially outside the United States. Inclusion of possible and speculative resources outside the United States—not now calculated or reported—might add several million tons. Table 2-1 shows that in the United States the ratio between potential resources and reserves, at the $30 level, is nearly five to one. If this ratio were valid outside the United States, potential resources there would amount to over 8 million rather than 1.4 million tons. This highlights the state of affairs in foreign resource estimation, which also appears to underestimate supplies substantially, questions of statistical comparability aside (see Table 2-3).

A recent ERDA paper, presented by Robert J. Wright at the 1976 Grand Junction Seminar in October 1976, puts feasible non-U.S. production capacity at nearly 80,000 tons by the mid-1980s. Such a level would be in the neighborhood of ERDA-projected requirements for "the rest of the world." However, the implied step-up has rather heroic proportions. For example, Australia, which has the largest reserves, has no production at this time. The ERDA speculation includes Australian productive capacity in the mid-1980s of 25,000 tons, African at 26,000+ tons, and Canadian at 15,000 tons. Canada's potential is uncertain because of rules requiring setting aside thirty-year fuel requirements for all reactors scheduled to be in operation ten years ahead. On the other hand, the Canadian government sees large future export opportunities.

[f]The ERDA estimate is the low-growth scenario and makes the following assumptions: (1) breeder commercially available in 1988; (2) capacity factor: 72 percent for developed, and 60 percent for developing countries with an average of just below 70 percent; (3) fuel reprocessing assumed to develop so as to work down backlog of unreprocessed spent fuel by the mid-1980s; (4) plutonium recycling starting in 1978 but constrained by processing capacity as under (3) above.

OECD estimates assume breeder introduction in the mid-1980s, but with no significant impact until the mid-1990s. Lifetime average capacity factor is 70 percent, and tails assay 0.25 percent. Both plutonium and uranium recycle to start in 1981.

In recognition of the uncertainties in these figures, Table 2-5 brings together estimates of resources vs. requirements to the year 2000. Much rides on the difference between the ERDA and OECD requirement estimates. If ERDA's lower requirement, even at the 0.30 tails assay level, is more nearly correct—as we believe is likely—then U.S. needs could be substantially aided by foreign supplies. If the OECD's estimate is accepted, the United States and the rest of the non-Communist world exhibit nearly the same relation between requirements and the sum of reserves and "limited potential" resources, both up to the $30 cutoff. However, at the "full potential" level, which admittedly is highly conjectural, the rest of the world is far better off, suggesting a large export potential.

Resource estimates are beginning to rise. An announcement in February 1976, following recent price increases, stated that the reserve estimate for Australia's Jubileko area, one of the world's largest uranium deposits, had been boosted from 115,000 to 228,000 tons. Another indication is the most recent authoritative reevaluation of the Canadian situation: after taking care of domestic needs and export commitments, about half of Canada's reserves are ". . . uncommitted to meet future export or domestic needs."

Freed only recently from a long-standing import ban, U.S. uranium import commitments in less than two years have reached about 41,000 tons through 1990, as against 11,000 tons of exports through 1979, with both amounts spaced quite evenly throughout the periods. In 1974 alone, contracts were made for future imports of 33,000 tons. This was more than twice the increase in purchase commitments from domestic suppliers in that year. Prospective imports have been boosted by the government's decision to phase out restrictions on U.S. enrichment of foreign-origin uranium for domestic use, to begin in 1977 and to continue until restrictions are fully removed in 1984. The United States can greatly expand its imports before falling into the kind of dependency that it is trying to avoid in the case of oil.

Summary and Conclusions

Uranium mining is an infant industry. Evaluated in the context of history and with an awareness of the peculiarities of the data, existing uranium resource estimates emerge as biased significantly on the low side. It is our judgment that, if there are any grounds for fear of shortages, it is not a lack of resources that gives rise to them, but rather a lack of incentives for exploration and development. The combination of a conservative industry and customers anxious for protective long-term contracts does not make for a broad-based effort. The constraints on uranium supply are not physical but economic constraints, amenable to resolution once the scope of the necessary effort is perceived.

We see no reason why uranium resources could not be developed to supply fuel for LWRs, up to and beyond the end of the century, at costs that would not exceed by much the spot prices at which small quantities have been sold

Table 2-5. Resources vs. Requirements to 2000 (Thousand Tons U_3O_8)

| | Resources | | | | | Requirements[b] |
| | Reserves | | Limited Potential | | Full Potential[a] | |
	$15	$30	$15	$30	$30	
United States	430	640	655	1,060	2,920	1,200
Rest of Non-Communist World	1,100	1,780	630	1,390	8,120	1,660/1,980 2,650

[a] From Table 2-1, for U.S., and estimated for Rest of World on assumption that total potential (all three U.S. categories) bear the same relation to reserves as in the U.S.

[b] U.S. is ERDA MID scenario; Rest of World low is drawn from ERDA, Edward J. Hanrahan, Richard H. Williamson, and Robert H. Brown, *World Requirement and Supply of Uranium*, September 14, 1976, AIF International Conference on Uranium, Geneva, Switzerland, depending on tails assay assumption; and Rest of World high is OECD, *Uranium: Resources, Production and Demand*, A joint report by the OECD Nuclear Energy Agency and the International Atomic Energy Agency (Paris, December 1975).

in 1976 (the $40 range). The panicked market that developed in the wake of Westinghouse's refusal to deliver on past contracts does not reflect the long-run situation. It does, however, point to the need to build up our information on uranium occurrences across the nation and abroad and their likely cost. The NURE program is a valuable contribution to this effort. Its findings could play a key role in policy decisions. NURE should be strengthened, however, by a closer relationship with the U.S. Geological Survey and by better arrangements for industry follow-up to make certain that geological findings are speedily translated into specific and concrete resource knowledge.

Foreign sources of supply are large and growing. The door has been opened to imports, and much larger imports may be desirable, especially in the light of the slippage in foreign requirements. The slow schedule for full removal of import restrictions could be revised in light of these considerations.

OIL AND GAS SUPPLIES

While the last word on oil and gas has not been spoken, these two energy sources are unlikely to be available in such quantities as to meet growing demands for energy. In the long perspective of history, natural oil and gas will have been short-lived, important only from the mid-nineteenth into the early twenty-first century.

Efforts to determine oil resources, which have been found as much as five miles below the earth's surface, are fraught with uncertainties and frustration. Estimates are made only to be criticized, revised, buried, exhumed, and generally held in low esteem, especially by those who do not like the results. A large new discovery discredits low projections, and declining reserves prove high estimates wrong.

There are even fashions. Since early 1975, those predicting shortages have taken over, heralding the disappearance of commercial oil and gas in as little as a couple of decades. This view is countered by others who point to a long string of earlier predictions of shortage, each of which has so far turned out to have been premature. While we have not made an independent judgment of reserves and resources, it seems clear that even on optimistic estimates, oil and gas cannot support a moderately growing energy demand for an extended period, either in the United States or in the rest of the world.

For the United States, projections in the 1950s and early 1960s, when oil and gas production was still rising and resource estimates were large, suggested that oil and gas would become a diminishing segment of the U.S. energy mix toward the end of the century. This judgment was based on estimates of the U.S. Geological Survey that discoverable and recoverable oil resources might measure 200 billion barrels and possibly twice that. In the meantime, the country has continued to draw ever more heavily on oil and gas and less on coal, and there has been a swing toward radically reduced estimates of remaining discoverable, recoverable oil and gas. In mid-1975, the U.S. Geological Survey revised its U.S. oil estimates sharply downward. Instead of a 200–400 billion barrel range, the USGS saw a 95 percent chance that only 50 billion barrels of oil would be discovered and recovered and only a 5 percent chance that the figure would be as much as 127 billion barrels. Taking the mean, computed by the USGS at 82 billion barrels of remaining undiscovered recoverable resources, and adding the measured, indicated and inferred reserves of about 62 billion barrels, one obtains a total of 144 billion barrels of potentially available oil. This potential supply could be augmented by greater success in discovery and by greater efficiency in recovery stimulated by higher prices.

U.S. oil consumption has been running at just under six billion barrels per year. The "life index" of reserves and resources at current rates of consumption, disregarding imports, is thus about twenty-five years. Holding imports to a level no higher than the current rate of 40 percent of consumption, and holding consumption constant, would extend the life of U.S. reserves to about forty years. Two-and-one-half billion barrels a year of non-U.S. oil would have to be imported.

The natural gas situation in the United States is similar, except that U.S. imports are a much smaller component of consumption, now and probably in the future. The time horizon has shortened rapidly; the ratio of reserves to production has declined from over twenty to one to about ten to one over the past fifteen years.

These are simplistic calculations: dividing resources or reserves by the annual consumption to obtain a "life index" is equivalent to assuming output at a given rate followed by an instantaneous decline to zero. Nevertheless, the figures do show that domestic oil production is unlikely to maintain its share of even constant per capita energy consumption to the end of the century and surely will not maintain it much beyond that. Any increase in per capita oil or aggregate energy consumption will have to be met by oil imports or by other energy sources. The prediction for natural gas is worse. Shale oil continues to be beset by technological and environmental problems, and its prospects are receding rather than advancing. Thus, inevitably, attention must turn to coal and nuclear power or other non-fossil fuel alternatives.

This is not to say that one must write off oil and gas. On the contrary, these two fuels still furnish together 75 percent of total primary energy. While that share is bound to decline, they will still be the largest source of energy through this century. For example, a simple calculation suggests that, if total annual energy consumption rose 50 percent by the end of the century and oil imports were not allowed to rise above the current share of about 40 percent of oil consumption, oil and gas would still furnish about half of the total. Such an estimate assumes, however, that the declining output of domestic oil and gas, in evidence since the early 1970s, will be halted by new, successful exploration and by higher rates of recovery from known fields.

An assessment of the world oil and gas situation is much more complex. Many areas have been poorly investigated, such as large portions of Latin America. Other areas, especially in the Middle East, are so rich in reserves that little effort has been made to evaluate the potential or to be concerned with higher rates of recovery. Demand in most industrialized countries since the early 1950s has shifted so radically toward oil that one must ask how rapidly this dependence can be reoriented to alternative energy sources.

As of 1975, non-Communist world oil reserves, other than those of the United States, stood at about 450 billion barrels. Discoverable, recoverable oil resources might be put at about two-and-one-half times that amount. Annual world oil consumption in the non-Communist, non-U.S. world totaled about 11 billion barrels, and demand is likely to see a much greater rise than is projected in the United States. U.S. imports, now near 2.5 billion barrels a year, will constitute a drain on the world's oil. Assuming that the potential resources will be discovered and drawn upon, production of oil at moderate growth rates of, say, 5 percent per year, would peak within the first quarter of the next century.

These general observations suggest that oil and gas will be declining in significance over the next fifty years. Absolute declines in worldwide output are likely to occur not too far into the twenty-first century, if not sooner.

COAL RESOURCES

Energy for the United States in the period after 1990 will be characterized by a much greater reliance on coal. This will be true even if there is a rapid expansion in nuclear power and whether or not the United States becomes less dependent on foreign oil. Unless much larger new reserves of world oil and gas are discovered, depletion of these fuels will force a transition to other energy supplies early in the next century or sooner.

The relative future roles of coal and nuclear power remain to be determined. While coal reserves are very large, extracting and using coal in an environmentally acceptable way presents many difficulties. Moreover, the orderly transition from an oil- and gas-based economy to one based on coal or nuclear power will require long-range planning.

The histories and structures of the coal and nuclear industries are entirely different. The nuclear industry is new, has been created mostly through government efforts, and has always been closely regulated. The coal industry is over one hundred years old in the United States and much older elsewhere, has been developed entirely by the private sector, and has traditionally been subject to very little government control.

During its long history, the coal industry has passed through a variety of phases. Until well into this century, coal was the primary domestic source of energy for both stationary and mobile power. As early as 1885, the coal industry was producing over 100 million tons per year. Production continued to increase steadily, reaching a level of about 600 million tons per year in the early 1920s. As a result of cheap oil and the depression, coal production declined. At the outbreak of World War II it was less than 400 million tons per year. After returning to 600 million during World War II and the immediate postwar period, output declined again during the 1950s with the appearance of very cheap Middle East oil, natural gas, and the conversion of the railroads to diesel locomotives. Over the past twenty years production has averaged between 500 and 550 million tons annually. In this process, coal has declined from 50 percent of U.S. energy supply in the mid-1940s to less than 20 percent in the early 1970s. As oil prices have increased, coal is again becoming an economically attractive option.

The Nature of Coal

Coal is a complex and varied mixture of solid organic compounds. Each coal bed differs from others in both physical and chemical properties. Even within the same seam there are variations. The major chemical variations are the amounts of carbon, hydrogen, oxygen, sulfur, water, and nitrogen, the percentages of volatile matter and fixed carbon, and the amount of ash and its chemical constitution. The major variations in the physical properties of coal are in weathering characteristics and in the nature of coking properties. Because of these great variations coal is classified according to its fixed carbon content, heat content, volatile matter, and agglomerating character.

The wide variety of the properties of different coal deposits requires that new processes using coal be tested for their compatibility with coal from specific seams. For a particular process, the environmental residues depend on the coal that is used. These widely differing properties require caution in generalizing about coal as an energy resource for different uses. Heating value, sulfur content, and ash properties are important in selecting coals for combustion to supply space and process heating and to generate electricity; coal reactivity and heating value affect processes for conversion to liquid or gaseous fuels; and coking properties and ash and sulfur content determine which coals should be used in the production of coke.

The International Coal Situation

As in the United States, coal resources in the rest of the world—considered collectively—are much larger than those of oil and gas. Estimates range from six to twelve trillion tons or approximately 600 to 1,200 years of supply at the 1972 world level of total energy use and for 1,600 to 3,200 years at the world level of coal use. The major known deposits are in the United States, the Soviet Union and the People's Republic of China, which together are estimated to have between 70 and 90 percent of the world total, depending on which reserve estimate is used.

The estimates of total remaining recoverable reserves of coal are thought to be reasonably accurate for industrialized countries which have had a long history of coal use. Exploration in these countries has been extensive, and it is unlikely that large new coal deposits will be discovered. The estimates for the People's Republic of China are uncertain, but the consensus is that the coal potential is huge.

On the other hand, little exploration has been carried out in Central and South America and in much of Africa, South Asia, and Oceania. The published estimates almost certainly understate the reserves. In some countries, such as Colombia, the understatement may be tenfold.

Even though Western Europe's reserves are only 10 percent of the world total, they are relatively large compared to the region's energy demand. With the fourfold increase in the price of oil, some countries in Western Europe may use more indigenous coal, although national energy policies do not so far reflect this. Both the United Kingdom and West Germany have large reserves that can be substituted for oil for electricity and at large industrial plants.

Mining research and development has received relatively little attention. In the United States the government's role in mining R&D has been greatly expanded in the past several years. Successful development of new, more economic mining methods would permit countries with coal resources to make wider use of them.

In 1975, world anthracite and bituminous coal production was 2.64 billion tons and lignite production was 950 million tons for a total of 3.6 billion tons. The Soviet Union, the United States, and the People's Republic of China together produced 53 percent of the total. Other major producing nations were Poland (187 million tons), the United Kingdom (139 million tons), West Germany (101 million tons), India (94 million tons), South Africa (76 million tons), and Australia (73 million tons). Major producers of lignite were East Germany (270 million tons), Czechoslovakia (91 million tons), Poland (44 million tons), and Yugoslavia (38 million tons). Along different lines of division, the developing countries used 5 percent of the coal, with the rest of the output divided about equally between the developed market and Communist economies.

International trade in coal has remained in the narrow range of 120 to 170 million tons per year between 1925 and 1970. In 1974, international coal trade was 210 million tons, with the United States, Poland, Australia, Canada, West Germany, and the Soviet Union being the major exporters. The major importers were Japan, France, Italy, Canada, and Belgium. U.S. exports in 1975 were 65.6 million short tons, or 31 percent of the world coal trade, the major part consisting of metallurgical coal.

The Reserve Base of the United States

The coal resources and reserves of the United States are very large, but most of the deposits have not been sufficiently explored to be classified as reserves.

The U.S. Geological Survey has identified deposits, at depths of less than 3,000 feet, of nearly 1,600 billion tons, and total resources of 4 trillion tons are surmised to exist. The demonstrated reserve base of coal in place[g] is 436 billion tons. The coal included in this reserve base is known with a high degree of precision and can be mined using current techniques. In making these estimates, the Bureau of Mines states that this amount of coal could be produced at current costs using current technology but also that "no consideration was given to marketability." In underground mining, about 50 percent of the coal can be recovered, and in strip mining as much as 80 percent can often be recovered.

Table 2-6 gives estimates of the demonstrated coal reserve base by geographic area and potential method of mining. Table 2-7 presents the information on coal rank and method of mining. As these tables indicate, about 54 percent of the total reserve base tonnage is found west of the Mississippi, but, because of their generally lower heating values, the western coals contain less than 50 percent of the total heat content of all U.S. coals. About half (44 percent) of the western coals can be strip mined, while only 17 percent of the eastern coals can be mined by surface mining methods. Nearly 85 percent of the coal tonnage with 1 percent sulfur content or less is found west of the Mississippi.

The concept of the demonstrated reserve base has several limitations. It no longer includes any estimate of the marketability of the coal, although the cost of coal will depend on many other factors besides the two that have been used to define the demonstrated reserve base—seam thickness and depth. Mining costs are also affected by the characteristics of the floor and roof, the presence or absence of methane gas, the pitch of the seam, the amount of water encountered in mining, the hardness of the coal and the type and amount of impurities, and, for strip mining, the characteristics of the overburden. Several recent studies[h] have concluded that much of the demonstrated reserve, especially underground eastern coal reserves, could be mined only at significantly greater costs. These studies make a large number of simplifying assumptions that cause the reserve base to appear smaller than it is; nevertheless, it is probably true that the reserve base estimates of the Bureau of Mines overstate the coal that can be produced with current technology at current costs.

Despite these reservations, the reserve base contains sufficient coal so that, at least until 2000, mining costs need not rise even if coal production should increase at 4.5 percent per year, to two billion tons per year in 2000. Some of the identified or estimated resources outside the reserve base can also be expected to be minable with present technology at present costs.

---

[g]As defined by the Bureau of Mines, seams that are 28 inches or more thick (for bituminous coal) and 60 inches or more thick (for subbituminous coal and lignite) to depths of 1,000 feet (120 feet for lignite).

[h]Such as the "PIES Coal Supply Methodology" conducted by ICF, Inc., and published in January 1976, and "The Supply of Coal in the Long Run: The Case of Eastern Deep Coal," by Martin Zimmerman, MIT, January 1975.

Table 2-6. Demonstrated Coal Reserve Base in the United States by Geographic Area

Location	Millions of Tons
East of Mississippi	
Underground	169,000
Surface	34,000
Total	203,000
West of Mississippi	
Underground	131,000
Surface	103,000
Total	234,000
Total	437,000
Total Underground	300,000
Total Surface	137,000

Source: U.S. Bureau of Mines, *Demonstrated Coal Reserve Base of the United States, By Sulfur Category, on January 1, 1974,* (Washington, May 1975).

Table 2-7. Demonstrated Coal Reserve Base in the United States by Potential Method of Mining and by Coal Rank

Coal Rank	Millions of Tons	
	Underground	Surface
Bituminous	193,000	40,000
Anthracite	7,000	1,000
Subbituminous Coal	100,000	68,000
Lignite		28,000
Total	300,000	137,000

Source: U.S. Bureau of Mines, *Demonstrated Coal Reserve Base of the United States, By Sulfur Category, on January 1, 1974,* (Washington, May 1975).

The development of improved coal mining technology could result in large increases in the size of the demonstrated reserve base, minable at current costs. Major improvements in coal mining methods have been introduced by the industry over the past fifty years, such as mechanized loading and underground transport, continuous underground mining machines, large draglines for removing overburden, and the ever expanding size of strip mining equipment. All of these developments, however, were largely evolutionary, and little has been done to apply scientific principles through specially designed R&D programs to improve coal mining practices. A large-scale, comprehensive R&D program could develop entirely new mining methods that could result in reduced mining costs. Such a program has been initiated by the U.S. government and includes studies on health and safety and advanced mining technology. As the coal mining industry expands, it should be able to supplement this government-sponsored program and accelerate the introduction of new technology.

Trends in Coal Industry Performance

Table 2-8 presents salient statistics of the U.S. coal industry, from 1970 to 1975. After fluctuating around 600 million tons per year, production rose to 640 million tons in 1975, about equal to production in the peak year of the 1940s. The price of coal, f.o.b. the mines, increased sharply during this period from $6.26 per ton to $15.75 per ton. Hand loading of coal underground virtually stopped and the percentage of coal produced by strip mining methods continued its long-term increase, reaching 54 percent in 1974. Productivity in underground mines continued to decrease as a result of the stringent provisions of the 1969 Coal Mine Health and Safety Act and the new United Mine Workers agreement signed in late 1974. These developments also slowed the long-term trend of rapidly rising productivity in strip mines.

Consumption of coal has been concentrated in two large markets—electric power and steel. Together these consume 85 percent of total coal production. The only growing market has been electric power, and this market is subject to competition from nuclear fuels and is affected by environmental regulations.

The Structure of the Industry

For a long period, the coal industry was characterized by a large number of small individual firms. The barrier to entry into the industry was low because coal deposits were large and widely dispersed and only a small amount of in-

Table 2-8. Bituminous Coal and Lignite Industry in the United States

Item	1970	1972	1974	1975 (Preliminary)
Production (Thousand Tons)	602,932	595,386	603,406	640,000
Exports[a] (Thousand Tons)	70,944	55,997	59,926	65,669
Price per Ton f.o.b. Mines[b]	$6.26	$7.66	$15.75	$18.75
Percentage Surface Mined	43.8	48.9	54.0	54.7
Production (Tons per Man per Day)[c]				
Underground Mines	13.76	11.91	11.31	9.50
Surface Mines	36.28	36.33	37.07	30.00
Average All Mines	18.84	17.74	18.68	15.15

Source: U.S. Bureau of Mines, *Coal—Bituminous and Lignite in 1974, Mineral Industry Surveys,* Annual, January 1976.
[a]Bureau of the Census, U.S. Department of Commerce.
[b]Interstate Commerce Commission
[c]Estimates based on data supplied by the Health and Safety Analysis Center, Mining Enforcement and Safety Administration.

vestment was required to open a new, small mine. As mechanization became more widespread, an increasing share was produced by a few large companies which could afford the larger investments in equipment. In spite of this increasing concentration, the coal industry remains highly competitive. The percentages of total production by the four, eight, and twenty largest producers for selected years are:

Number of Largest Producers	Percentage of Total Production		
	1955	*1970*	*1974*
4	18	30	26
8	25	40	37
20	40	56	51

Concentration actually declined between 1970 and 1974, presumably as a result of a scramble by smaller mines to profit from the high prices. Rapid short-term expansion of production comes from an increase in the small mines that require only simple equipment and a minimum of preparatory work.

Even though competition remains high, as measured by the customary indicators, the structure of the industry has changed radically in the past ten years. In 1964, only three of the fifteen largest coal companies were owned by other interests; by 1974, only four of the fifteen largest were entirely devoted to coal mining, the other eleven companies being subsidiaries of oil and metal producing companies. This shift in ownership has caused concern that the trend toward energy conglomerates will reduce competition among different energy forms. The issue remains active in the attempt to force divestiture in the oil industry.

Attempts to restrict ownership of coal companies could have an adverse effect on future coal developments. If coal is to supply an increasing share of domestic energy, large investments to open up new mines will be required. These investments will be much larger than those needed in the past, and even the largest companies may have to strain to raise the necessary capital.

Coal Transportation

Unlike uranium production, in which transportation represents only a minor share of the delivered price of fuel, the transportation capacity and cost are major factors in the timely and competitive supply of coal. Approximately 65 percent of all coal is moved either entirely or part by rail, much of it over long distances; 11 percent by barge, and 11 percent by truck; 10 percent is used at the mine mouth. Although the lowest cost is by barge, use of this mode is largely confined to the Midwest along the Mississippi, Ohio, and Missouri rivers. Most truck transport is short-range, up to one hundred miles; while flexible, this is expensive. As recently as 1970, transport costs represented 35 percent of the delivered cost of coal. By 1974, the rise in prices reduced the transport share to less than 25 percent.

Regional differences are significant in rail rates for coal. Because of generally better management and operating efficiency, costs per ton mile have been less for railroads west of the Mississippi than on eastern railroads. Equipment and roadbeds for eastern railroads are not in as satisfactory condition and, in some areas, trains must be operated at reduced speeds to prevent further deterioration of the rail system. Expansion of production to meet potential increased demands for coal, perhaps doubled by 1985, could be constrained by transportation. In railroads, the constraint would be the deteriorated status of equipment, while barge movements could be limited by lock capacity on the Mississippi River. Since coal output now has reached its highest level, any increase would add to the strains.

In the early 1960s, a new method of transporting coal was introduced commercially, the coal slurry pipeline. In this method, coal is ground to a specified size and mixed with water. The "slurry" can be pumped by pipelines over long distances. One such pipeline was operated successfully for several years in Ohio over a distance of one hundred miles. This pipeline was ultimately shut down with the development of unit trains devoted exclusively to the transport of coal, since for the particular circumstances surrounding the Ohio slurry pipeline, unit train transport could be accomplished at lower cost. A second slurry pipeline is now in operation between a coal mine in Arizona and a power plant in Nevada. Generally, slurry pipelines are economical when new rail facilities would otherwise have to be constructed and when large volumes of coal are moved from a limited geographic area to a small number of receiving areas.

The main constraints on slurry pipelines are water and the legal barriers that currently exist because such pipelines do not have the right of eminent domain. Widespread use of pipelines would require the enactment of new federal legislation to confer this right; legislation now under consideration by the Congress to give those pipelines the right of eminent domain is opposed by the railroads.

The availability of water for slurry pipelines depends on where the coal is mined. Unfortunately, the low-sulfur, low-cost strippable coal in the Northern Great Plains, where much of the new productive capacity is expected to be installed, is in a water-short area, and local opposition to shipping water from the region in the coal slurry is already apparent. A return pipeline to recycle water would reduce water requirements by 70 to 85 percent and eliminate water cleaning problems, since the returned water would be used to reslurry.

Adequacy of Coal Resources

Long-range reliance on coal as a primary source of energy is not basically resource-constrained. As noted earlier, if coal production in 2000 were triple current production, or two billion tons per year, there would be sufficient coal at today's costs to satisfy that demand easily; the annual growth rate would be 4.5 percent. It is conceivable that by 2025 annual coal demand could be six billion tons or more. This would be a tenfold increase over 1975 but would still require only a 4.7 percent annual growth rate. While this rate of growth would be unusual for a mature industry to sustain over an extended period, we believe that it would be technically possible to achieve, though real coal costs might start to rise slowly early in the next century.

Constraints on Coal Production and Use

Although it has been obvious for many years that domestic coal resources could play a major role in providing a secure supply of energy for the foreseeable future, coal consumption did not increase significantly even after the Arab embargo. The explanation for this lack of response by the coal industry is found, not in resource constraints but in problems and uncertainties, discussed below, that increase risks and make development and investment decisions difficult.

Environmental Controls

Recent environmental protection measures have had an inhibiting effect on the growth of the coal industry. Possible new federal and local constraints on strip mining, and increasing local controls on water usage, have introduced uncertainties into prospective coal expansion. Public opposition to the siting of power plants (such as the Kaiparowits case in Utah) have added to this uncertainty.

Rising standards on emissions of particulates and pollution from the oxides of sulfur and nitrogen have also presented problems for the industry, not only by holding down demand for coal but also by adding requirements for expensive controls at power plants and by presenting complex engineering problems that have not yet been completely solved. Longer-term environmental concerns involve carbon dioxide and its potential effect on climate.

Health and Safety Constraints. The possibility of increasingly strict health and safety regulations also discourages new mining ventures. The Coal Mine Health and Safety Act of 1969, while providing safer and healthier working conditions, has reduced the productivity of miners and introduced significant costs in the form of mandated disability payments for black lung disease. In 1975, these payments amounted to more than $1 billion. Beginning in 1976, this liability became the responsibility of the coal industry.

Uncertainty in the Supply and Price of Competitive Fuels. Prospects for the future profitability of coal depend to a large extent on its competitive position compared to oil and gas and nuclear power alternatives. Uncertainty about the future prices of world oil and domestic natural gas inhibits the growth of coal production. So also does uncertainty about the relative costs of generating electricity from nuclear power and coal. The possibility of falling oil prices and the prospect of an increasing cost differential in favor of nuclear power act as real constraints on increased coal production.

Lack of Infrastructure to Support Coal Expansion. Besides transportation, another possible constraint, at least for short periods during rapid expansion of the industry, is a lack of trained manpower, especially in underground mines. An expansion of production from the 1975 level of 640 million tons per year to even 2 billion tons would require between 100,000 and 160,000 more miners underground in the year 2000 in addition to the 140,000 currently employed. Siting new mines in relatively sparsely settled areas might also be inhibited by local restrictions on the influx of the necessary work force.

Availability of Investment Capital. Investment per yearly ton of underground capacity has increased from $10 per ton to $30-$40. To double production in ten years would require $11 billion to $25 billion, depending on the mix of underground and strip mine capacity. Except for 1974 and 1975, profitability in the industry has historically been too low to generate internal funds or to raise outside funds at the rate projected. Only large companies with outside sources of capital could provide for the expansion in production thought to be needed.

All the issues cited above make it difficult to generate capital required for expansion. To add to this problem, Congress may divest oil companies of their holdings in other energy industries, thus depriving some coal enterprises now owned by oil companies of a valuable source of expansion capital. Federal and state tax policies, such as state severance taxes on coal shipped outside the state, may further discourage investment.

Institutional Problems. Development of the western coal regions poses special problems. It must be acceptable to the resident population. Financial help for state and local governments will be needed so that the required infrastructure can evolve on schedule. Competition for the limited supplies of water will have to be resolved.

Such issues are part of a growing conflict between national and local interests on a variety of energy matters. In many instances, what is in the national interest can be achieved only at the cost of burdens on local communities. A method for adjudicating these conflicts needs to be found.

Another problem is financing the R&D required to develop new technologies to utilize coal more efficiently and with less environmental effect. New mechanisms for government/industry cooperation need to be devised to support R&D and bring promising technologies to commercial use.

In summary, coal resources are not a constraint; but there are institutional, economic, and social problems to be solved if there is to be major expansion of U.S. coal production.

Future Uses for Coal in Conventional Markets

The major uses for coal in the next ten to twenty years will be electricity, steel, and industrial process heat. Use for electricity will depend on the relative costs of nuclear, oil, and coal fuels. The comparative economics of coal and nuclear power are discussed in detail in the next chapter.

Factors affecting consumption of coal by the steel industry include the general state of the economy, the level of steel imports, and competition from steel substitutes. The use of coal by the industrial sector will be greatly influenced by future environmental standards. Because fuel consumption at individual industrial plants is generally small, the cost of environmental controls can be high. With present technology, meeting strict emission standards significantly increases the cost of using coal for industrial plants and may induce many plants to use other fuels.

Fluidized Bed Combustion

Encouraging prospects for the environmental acceptability of coal come primarily from the technology of fluidized bed combustion (FBC). In principle, FBC should meet not only the emission limits on sulfur oxides, but those on nitrogen oxides as well, at less cost than conventional methods. The new technology has, however, been tested only on a small scale. It appears that this method of combustion would allow coal to compete with oil in a wide range of industrial uses, including cogeneration (the production of electricity along with industrial heat or steam), and central station electric power.

A fluidized bed consists of a container partially filled with coarsely ground solids, which is "fluidized" by the injection of gas through an apertured plate supporting the solids. The fluidized bed technique is widely used to obtain intimate contact between gas and solids in catalysis, heat treatment, and burning materials such as garbage. When the solids are a mixture of ground coal and limestone or dolomite (limestone containing magnesium) and the fluidizing gas is air, the coal can be burned efficiently and the sulfur removed at the same time.

Fluidized bed combustion can operate at atmospheric pressure to provide steam in water-tube boilers or at elevated pressure (four to ten atmospheres) to provide hot gas for gas turbine or combined-cycle (gas turbine followed by steam) electrical generation. A 20 MWe plant is under test now, and designs are complete for another plant to drive a 70 MWe gas turbine. If these tests are successful, ERDA plans to build a 250 MWe demonstration plant for testing on a larger scale.

Synthetic Oils and Gas

Oil and gas, which currently provide more than 75 percent of total U.S. energy supplies, are convenient for many energy uses and, with current technology, indispensable for others. Gas provides a low-cost, convenient, and environmentally acceptable fuel for residential use, and liquid fuels are almost the sole source of energy for transportation. The large stock of existing equipment which requires these fuels will not be replaced for many years.

When supplemental sources of oil and gas are required in the future, they can be produced from coal. Coal gasification is an old art that was widely used before it was displaced by natural gas. Coal gasification continues in many countries that do not have natural gas. In this country, however, gas made from coal would currently be considerably more expensive than natural gas, even where the price is not regulated.

Industrial scale coal liquefaction plants supported the German war effort during World War II. Costs of producing synthetic liquid fuels from coal, however, are very high. It would be impossible in this country for such fuels to compete with other liquid fuels unless the price of oil reaches about twice the 1976 OPEC price, or unless new federal policies permit higher cost fuel to find a market. Because of the lead times in planning, licensing, and constructing plants, the introduction of synthetics into the fuel market would require government action from the start.

In the absence of an unexpected technological breakthrough, R&D on new gasification processes will probably not reduce the present estimated cost of the synthetic gas (about $4.50 per million BTU at the plant) by more than 25 percent, and possibly by less. Therefore, consumers who require energy in this form will pay much higher prices than they have in the past, although the increase may occur slowly if the high-cost synthetic gas is mixed with natural gas and "rolled-in" prices are permitted.

One R&D proposal for production of synthetics which continues to attract attention is the underground gasification of coal. This method, which was first proposed more than a hundred years ago, has attractive features. In most underground gasification processes, no mining is required. Seams of such low quality that they could not be mined might be utilized. Underground gasification has been under development in the Soviet Union for more than fifty years, and several commercial-sized projects are said to be in operation. Following World War II, a number of other countries experimented with underground processes, but eventually these experiments were abandoned. For two years, ERDA has been experimenting at a site in Wyoming, and a second experiment in a deep bituminous coal seam is in the planning stage. Whatever its prospects, however, it will probably be a long time before this technology will be widely used.

In 1975, the Administration proposed a large-scale demonstration program for synthetic fuels. The total program would have produced the equivalent of

350,000 barrels of oil per day (compared to consumption of about 17 million barrels per day in 1976). Congress rejected the program in late 1976.

SUMMARY AND CONCLUSIONS

While estimates of the size of both U.S. and world coal reserves vary, it is clear that reserves are adequate for coal to become once again the primary source of energy in the industrialized countries, even if total energy consumption grows at 3 to 5 percent annually for the next several decades. Real costs would not have to rise until the next century. Expanded use of coal is developing in response to higher oil prices and national policies to create wider energy options. Coal should thus become more prominent in international trade.

Concern about the environmental effects and safety of coal mining and use have led to local and federal regulations on mining, use of water resources, and enforcement of ambient air standards. Development of the coal industry will be constrained until these social concerns are satisfied by technical and management measures.

Availability and use of coal may be increased by new technologies (such as fluidized bed combustion) for burning coal more cleanly and efficiently, for transporting it more cheaply, and for transforming it into synthetic substitutes for oil and gas. Substantial investment will be required to develop these technologies; to build railroads, pipelines, transmission lines, and facilities and equipment; and to establish communities in the mining areas. While much of the investment will be private, high-cost research and development will require government participation in funding.

※ *Chapter 3*

Economics of Nuclear Power

The significance of nuclear power to the overall economy has been discussed in general terms in Chapter 1. In this chapter, the costs of nuclear power are compared with the costs of coal, the competitive energy source for the generation of electricity. Although oil or gas may be preferable to coal or nuclear for meeting peak loads, for the next few decades, the choice for the generation of baseload power will be between coal and nuclear power from light-water reactors (LWRs). The prospects of alternative technologies such as solar, geothermal, and fusion energy, as well as the potential of energy conservation, are considered in Chapter 4.

The basic conclusion of this chapter is that, despite large uncertainties, nuclear power will on the average probably be somewhat less costly than coal-generated power in most of the United States, or, to be more precise, in areas that contain most of the country's population. The advantage for nuclear power is likely to be largest in New England and in parts of the South. In large areas of the West, containing a small fraction of the country's population, coal-generated power is likely to be less costly than nuclear power. In some areas, the costs can be expected to be very similar. The cost factors are sufficiently uncertain, however, that the balance could shift to increase or eliminate the small advantage that nuclear power appears to have on the average in present estimates. In view of the large regional and even local variations in costs, a mix of coal and nuclear plants will probably be built in most parts of the country during the rest of the century. At present such a mix is desirable in view of the many uncertainties in the economics and social costs associated with these sources of energy and the unforeseen problems that may emerge during the balance of this century.

SOURCES OF UNCERTAINTY

In order to make predictions about comparative generating costs, it is necessary to attempt to understand the factors which have caused costs to differ or to change in the past, and then to make judgments about how these factors and others will operate in the future. Uncertainties in some of the factors that enter into overall costs make comparative estimates on the coal and nuclear alternatives complex and imprecise. It is convenient to partition the problem into two categories: (1) simple escalation in costs, i.e., the changes with time in the cost of a piece of equipment, an hour of craft labor, the transport of coal from a mine to a power plant, etc.; and (2) the effects of legislation, regulation, safety standards, taxes, etc., on the costs of plant construction, operation, and the mining and processing of fuels.

Cost escalation has been substantial in recent years, with many cost elements for power plants increasing more rapidly than the general rate of inflation. In a few instances it is possible to predict with some confidence that this will continue for several years, as with coal miners' pay and benefits where contracts involving escalation have been already negotiated. In the intermediate term at least, the difference between the escalation in various factor costs and in the general price level should moderate.

The principal causes for dramatic recent increases in the cost of nuclear power generation have been such factors as the lengthening of construction and licensing time and changes in plant design that have had their basis, at least in part, in regulations directed at safety and environmental concerns. In general, we expect these types of changes to moderate and to be offset by other improvements, but the uncertainties remain large and are greater the further in the future one attempts to project costs.

The situation for coal plants shows notable differences but is no less complex. There is much more experience with construction of coal plants, but not a great deal with very large ones. Until recently, coal capital costs have been both lower and less variable than nuclear. As a result, uncertainty in capital costs has been less of a factor in estimates of coal-generated power costs than in the nuclear case. The major uncertainty has been in fuel costs. Depending on the source of coal and the location of the power plant, coal costs at the mine can dominate transportation costs or vice versa. Uncertainties in the cost of either can be important.

The situation with respect to coal is changing significantly with the imposition of constraints to limit emissions of sulfur oxides and other pollutants. Where sulfur removal equipment has been added, capital costs have been substantially increased and plant reliability reduced. Of particular significance is the possibility that desulfurization may be required even for low-sulfur western coal in order to meet more stringent air quality standards. Should this occur, the cost would be substantial and the geographical area within which western coal could compete with nuclear power (or medium- and high-sulfur coal) would diminish greatly. These cost increases could be partially offset by developments reducing cost for coal transportation such as coal slurry pipelines or electrical transmission. Another technological development—fluidized bed combustion—may prove to be effective in removing SO_2 and other pollutants as well as in improving efficiency of fuel utilization.

Thus, in projecting the future of coal for the rest of this century, uncertainties with respect to both the regulatory environment and technology are probably greater at this time than in the case of nuclear power. As a result, in our tables and discussions, we have presented two variants for the coal alternative, one with and one without scrubbers to remove pollutants.

THE COST OF NUCLEAR POWER

The cost of nuclear power is dominated by the capital charges.[a] These charges per kilowatt hour are directly proportional to the cost of plant construction and inversely proportional to the plant "capacity factor." The capacity factor is the ratio of the electricity actually produced during a fixed time period compared to the amount which would be produced if the plant operated continuously at full capacity. Other cost factors about which there is also great uncertainty are the price of uranium, the cost of enrichment services, and the cost of disposing of spent fuel.

Construction Cost

The actual cost of constructing nuclear power plants has greatly exceeded original estimates. The cost per kilowatt for plants becoming operational in 1976 is double those that became operational in 1970. Indeed, the increase in costs with time, the variation in plants built at the same time, and the fact that many estimates of individual plant costs have been low by a factor of two or more combine to cast grave doubt on the ability of the industry to predict or control costs for future plants. Despite this disconcerting record, there seems to be convergence on estimates of future capital costs and some reason for optimism that cost escalation and uncertainty can be alleviated.

Construction costs for nuclear plants are difficult to estimate because of

[a]Capital charges reflect not only the cost of borrowing money but also amortization of investment and allowances for taxes. Because amortization is generally accelerated and because of taxes, the fixed charge rates used by utilities in making investment decisions are high relative to those appropriate from the perspective of society as a whole. Thus, there is no inconsistency in a higher fixed charge rate in this chapter than that used in Chapter 1. In recent years, fixed charge rates for baseload power plants have been 15 to 20 percent. The rates have been as high as they are in part because of high interest rates. In an inflation-free economy, or in an analysis such as that of this chapter, which is based on constant dollars, a lower rate is appropriate. We have used 13 percent, which would be equivalent to 17 percent in current dollars assuming a long-term inflation rate of 4 or 5 percent per year.

the following complicating factors that have to be taken into account in cal-culations.

"Turn-key" Plants. During the period 1963–72, thirteen plants in the United States were built by General Electric and Westinghouse on a fixed-price turn-key basis. The cost to the utilities of these plants was only about two-thirds that of comparable plants during the same period. General Electric and Westinghouse apparently absorbed substantial losses on those plants, presumably to gain an advantageous position for future business.

Scaling. Since there are economies of scale in the construction of power plants, comparing costs using $/kw of generating capacity can be misleading unless adjustments are made. Most of the units that became operational up to 1972 were in the 400 to 800 MWe range. Since then, new plants have been in the 800–1,100 MWe range. It now seems probable that 1,300 MWe will become something of a standard, in part because the Nuclear Regulatory Commission (NRC) is unwilling to license plants greater than 3,800 MWt or about 1,300 MWe capacity. Both the government and industry estimate that a 1,300 MWe plant will cost only 50 to 75 percent more than a plant half that size built at the same time.

Dual Versus Single Units. Significant savings can be realized by siting two plants at the same location, particularly if they are identical and if the second follows the first by a year or two. The cost of the second unit will probably be about 85 to 90 percent of the first.

Labor Productivity. The regional variability in labor productivity is surprisingly large. Construction of plants now being started in New England can be expected to require about 10.5 man-hours of craft labor per kilowatt of capacity. The corresponding figure for the South is apparently about 8 man-hours. Such differences provide a partial explanation of the difference in construction costs between the Middle Atlantic-New England states and the South. WASH-1345 suggests that the latter are about 85 percent of the former.[b] There are even more extreme variations. For example, construction costs for Duke Power, which employs nonunion labor and which also serves as its own architect-engineer (possibly a significant factor), have been about half the costs experienced elsewhere. Nonunion labor costs are lower primarily because the same individuals work in several craft areas, increasing productivity, not because of lower wages.

[b]United States Atomic Energy Commission, "Power Plant Capital Costs Current Trends and Sensitivity to Economic Parameters," WASH-1345 (Washington, D.C.: U.S. Government Printing Office, October 1974).

Changes in the "Scope" of Work. Figure 3-1 shows how manpower requirements per kw of generating capacity have changed. The trend is not due primarily to changes in labor productivity but results from changes in the "scope" of a project. These changes are a consequence of more stringent safety and environmental standards. Additional quality assurance, inspection, and documentation requirements in recent years have also contributed to the increase in man-hours per kilowatt installed. Other indices reflect the same kind of change, such as increases in engineering manpower, increases in materials—steel, concrete, wire per kw of plant, and increasing plant volume per kw.

Design changes, particularly changes made after engineering is well underway or retrofitting after construction is underway, have been very costly. Studies have been carried out on three plants—Sequoyah and Belefonte, both TVA plants, and Millstone 2, a Northeast Utilities plant—to try to relate increase in plant cost to such changes. Generalizing from these is difficult; however, in the case of the Sequoyah plant, TVA identified "90 significant engineering changes" made after the original cost estimate, and attributed them as follows:

NRC requirements	35
TVA redesigns	28
TVA decisions because of improved technology or cost savings	13
EPA requirements	4
Equipment manufacturer redesigns	4
Equipment manufacturer modifications because of safety criteria changes	4
Other requirements	5

There were twenty-three cases where a structure or component had to be torn out and rebuilt or added. Construction labor estimates had increased from an original 8.2 million to 16.4 million hours as of March 1975, and the plant was still not complete at the end of 1976. A particularly troublesome problem in this case appears to have been TVA's inexperience in nuclear plant design (it served as its own architect-engineer), and the fact that it began construction when engineering was only about 2 percent complete.

Lengthening of Construction Period. The time required for the construction of a nuclear power plant from award of a contract to startup of commercial operations has increased from five to seven years in the mid-1960s to about ten years. As a result, interest during construction has become a larger fraction of total plant cost, rising from about 8 percent of total capital costs for plants going into commercial service in 1972 to about 20 percent for those going into service in 1983.

Inflation. Much of the increase in plant costs can be attributed to the abnormal inflation of the last several years, and in particular to the fact that inflation in the costs of labor, materials, and components has greatly exceeded the general rate of inflation.

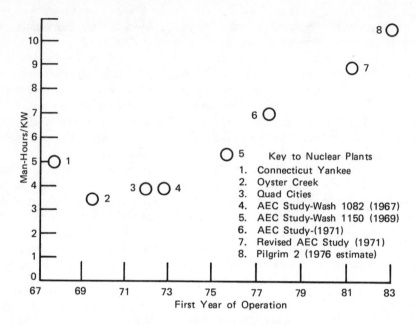

Figure 3-1. Trends of Manpower Requirements in Nuclear Plants Construction

Source: "Trends in Light Water Reactor Capital Costs in the U.S.: Causes and Consequences." Center for Policy Alternatives, MIT, CPA 74-8, December 18, 1974 (with estimate for Pilgrim 2 added).

Site-specific Differences. Two nuclear plants of the same capacity and design built by the same utility, architect-engineer firm, and labor force in the same time period may experience significant cost differences due to site-specific variations. Some example are:

Item	Cost Variation (Millions of 1976 Dollars) for a 1,000 MWe Plant[c]
Cooling tower rather than offshore diffuser	50[d]
Seismic design in sensitive geologic zones	10–30
Flood protection depending on nature of flood plain	10
Aircraft hardening if near airport	20–40
Transportation of major equipment depending on access	40

[c]An increase of $10 million in capital cost will result in an increase of about 0.25 mills/kwh in the cost of generating electricity.

[d]The typical power output decreases about 3 percent due to cooling towers, adding about 0.7 mill/kw hour to the power cost. When the cooling tower's capitalized costs are added in, the total power cost increase is about 2.0 mills/kw hour.

Other Sources of Variation. Estimates for different plants often involve different assumptions about reserves for contingencies, or differences in what is included in capital cost estimates, e.g., whether or not real estate, transformers, and switchboards are included. Although it is not normal industry practice, there are cases where nuclear steam supply system costs include a first fuel loading. If so, this last factor alone would make a difference in capital cost of the order of 15 percent.

Table 3-1 presents a sampling of actual estimates of construction costs for large nuclear plants scheduled for the mid-1980s. The following observations about some of the entries in Table 3-1 may help to illuminate the problems involved in interpreting such estimates. The first estimate for New England Electric includes contingency allowances that are substantially larger than those included in other cases, such as the Commonwealth Edison estimate. If an adjustment is made on the assumption that the New England Electric reserve is excessive, the cost estimate would come down somewhat, perhaps $40/kw. The Commonwealth Edison figure is based on a 7 percent interest rate, whereas the New England Electric figure is based on 9 percent, a more representative figure. Adjustment of the Commonwealth Edison figure to 9 percent would bring it up, probably about $40/kw. The Boston Edison figure (which like the New England Electric figure is based on an Arthur D. Little/Stoller study) appears to be at least as generous in its treatment of contingency reserves as the New England Electric estimate and is based on a 9½ percent interest rate. This may be a partial explanation of why it is as high as it is. Although it was Northeast Utilities' plan, as of last year, to bring Millstone 3 on stream in 1980, the date has now been deferred until 1982. Had the unit been started two years later to meet the 1982 operational dates, the interest cost on funds spent during construction would have been less, but the escalation in costs on some items would have been greater. The net effect would probably have been an increase since the inflation rate exceeded the interest rate on funds used during construction. The WASH-1345 figures and the Union Electric figure for its Callaway plant were derived using the Oak Ridge National Laboratory computer model code CONCEPT. There is some feeling that the code underestimates costs. CONCEPT generated an estimate of $780 million, or $660/kw for the Boston Edison Pilgrim 2 plant compared with the utility's $1,034/kw figure of Table 3-1. The estimates for Duke Power and Commonwealth Edison are not necessarily anomalously low. Duke Power achieves construction costs that are low relative to the rest of the industry, in part because of its location and use of nonunion labor. Commonwealth Edison also generally does better than most of the industry. In both cases lower costs are partly a consequence of large company size and may reflect superior management as well. Both companies have added nuclear plants more or less continuously, permitting better use of manpower and negotiation of more favorable contracts with equipment suppliers. There seems to be widespread agreement that costs in

Table 3-1. Capital Cost Estimates for Future Nuclear Plants

Size	Utility or Source of Estimate	Operational Date	Cost $/kw as Quoted in Current $s (Except as Noted)	Cost $/kw (1985 Constant Dollars)*
2 × 1150 MWe	New England Electric	1983, 1985	863	1,007
2 × (1100-1200)	Commonwealth Edison	1985	463 (mid-1974 $s)	821
2 × 1100	Bechtel	1983, 1985	1,030	1,139
1 × 1180	Boston Edison	Late 1982	1,034	1,205
1 × 1150	Northeast Utilities	1982	878	1,098
		1984	1,015	1,089
1 × 1100	Sargent & Lundy	1984	868	855
2 × 1153	Duke Power	1981–1982	477	624
2 × 1150	Union Electric	1982	757	954
2 × 1200	Ebasco	1/86	1,005	980
2 × 1300	WASH-1345 ("site favorable in all respects")	1/81, 1/82	551	779
		7/84, 7/85	660	756

"BEST" ESTIMATE: $1,000
($667 in 1976 $s)

*All costs are normalized to a plant with a mid-1985 operational date with two 1,150 MWe units using cooling towers. Adjustments include the following: (1) a reduction of single unit costs by 7½ percent; (2) adjustment for economy of scale to 1,150 Mw size, assuming that the cost/kw is proportional to (unit size) $^{-0.32}$, based on WASH-1345; (3) escalation of capital costs in current dollars at 8.0 percent per year to mid-1985; and (4) addition of $75/kw (1985 dollars) for cooling towers to estimates that are based on once-through cooling. Escalation of Commonwealth Edison costs is based on GNP deflators of 10.65 percent, 8.70 percent, and 6.8 percent for 1974, 1975, and 1976, respectively. An average of 4.0 percent/year is used for subsequent years to reduce the $1,000/kw "best estimate" to $667 in 1976 dollars.

the Northeast are high, mainly because of high labor costs and possibly low productivity. WASH-1345 estimates 7.7 percent higher cost than the U.S. average, but the figures in Table 3-1 suggest the differential is closer to 15 percent.

Taking all these points into account and recognizing that the WASH-1345 estimates are for "a site favorable in all respects," we conclude, as indicated in Table 3-1, that a reasonable national average cost/kw figure for dual 1,150 MWe units becoming operational in 1985 is about $1,000 (1985 dollars) or about $667 (1976 dollars).

How much confidence can one have in such estimates in light of the record of the last decade or so? Is there any reason to believe that there will be an end, or at least a moderation, in the escalation of costs?[e] We have concluded that the differential between the escalation in construction costs and the general rate of inflation will probably begin to diminish for plants ordered in the next few years to begin generating power in the late 1980s. In considering how likely this is, it is important to bear in mind that indirect costs—escalation and interest on funds used during construction—have become an increasingly large component of total costs. Major equipment items are now a relatively small component. For example, the nuclear steam supply system and the turbine generator together amount to only about 15 percent of total cost for plants to be operational in the early 1980s.

In these circumstances the critical question is whether there will be further delays in the construction and licensing process, now typically ten years. There are reasons to believe that this period will not lengthen and may even be reduced. First, it has been substantially less in some other countries that are building reactors of the same size and complexity. Second, there has been some progress in standardization in the United States. This will presumably result in some savings in direct costs and more importantly should reduce the time and effort required in the licensing process.

The following developments in particular are of interest. The nuclear steam supply system vendors appear to be settling on standard designs of 1,300 MWe, and the architect-engineers are also moving toward standardized designs. With advance NRC approval, which has already occurred in the case of Stone and Webster standard designs, the licensing procedures for individual plants will be simplified and shortened. There are also moves toward advance approval of sites. With recent plant cancellations, stretch-outs, and diminished demand, it seems less likely than in recent years that there will be delays in construction due to shortages of critical manpower, strikes, and delays in delivery of equipment. Third, there has been a trend toward less concurrency in design and construction, and this should lead to more accurate cost estimation and less

[e]In this connection, it should be noted that some of the studies we have used assume cost escalation of about 8 percent per year (in current dollars) for plants coming on-stream in the mid-1980s. The adjustments in Table 3-1 reflect this.

retrofitting. The trend is illustrated by TVA experience: its first, second, and third plants had 2 percent, 11 percent, and 24 percent, respectively, of design complete when construction began; the fourth had 30 percent design completion as of December 1975, with construction not yet begun. Finally, with the slowing of utility growth and with excess manufacturing capacity, the inflation in wages, materials, and equipment in the industry is not likely to exceed the general price index as much as it has in recent years. For all of these reasons, we believe that the trend of cost increases for nuclear electric power (in constant dollars) will be moderated.

The question remains as to whether there may be major changes in Nuclear Regulatory Commission or Environmental Protection Agency (EPA) requirements that could lead to backfitting and expensive delays in construction and licensing. Two such developments merit comment. Reconsideration of seismic criteria is now underway and insistence on more demanding standards could result in increases in costs, especially in plants already under construction. Additional costs are likely to be 3 to 5 percent. The other issue is cooling. Technically, "once-through" cooling will not be permitted by EPA in the future, but variances have been granted (e.g., Seabrook) and planning is going ahead on the assumption that such variances will be allowed (e.g., Millstone 3).

With all of these considerations in mind, we estimate that the uncertainty attached to the capital cost figure cited above ($667 per kwe in 1976 dollars for and *average* dual 1,150 MWe plant operational in 1985) is about $^{+25}_{-15}$ percent and that the increase in cost of plants coming on-stream in the years thereafter will be less than 5 percent/year in constant dollars.

Capacity Factors

The cost of nuclear generation of electricity is highly sensitive to the capacity factor, which is the total amount of electricity actually produced by a unit in a year divided by amount of electricity the unit could produce running at full capacity for the entire year. There is considerable controversy as to the appropriate value to use in estimates. A major problem is that there are virtually no data on large, mature units in the 1,000 MWe range that have been operating for several years. There seems to be agreement that performance during the first year or two is atypically low. There is also some contention that after about the fifth or sixth year the capacity factor begins to decline as a consequence of wear and corrosion. However, there is not enough experience to support or refute this assertion. Estimates must be made based largely on experience with units smaller than those now being built, that are in their second through fifth or sixth years of operation.

One would think that there would be agreement on capacity factors calculated on the basis of experience to date, even if there is little about what factors might be appropriate for the future. But there is not. This reflects disagreement about exclusion of certain units from the data base used, ambiguity about the capacity factors of individual units, and different methods used in computing averages. When considering future performance, it is reasonable to exclude certain units from the data base (e.g., those that are very small and those that are atypical in their design); however, this leads to a certain arbitrariness in selection.

In calculating the capacity factor for individual units, there is the question of the appropriate figure to use for capacity, since the maximum capacity of some units will vary seasonally because of differences in cooling water temperature. In other instances, units may be derated because of environmental or safety considerations. For example, Zion 1 has a design rating of 1,050 MWe, a restriction to 880 MWe, and a maximum dependable capacity of 850 MWe. Since the capital charge for generating electricity is proportional to cost per kw of capacity for an installed unit divided by the capacity factor, it makes no difference which definition of unit rating is used in estimating the cost of power generation provided the same definition is used in calculating unit cost. Since this is usually based on design rating, that would seem to be the appropriate figure to use in calculating capacity factors.

Then, there is a question of averaging when one is using data from a number of units. The most obvious approach is to weight each unit in proportion to its design rating. An alternative is to weight all units equally regardless of size. The latter method produces a higher average capacity factor since to date small units generally have a better record than large ones. This kind of averaging is used by Westinghouse in calculating a capacity factor of 62.8 percent for units of all kinds and 65.8 percent for its own. The procedure is dubious since for most other purposes (expressions of demand, investment, etc.) units would be weighted according to size, and in making decisions about future investment the interest is in large units. A less defensible method is to weight units according to the energy they actually generate. The problem with this is that a unit not operating, i.e., with a zero capacity factor, simply drops out of the calculation. The Atomic Industrial Forum (AIF) used this approach to calculate a capacity factor of 64.4 percent based, presumably, on units operating in 1975. Adjusting the AIF figure to a weighted average of all units, one gets a capacity factor of 61.7 percent.

A number of other estimates of capacity factors are available. Those appearing in the NRC's "Grey Books" (Operating Units Status Reports) would generally be regarded as authoritative. However, there are errors and inconsistencies is some of the material in the Grey Books. For example, on occasion, restricted power rating is used in place of a design rating; yet, the results are presented as if the latter had been used. Thus, even the Grey Book figures have to be treated with some care. The Grey Book figures for 1975 for all nuclear units are 58.3 percent based on the design rating and 66.5 percent based on the maximum dependable capacity.

Several analyses suggest that performance is poorer for large units than for smaller ones, and the Grey Book data for 1975 are consistent with this. Thus, if one simply averages the capacity factors for all units of greater than 800 MWe size (without weighting them according to their rating), one gets a value of 51.1 percent, whereas for the units of less than 800 MWe the figure is 65.3 (62.9 if one includes Indian Point I, which has been completely shut down). This is also borne out by a study completed in late 1976 by the Council on Economic Priorities.[f] That study projects 1984 capacity factors for large units of 0.49 and 0.55 for BWRs and PWRs, respectively.

Virtually all new units are large. Is a capacity factor as low as about 50 percent, based on historical experience, appropriate for them? Is the experience figure low because the larger units are newer and still going through a "shakedown" period, or is it because there is an inherent inverse correlation between reliability and size? We do not believe there are persuasive reasons why larger units should be less reliable than small ones, once there is substantial experience with them. The experience with fossil fuel units and that with smaller nuclear units supports the view that initial capacity factors are likely to be lower than expected lifetime averages.

Taking all of the above into account, we have concluded that a figure of around 60 percent is realistic for near-term planning. It seems reasonable, but by no means certain, that there will be improvements beyond the 60 percent level as the industry matures.

One way of increasing the capacity factor slightly would be to cut the time required for refueling. With present designs, a reactor must be shut down for about thirty days per year for refueling. Westinghouse has developed designs that would permit reducing this to about a week, but there is little utility interest in the concept, since other planned maintenance requires the longer period.

It should be noted that many of the problems affecting the capacity factor of nuclear units occur in the nonnuclear part of the plant. Large coal units also have little historical data accumulated and may suffer similar problems affecting capacity factor. Comparable large fossil and nuclear units are compared below:

	Nuclear	*Fossil*
Miles of tubing	140	800
Primary working fluid temperatures	650° F	1,000°–1,300° F
Primary working fluid pressure	1,000–1,700 psi	3,000–4,000 psi
Flame temperatures	–	3,000° F

[f]Table 10.4 in "Power Plant Performance," published by the Council on Economic Priorities, 84 Fifth Avenue, New York, New York 10011.

It would seem that large coal units would have comparable or more severe operating stresses and temperatures than nuclear units. However, because of radiation hazard, nuclear units are more likely to be shut down when problems develop, and repairs on a nuclear plant are often more time-consuming.

Although the uncertainties for capital charges are by far the most important in estimates of the costs of nuclear power, there are also major uncertainties in other cost components that may also affect the future costs of nuclear power.

Uranium Fuel

Increases in the cost of uranium ore (U_3O_8) will obviously increase the cost of nuclear power (see Chapter 2). Uranium costs, however, are a sufficiently small element in the total cost of electricity so that uncertainty about U_3O_8 costs should not be a critical factor in determining the cost of nuclear power in this century. For our comparative calculations, we have estimated an average uranium cost of 2.5 mill/kwh for a reactor coming into operation in 1985. This is based on the assumption that U_3O_8 will cost $30 per pound (1976 dollars), and that the cost in constant dollars will not increase substantially over the period of interest, i.e., out to the turn of the century. Other analyses assume lower costs for the mid-1980s, but then have assumed escalation (in constant dollars). If the 1985 price (in 1976 dollars) is $20/pound increasing to $50/ pound in 2000 (i.e., at 6.3 percent/year in constant dollars), the levelized cost, using any reasonable discount rate, would also be about 2.5 mill/kwh.

Enrichment

In our projection (see Tables 3-3 and 3-4), we have estimated that future enrichment costs will remain constant at $80 per kg of separative work[g], which is equivalent to 2 mills/kwh. This is about 20 percent higher than the present Energy Research and Development Administration (ERDA) cost, but 20 percent lower than expected European costs from URENCO and Eurodif. The $80 figure is reasonably consistent with estimates of what the cost would be with new American plants based on demonstrated technology (see Chapter 13). It is likely that cost of enrichment using gaseous diffusion technology will increase over the next few years at a rate somewhat greater than the general rate of inflation because of increasing power costs and the cost of new plants that may be built.

Costs based on centrifuge technology are more uncertain, but there is a greater likelihood of cost reduction since the technology is less mature. Laser

[g]It should be noted that the costs in Tables 3-3 and 3-4 are based on the assumption that the tails from the enrichment process contain 0.20 percent uranium-235. With the prices assumed for U_3O_8 ($30/lb) and for enrichment ($80/SWU), this is not the optimal tails assay. It would pay to use more U_3O_8 and less separative work, stripping to about 0.22 percent uranium-235. The reduction in the cost of generating electricity would, however, amount to only 0.01 mills/kwh.

isotope separation may be realized during the lifetime of reactors coming into operation in 1985. While meaningful cost estimates cannot be made now, laser enrichment costs are likely to be much cheaper than other methods. Laser separation could also extend uranium resources by as much as 35 to 70 percent since it may be economically advantageous to recover most of the uranium-235 in the enrichment process instead of only 60 to 70 percent. It may also be advantageous to rework existing uranium tails to extract most of the 0.2–0.3 percent uranium-235 contained in them. Thus, laser enrichment may have the effect of substantially reducing the amount of U_3O_8 required per kwh as well as lowering enrichment costs.

Spent Fuel Storage, Transportation, and Disposal

The exact costs of the "back end" of the fuel cycle are not yet clearly defined. However, enough information exists to conclude that they will not contribute substantially to the cost of electricity. Without reprocessing or recycle we have estimated these costs at about 0.4 mills/kwh. This figure is based on storage of fuel for five years at $5/kg-year, on transportation at $15/kg, and on disposal at $100/kg (all in 1976 dollars), with a discount rate in constant dollars of 9 percent. This estimate is probably uncertain by ± 0.2 mills/kwh. The effect of reprocessing and recycle on the fuel cycle costs is discussed in detail in Chapter 11.

THE COST OF COAL-GENERATED ELECTRICITY

Capital Costs and Capacity Factors

Coal is the principal competitor with nuclear power for the generation of baseload electricity for the remainder of the century. Indeed, while overall coal production has changed little since 1970, consumption of coal by electric utilities has grown at the rate of nearly 5 percent per year. Nonetheless, the future of coal is clouded by uncertainties. The cost of coal production has risen significantly in recent years reflecting more stringent safety, health, and environmental standards. The effects of large-scale coal combustion on public health and the environment are poorly understood. Existing emission control technologies, e.g., flue gas desulfurization, increase the cost of coal plants and reduce their reliability. Advanced technologies, e.g., fluidized bed combustion of coal, while promising, are not fully proven. In addition, like nuclear power, the coal situation and its economic position varies considerably from region to region.

Estimates of capital cost ratios for coal and nuclear plants becoming operational in the mid-1980s are summarized in Table 3-2. Cost estimates for plants designed to burn low-sulfur coal without flue gas desulfurization range from $345/kw to $680/kw (1976 dollars); and those for plants designed to burn high-sulfur coal with SO_2 removal range from $465 to $743. Based on these

Table 3-2. Comparative Capital Costs
(Ratio of Estimated Coal and Nuclear Plant Capital Costs)

	High-Sulfur Coal With Scrubbers	Low-Sulfur Coal Without Scrubbers
Specific Sites		
Northeast Electric (Estimates by ADL/Stoller)	0.81	0.65
Pilgrim 2 (Boston Edison)	0.82	
Millstone 3 (Northeast Utilities)	1.01	0.93
Yadkin (Duke Power)	0.82	
Cherokee (Duke Power)	0.82	
Callaway (Union Electric)	0.81	0.67
Architect-Engineers		
Bechtel	0.83	0.74
Ebasco	0.91	0.69
Others		
Commonwealth Edison (Plants Not Specified)	1.00	0.70
WASH-1345	0.84	0.64
Komanoff (Council on Economic Priorities)	0.95	0.83
Westinghouse	0.83	0.71

estimates, we have applied capital cost ratios of 0.83 and 0.70 to a cost of $667 for the nuclear case to calculate 1985 capital costs (in 1976 dollars) of $555/kw and $465/kw respectively for coal plants with and without scrubbers. The difference between the two coal cases is consistent with an independent estimate of the added cost of scrubbers on new plants of $70–$100/kw (1975 dollars). It is interesting that the range in estimates for coal plants should be so large considering the vast experience with them and the fact that, aside from effluent-related problems, the impact of regulatory uncertainties on coal plants is less than in the nuclear case.

Capacity factors for coal plants operating without sulfur removal can probably be estimated with more confidence than for nuclear plants. We have estimated a capacity factor of 67 percent since they are likely to average 5 to 10 percent higher than nuclear plants. If sulfur removal is required, the figures become less certain. There is little experience with flue gas desulfurization in large plants. A question arises as to the operation of plants in the event of scrubber failure. If temporary variances are granted or a contingency reserve of low-sulfur coal is available, the effect of scrubbers on a plant's average capacity factor may be negligible; however, if plants must be shut down in the event of scrubber failure, substantial degradation in capacity factors will occur. We have estimated a capacity factor for coal plants with scrubbers of 60 percent, the same as in the nuclear case.

The foregoing suggests that improvement in the technology of desulfurization should have high priority in the allocation of R&D funds by ERDA. Looking further into the future, additional limits may be placed on the emission of nitric oxides, heavy trace elements, and toxic and carcinogenic organic compounds. There is little basis for estimating the costs of limiting these emissions, but the imposition of such limits will almost certainly increase the investment required to burn coal. Advanced technologies, e.g., fluidized beds, may be effective in limiting some of these emissions as well.

The Cost of Coal

The price of long-term contract coal doubled between early 1973 and mid-1975, largely reflecting coal's response to the 1973 quadrupling of oil prices and its short-term inelastic supply. In light of this, and with uncertainty in future policy with respect to mine safety, strip mining reclamation, state severance taxes, wages, and benefits for miners, future coal costs cannot be predicted with certainty. In the case of high-sulfur Appalachian coal, near-term cost escalation (in constant dollars) is inevitable, considering the fact that underground mining is labor-intensive and that contracts recently negotiated call for wage increases in excess of expected increases in general price indices. Improved safety standards may for a time continue the recent trend of declining productivity. With this in mind, it is likely that the price of coal will rise to somewhere in the range of $0.90 to $1.25 per million BTUs in the early or mid-1980s. Thereafter, stabilization is expected with modest improvements in productivity more or less offsetting any increase in wage and benefit costs that exceed the general rate of inflation. We have assumed a price of $1.00 per million BTUs in mid-1975 dollars (or $1.08 in mid-1976 dollars).

The situation is more uncertain for western low-sulfur strip mined coal. The costs of production are more predictable and less subject to escalation since strip mining is less labor-intensive than underground mining. But production costs may not be a very good guide to price. With uncertainty about leasing of federal lands and about state policies with respect to rapid expansion of the industry, it is by no means clear that supply will expand to match demand. In addition, state severance taxes can (and already do) increase price substantially. Taking these factors into account we have estimated a price of $0.43 per million BTUs (1976 dollars) for western coal.[h]

Transportation

There is substantial experience in moving coal with unit trains and by barge. The cost is generally higher in the East, and great increases in traffic could re-

[h]Richard Gordon has recently estimated $0.25 to $0.50 per million BTUs in 1975 dollars, "Historical Trends in Coal Utilization and Supply," Pennsylvania State University, University Park, Pennsylvania, August 1976.

quire additional investment to increase the capacity of the eastern railroads. The western railroads are in better shape. Transportation costs have increased at a rate greater than the general rate of inflation, and most analyses suggest that this will continue. We have assumed an increase in cost of 3 percent per year in constant dollars.

CONCLUSIONS

When the various uncertainties are combined, the uncertainty in the generating cost for coal is not much different than for nuclear power—the uncertainties in both surely overlap any difference between them.

In Table 3-3, we present our estimate of the costs of power generation in the Midwest, where nuclear and coal-generated power may be about equal. The estimates are for a plant coming on-stream in mid-1985 and draw on a variety of studies and extrapolations by government, utilities, and consulting firms. The uncertainties indicated in the last row of the table should *not* be regarded as independent, since some of the factors which might result in deviations from our central estimates would affect costs for nuclear and coal generation in the same way. For example, if craft labor costs increase more rapidly than we have assumed, this will increase both nuclear and coal-generating costs, with the effect being somewhat larger in the nuclear case since capital charges represent a larger fraction of total costs. Thus, it is likely that, if costs for the nuclear case are substantially higher than our central value, the costs for the coal cases will also be higher than the central values.

The effect of regional variations in costs can be demonstrated with the help of Table 3-3. For example, in New England additional transportation costs will increase the cost of coal-generated power by about 2 mills/kwh over that shown in the table. Also, construction costs for both nuclear and coal plants will be 10 to 20 percent greater in New England, the effect of the increase being larger in the nuclear case, where construction is a larger share of the total. On the other hand, costs of burning Montana-Wyoming coal in Kansas City will be about 5 mills/kwh less than the figure shown in Table 3-3. In the South, nuclear power may be more attractive than Table 3-3 suggests because of relatively low construction costs. In the case of Duke Power, this is likely to mean a differential in favor of nuclear power of perhaps 1 mill/kwh greater than Table 3-3 suggests. In some parts of the South, e.g., Florida, the long-distance transportation costs of coal will give nuclear power an additional advantage of several mills/kwh.

The effects of such geographical variations are summed up in Table 3-4. It will be seen that there is substantial overlap in costs for the coal and nuclear power. Thus, we conclude that the trade-off between coal and nuclear on direct economic grounds is a close one, with many uncertainties, and will vary with local conditions and expectations in each case.

Table 3-3.   Projected Cost of Generating Electricity in the Midwest in 1985[a] (mills/kwh in mid-1976 dollars)

		Nuclear	Coal With Scrubbers	Coal Without Scrubbers
Capital Charges[b]		16.5	13.7	10.3
Operations and Maintenance (O&M)		2.0	2.8	1.6
Fuel[c]		5.4		
U_3O_8	2.5	Coal at Mine	10.0^d	4.0^e
Conversion to UF_6	0.1	Transportation	2.0^f	11.3^g
Enrichment	2.0			
Fabrication	0.4			
Spent Fuel Storage and Disposal	0.4			
TOTAL[h]		23.9^{+5}_{-4}	$28.5^{\pm4}$	$27.2^{\pm3}$

[a]In computing these costs it is assumed that the O&M and fuel costs remain the same (in constant dollars) over time. However, capacity factors will change with time, which has the effect of changing capital charge costs per kwh with time.

[b]Based on costs for nuclear, coal with scrubbers, and coal without scrubbers of $1,000, $833 and $698/kw (in 1985 dollars), equivalent to $667, $555, and $465/kw (in 1976 dollars); on a capital charge rate of 13 percent; and on capacity factors of 60 percent for the nuclear and coal with scrubber cases and 67 percent for the coal without scrubber cases.

[c]Based on tails assay of 0.20 percent for uranium-235; no reprocessing of spent fuel; costs (1976 dollars) of $30/lb for U_3O_8, $3.33/kg for conversion, $80/kg/SWU for enrichment, and $90/kg (1976 dollars) for fabrication, respectively; and on carrying charges of 28 percent for U_3O_8 and conversion costs, and 24 percent for enrichment and fabrication costs. In the event of reprocessing, credits for recovered uranium and/or plutonium could offset part or all of the costs of the "back end" of the fuel cycle. With reprocessing and recycle, the figure for the back end of the fuel cycle is likely to range from a cost of 0.6 mills/kwh to a credit of 0.4 mills/kwh.

[d]Northern or Central Appalachian coal $1.08/$10^6$ BTU at the mine (1976 dollars).

[e]Montana-Wyoming coal at $0.43/$10^6$ BTU at the mine (1976 dollars).

[f]Movement of 300 miles in the East.

[g]Movement of 1,400 miles from Montana to the Midwest. These figures may suggest that movement of western coal is more expensive than for eastern coal. Since the heat content of western coal is about 75 percent of Appalachian coal, more western coal is consumed per kwh. Cost of transportation, by rail, measured in *dollars per ton*, is less in the West than in the East.

[h]These are bus bar costs at baseload generating plants. The charge to the consumer would be 50 to 200 percent higher because of distribution costs, the need for high-cost peak power, etc.

Table 3-4. Ranges in the Projected Cost for Generating Electricity in 1985[a]
(mills/kwh in mid-1976 dollars)

	Nuclear	Coal With Scrubbers	Coal Without Scrubbers
Central Values for the Midwest	23.9	28.5	27.2
Range Considering Geographical Variations Variations[b]	18–28	25–34	16–31

[a]These are bus bar costs at baseload generating plants. The charge to the consumer would be 50 to 200 percent higher because of distribution costs, the need for high-cost peak power, etc.
[b]This range does not include uncertainties and variables other than location.

Chapter 4

Alternative Energy Sources

An assessment of the need for nuclear power should consider potential alternatives that might be economically competitive or might avoid specific serious problems associated with nuclear power and fossil fuels. Three such energy sources which may offer prospects for successful market penetration and broad applicability are solar energy, geothermal energy, and fusion. A fourth source of energy is more efficient use of existing sources through conservation, cogeneration (of electric power and process heat), and improved distribution.

Critics of nuclear power frequently claim that these sources of energy can be successfully exploited in the relatively near future. Whether or not these sources can in fact contribute in the near term, they will probably be needed eventually. The potential availability of these alternative energy sources is indeed central to our conclusion that mankind faces no long-term shortage of energy.

In evaluating each potential resource, consideration must be given to technical feasibility, capital cost per kilowatt of capacity created or saved, fuel or resource cost, health or environmental impact, and impediments to large-scale deployment of the technology.

Our evaluation fixes attention not on the present or near-term prospects of advanced systems but on both their ultimate potential and their inherent limitations. For example, the economic aspects of fusion are assessed as they would exist if one of the fusion technologies presently being considered were to prove technically successful. A given fusion electric plant will incorporate some conventional elements whose costs can be determined, and other more novel aspects where cost estimates are difficult. If identifiable cost elements exceed total costs for nuclear or coal plants, it will be impossible for the new technology to compete. Thus, a particular technical approach may be eliminated as a competitor of coal or nuclear power on some time scale because simple considerations of capital cost and capacity factor of the identifiable portions of the system leave no room for the probable costs, quite aside from the technical uncertainties of the novel aspects of the system.

If an alternative system is not determined to be noncompetitive, we must then look at technological details—present status, plans, schedules, probability of success. Experience shows that, while unsuspected difficulties may be encountered in development of any new technology, unexpected opportunities may also arise, especially in the exploratory phases of the work. A consequence of this is that it is a waste of funds to have a demonstration of a commercial-scale plant using a technology that might be important a century hence but is predictably noncompetitive now. Too rapid a move toward demonstration also tends unnecessarily to impede progress on a long-term resource; focusing the technical effort prematurely may eliminate unexpected opportunities with no assurance of avoiding difficulties. Instead, the early phases of the development process should be sufficiently broad that problems and opportunities can be perceived over the widest possible field. At the same time, tools of research must be improved and understanding of requirements for alternative energy sources must be increased.

The goal of research and development efforts is to have moderate cost energy supplies available when the present low-cost—but, like oil, not always low-priced—energy resources are gone. In view of this goal, it is inefficient and unnecessarily limiting to try to make alternative sources available prematurely, particularly if the present costs of these sources is so high that they would have essentially no market share and could play no immediate part in supplying current energy needs.

SOLAR ENERGY

The amount of solar energy falling on the United States is enormous: 44,000 quads per year. The present annual U.S. consumption of electrical energy could be supplied by 0.15 percent of this solar energy, if it could be used at 10 percent efficiency.

There is no doubt that solar energy can be used to generate heat, electricity, and biomass or other fuel, but it is not certain how much it could cost to collect and use it for such purposes. The energy intensity of sunlight is much lower than that in the heat-generation portion of a nuclear or fossil fuel plant. In the United States, sunlight provides an average of 190 watts of energy per square meter throughout the year, compared with nuclear or fossil fuel plants operating with energy intensities of hundreds of kilowatts per square meter. The problem is to find a way of utilizing the low-density energy of sunlight at a capital cost low enough not to outweigh the benefits of "fuel" which is essentially free.

Methods of using solar energy directly include water heating, space heating and cooling, heating the working fluid of a heat engine, or using the photovoltaic effect to generate electricity. All of these suffer from the variability with which solar energy arrives at the earth's surface and the mismatch between time of production and time of use. Solar energy is available during only a portion of the day, and weather and seasonal variations greatly affect the amount available. Both contribute to a low load factor (fraction of time plant can operate) for solar plants. The fact that consumption is not limited to times when direct solar energy is available implies a need for large-scale storage.

When the bulk of a system cost is in capital expenditure, as it would be for solar-electric plants, the cost per kilowatt hour of energy produced is almost inversely proportional to the percentage of the time that the system is putting out power (except to the extent that components wear out with energy throughput rather than with time). Thus, a hypothetical energy plant with a capital cost of $1,500 per kilowatt peak electrical output (and a 15 percent annual capital charge rate) would have a capital charge of 26 mill/kwh at a 100 percent load factor, 43 mill/kwh at the 0.6 load factor typical of coal or nuclear plants, and 86 mill/kwh at the 0.3 load factor enforced by daily variation of insolation (the rate at which the sun's radiation is received at the surface) even at a favorable site.

Even apart from variations in power generation throughout the year (owing to changing lengths of days, sun angles, and long cloudy periods), the variation of insolation during each day even at a cloudless tropical generating site would make energy storage necessary for solar energy systems to compete in supplying power for normal uses. Peak electrical consumption, summer or winter, comes in the late afternoon or early evening when solar flux is low or even zero. Thus, without storage, solar energy cannot provide for peak load periods so as to save the fuel otherwise consumed in low capital cost plants such as those fueled by gas turbines, which might be used during times of high demand. Without energy storage, the generation of off-peak electricity by solar energy would have to compete in cost with nuclear or coal-fired plants, which however have a great advantage in not being limited to periods of sunlight.

Solar energy could be more widely applicable if energy storage was provided, and several techniques for storage are in various stages of development. Examples are thermal storage in hot oil or hot rocks, electrical storage in batteries, and mechanical storage using pumped water or compressed air. Storage systems thus consist of a reservoir, the capital cost of which is proportional to energy stored (in the case of electrical energy, a cost per kilowatt hour), together with devices for transferring the energy into and out of the store (the capital cost of which is stated per kilowatt for electricity). Up to one-third of the energy is lost in the transfer to and from storage. A reasonable figure for capital cost for storing six hours of generated electric power might be $300–$500/kw, plus an additional 50 percent capital cost increment on the primary solar thermal-electric system to make up for losses. Thus a plant costing $1,500 per kilowatt of peak capacity could have a capital cost element of 86 mill/kwh if the electricity were used currently; if six hours of generated energy were all stored for later use, the capital cost element would rise to 150 mill/kwh.

An alternative to storage is to use another energy system as standby capacity to augment the solar plant. However, this alternative would increase the effective cost of solar power by the additional capital cost of the auxiliary system plus the value of the fuel it used. Storage (thermal or electrical) could also be used with nuclear plants, another low-fuel-cost system, to allow load-following operation without consumption of fossil fuel. Thus the availability of storage may not improve the competitive position of central-station solar-electric plants versus nuclear reactors. It may do so, however, in small-scale systems.

Indirect uses of solar energy avoid the storage problem by utilizing a fuel which has stored the sun's energy. Examples are the growth of plant matter for fuel (biomass), or the more hypothetical artificial photosynthesis and the solar photodissociation of water to produce hydrogen. Alternatively, solar energy can be extracted from the ocean (where, in the tropics, the temperature varies by 20° C between the surface and a depth of 1,500 meters and where this temperature difference can be used to run a heat engine), from waves, or from the atmosphere, where solar energy is stored in the wind. In each of these cases, the energy reservoir is independent of whether the sun is shining, though the energy available may be periodic or otherwise variable for other reasons, as in the case of wind and waves.

Transmission of energy may also be a more severe problem with some forms of solar energy than with nuclear or even fossil fuel energy. Energy is consumed for the most part where people live and work. Insolation is about 40 percent higher in southern Arizona and New Mexico than the average over the continental United States, but the bulk of the nation's population is 1,000 to 2,000 miles away. For solar-electric plants, for which the contribution of energy-collector capital cost to the cost per kilowatt hour is inversely proportional to insolation, the cost (perhaps a cent or so per kilowatt hour) required to transport electrical energy will probably prove economical, but there will be a substantial incentive to reduce electrical transmission costs for very large blocks of power (10,000 MWe and up) by developing devices such as cryogenic superconducting underground cables. Converting and transporting energy from regions of high insolation in chemical form (e.g., as hydrogen) is likely to prove more expensive unless the process is such that energy is generated originally in such form. Small-scale applications of solar energy, such as for space heating and cooling, have an advantage over other sources in requiring no transportation or transmission.

Comparisons of solar-thermal, solar-electric, or ocean-gradient energy costs with those of nuclear power rest on capital cost judgments, since fuel costs are nonexistent in one case and low in the other. Cost comparisons between biomass and fossil fuel plants are even more straightforward; since the generating plants are similar, the comparison is a question of fuel costs. Cross-comparisons between capital-intensive and fuel-cost-intensive methods are more difficult since they depend on the assumed cost of money, amortization times, construction times, the differential rate of inflation between the cost of construction and that of fuel, and load factors.

Water Heating and Space Conditioning

Water and space heating by solar energy is currently being demonstrated in a number of projects as a substitute for direct electrical heat in residential use. At 40 percent solar absorption efficiency and a (naked) collector cost of $3–$5 per square foot, which is approximately the present situation, costs are competitive with direct (resistive) electric heating. (In contrast, installed system costs are about $20 per square foot of collector, including profit.) Capital costs are about $680 per kilowatt (average) of heat delivered at a location of insolation averaging 1,200 $BTU/ft^2/day$. At a capital charge rate of 15 percent per year, this cost would come to $3.40 per million BTU *if* there were a use for this much energy all year round; otherwise, the cost per million BTU produced would be higher. These solar costs, however, are several times those of heating systems using oil or unregulated gas, where costs are due principally to fuel.

Solar cooling is not currently competitive even with electrical cooling; commercial use of solar energy for space conditioning is at present ten times as costly for cooling as heating. Although better design could improve the economic position of solar cooling systems, their large-scale utilization awaits a further rise in oil and gas prices. Since space conditioning comprises about 17 percent of total U.S. energy consumption, solar heating and even cooling for all new construction to the year 2000 could provide for only a small percentage of total U.S. energy consumption in year 2000. Retrofit of existing construction, requiring much larger capital investment per kilowatt, would produce larger effects and sooner. Nonetheless, for the long term, solar energy may well be used widely for space conditioning future structures which will be built with much better insulation than at present so as to require less energy for space conditioning, whether by solar energy or by fossil fuel or electric power.

Solar Thermal-Electric Systems

The solar thermal-electric system is much more hypothetical than solar space heating. It uses fixed or movable arrays of flat or parabolic mirrors to reflect the sun's energy onto a central tower or onto multiple tubes containing a working fluid. The heated fluid is then used, directly or by making steam, to drive turbines and produce electricity. For plants expected to be ready in the early 1980s, present estimates of the capital cost in (present) dollars per kilowatt of peak electrical capacity appear to be about $2,000, two to three times the capital costs for nuclear plants, but, because of diurnal variation of insolation, some six to nine times as capital-intensive per kilowatt hour. The relative total energy costs are little changed by the inclusion of nuclear fuel costs since the latter are equivalent to at most a few hundred dollars of capital cost difference.

Estimates of the capital cost ratio of solar thermal-electric plants to nuclear power plants are more likely to rise than to fall, even though nuclear costs are likely to rise, since solar thermal-electric plants are much less well defined than nuclear ones and since we have had no experience with them. Considering relative load-factor and energy storage costs, a generating cost of 100 mills per kilowatt hour can be estimated for solar plants, while the range for nuclear and coal plants is 20 to 40 mills. Such solar systems thus appear to be a possible, but very expensive, alternative to large-scale nuclear or coal plants at some time well into the twenty-first century, if real costs of coal or uranium become very high. They would be used before then only if noneconomic factors such as nuclear hazards or CO_2 accumulation prevented the use of fission energy or coal. Nevertheless, they are worthy of modest research and development. Demonstration should be delayed until there is some prospect for competitive entry into the market. Primary areas of high and uncertain cost are the collectors, tracking mechanisms, and especially storage.

Solar Photovoltaic and Mixed Systems

Solar photovoltaic systems using silicon, gallium arsenide, or other materials could dispense with thermal conversion equipment and generate electricity directly. Efficiencies of about 10 percent have been achieved in direct sunlight. Even if the theoretical limit of 22 percent (for single silicon crystals) for photoelectric conversion could be reached, current collector costs are about $200,000 per kilowatt of peak electrical capacity, several hundred to one thousand times what they have to be (about $1 per square foot of collector) in order to compete with nuclear or fossil plants. Polycrystalline substances are much less expensive than silicon crystals, but their efficiency is also lower. Moreover, the lifetime of polycrystalline collectors must be increased substantially to equal the expected lifetimes of nuclear or fossil plants. An efficiency near 40 percent might in principle be achieved by stacking heterojunction cells of different composition to cover the sun's spectrum more broadly. Since a change of a factor of a hundred or even a thousand in cost per square foot of photoelectric surface is conceivable, the search for new techniques is worthy of intensive research and development.

Another possible system concentrates sunlight onto photovoltaic surfaces through the use of mirrors or lenses. This method reduces the area of photovoltaic crystal needed to about 1/1,000 the mirror area, so that its cost becomes less dominant. Gallium arsenide had demonstrated 20 percent efficiency of conversion of concentrated sunlight to electrical energy in such a configuration, and the opportunity to achieve higher efficiency is even better for stacking junctions. Since gallium arsenide crystals can stand temperatures only up to 350° C, they would have to be cooled, but the coolant might be used for a secondary energy recovery process. The photoconversion efficiency for such "hot photocells" may be only one-half that of a solar thermal-electric cycle (20 percent versus 40 percent), although economy of scale should be available at much smaller sizes with the photovoltaic approach. Mirror costs would have to be brought down to the neighborhood of $1 per square foot in order for the hot photocells to compete with the thermal-electric approach for central-station power plants; the competition with nuclear or coal plants would depend upon the existence of storage. Thus, it seems possible that concentrated photovoltaic systems may have some application (with storage), in sizes of kilowatts to megawatts, in situations where fuel costs are very high.

Wind and Sea

Present capital costs per kwe of rotor-style windmills are substantially higher than for fossil or nuclear plants. Advanced designs, if they are successful, could have capital costs comparable to those of nuclear or fossil plants, with present estimates ranging as low as $1,000/peak kilowatt of electrical capacity. Component lifetime, particularly of blades and rotors, is a major source of cost uncertainty since designs have not been tested. If these uncertainties are resolved favorably, there will still be an extra cost factor of three per average kw due to the relatively low load-factor and intermittency problems or due to storage to remedy these problems. Nevertheless, wind power has the potential of meeting a significant portion of electrical energy requirements in some areas, albeit at a cost which may be three to five times that of nuclear electrical power. One estimate of availability would permit generation of an average of 40 MWe per square mile in rural districts where construction would not interfere with human activities.

The various oceanic schemes appear much further off in time and economic feasibility. Most involve an extrapolation of existing technologies to a severe environment, and their proponents usually make the questionable assumption that the technology transfer to such larger units is accompanied by a large reduction in cost per unit volume of flow or unit area of heat transfer. Making the ocean-gradient scheme work requires development of heat exchangers, pumps, compressors, and other equipment. Only currently unrealistic assumptions about efficiencies, pressure differences, and equipment survivability can make the energy cost competitive, even ignoring the problem of transporting the energy from sea to land. Careful existing analysis gives a capital cost in the neighborhood of $2,000 per kilowatt of capacity, a cost that does not take into account the possibility of low load-factor and transmission needs. Theoretically, the maximum efficiency for operating a heat engine on the 20° C ocean thermal-gradients is only 7 percent; in practice 2 percent is probably the best that can be achieved. This low efficiency makes economical utilization of ocean-thermal energy very difficult.

Biomass

Despite its long history, the use of vegetable matter for fuel is still undergoing development and may benefit from advanced biochemistry and molecular biology. The theoretical limit for conversion efficiency from solar energy to fuel is about 5 percent; in practice, 0.5-1 percent conversion over the year is apparently achieved with certain high-yield crops such as eucalyptus, conifers, sorghum, and sugar cane. One pound of dry plant matter yields about 7,500 BTU, about two-thirds the energy derived from an equal weight of coal. Thus fifteen tons of annual dry crop per acre corresponds to about 220 million BTU per acre, a product equivalent to about 1 percent of the average incident sunlight. Pyrolytic heating of biomass can simultaneously produce oil, medium BTU gas, and char containing, respectively per ton of biomass, about 7 million BTU, 0.6 million BTU, and 8 million BTU and, altogether, about 85 percent of the initial fuel value of the biomass.

At 1 percent efficiency of conversion of solar energy to hydrocarbons and about 33 percent thermodynamic efficiency of conversion to electricity, it would take 4.6 percent of the area of the continental United States, or about 100 million acres, to provide biomass-generated energy for the present level of U.S. electrical consumption. This land requirement would pose severe problems, such as finding areas adequate from the standpoint of insolation, rainfall, nutrients in the soil, and topography suitable for crop or tree cultivation and harvesting. Fertilizer requirements could be a substantial element of cost unless genetic change can be accomplished which allows nitrogen fixation by bacteria in the plant roots, or unless more efficient use can be made of fertilizer. A typical input of 150 pounds of fertilizer for one year, yielding six tons of dry biomass (with fertilizer at $250 per ton) would cost $0.20 per million BTU of fuel value. Trace elements could be expensive as well if they had to be introduced or replaced. However, a 1,000 MWe plant could operate from the output of biomass raised by intensive cultivation within a radius of twenty to thirty miles of the plant, reducing transportation costs and allowing recovery of trace elements by plowing the ash back into the soil if combustion is controlled so as to permit recovery of those elements.

Burning currently grown plant matter rather than fossil plant material is one way to avoid the possibly serious consequences of a greenhouse effect (as discussed in Chapter 6) since the carbon dioxide put into the air by combustion of currently grown biomass would be offset by the removal of carbon dioxide in the photosynthesis that creates it.

Thos cost of such an operation would depend on the price of land, farm equipment, labor, and water. Estimated costs for biomass fall in the range of $15-$25 per ton ($1.00 to $1.60 per million BTU). This is perhaps 20 to 100 percent above the present average price of coal, but below the $2 per million BTU price of foreign oil. The higher hydrogen content as compared with coal would produce larger fractions of liquid and gas from pyrolytic processing. The sulfur content is below 0.1 percent. If the crop could be sold at $2 per million BTU, crop value added would be $5 to $15 per ton; at a high yield of fifteen to thirty tons per acre per year this would total some $75 to $450/acre/yr, comparable with the value added for wheat acreage in the Midwest. Conifers or other perennials have larger cost per BTU but avoid the otherwise difficult problem of scheduling harvest or storage, both of which influence the load-factor. There is, of course, no reason for the market to pay $2 per million BTU for biomass when coal is available at less than $1 per million BTU. Thus biomass use will be limited to local production of heat and electricity.

Assessment of the biomass potential must deal with many unresolved problems of which the most serious is probably assuring supplies of water and nutrients. If land is well supplied with these, it could, however, presumably be cultivated to produce other cash crops in preference to biomass for fuel. Kelp can be grown in the ocean, solving the problem of water but posing a rather large harvesting and drying cost. Using thin films of plastic over biomass crops to recover transpired water, direct drip irrigation of roots, recovery and recycling of trace elements in the combustion process may or may not be necessary or practical. Water projects of continental scale may be economical if water is used more sparingly for irrigation. Finally, the environmental impacts of biomass production would not be small. While one may think of biomass as sylvan or bucolic, each 1,000 MWe installation would be a factory extending over a thousand square miles. The proper analogy is a logging forest and pulp mill.

The extensive land requirement is a further economic bar to biomass fuel since the opportunity cost of farmland, water, transportation, etc., would often be greater for food or other cash crops. Thus, biomass fuel would not appear to be economically viable in the United States on a large scale for the foreseeable future. In some areas, biomass as a by-product of agricultural or wood-product operations could be used economically now on a limited scale. If CO_2 from coal eventually turns out to have an unavoidable deleterious effect on climate, biomass could provide, at higher cost, a substitute that avoids this problem, and serious research should continue in view of this possibility.

Prospects for Solar Energy

There is enough solar energy to meet any foreseeable needs for electrical energy. The question is one of price. In general, the capital cost per kilowatt of peak capacity for most methods of converting solar energy to electricity in central-station facilities is estimated to be more than double that of nuclear power, a figure that must be multiplied by another factor of 1.5 to 4 for differences in load-factor and possibly for additional costs for energy storage. While fuel costs are zero, this by no means compensates for capital cost differences in present cost comparisons with coal or nuclear energy. Moreover, solar technology (except for biomass) is not nearly so well developed as nuclear or fossil technology. In most cases, capital costs are first likely to rise as we learn more about the problems before they decline as we learn more about the solutions.

A small fraction of U.S. electric energy could nevertheless be generated by solar thermal-electric sources by 2000 in view of the varied circumstances over the United States. Specialty uses (water and space heating and perhaps cooling, remote-area windmill or solar-thermal generators) are nearer competitive feasibility and are much more likely in this century than solar central-station generation of electricity.

Photovoltaic methods are far less competitive at present as economic sources of energy but, since large improvements may be possible with new ideas, economically competitive designs might emerge within a few decades. Ocean thermal gradient methods are presently noncompetitive, and large improvements in them are unlikely. Tides and waves are also unrealistic prospects as sources of an appreciable fraction of U.S. or world energy.

The environmental, health, and safety advantages of using solar rather than fossil or nuclear energy are real, but environmental problems are not absent. Changes in land use, industrial, and other patterns of activity will be substantial but should be manageable.

Biomass methods are unwieldy and of low efficiency, but well-proven in small scale. Close to being economic on a small scale, especially as a by-product use, their feasibility and economy on a mass scale are much more doubtful (considering the demand for water, nutrients, land use patterns) but deserve some research and development. A limited demonstration program is justified for special biomass applications which have immediate practical value, such as crop drying or combined garbage disposal and energy generation.

For the longer term, one or more of the solar energy methods may provide a significant fraction of energy in the United States, but not until rather far into the twenty-first century, and with a price premium over nuclear or coal power. A cost factor of three or more, including storage and load-factor effects, may be incurred to pay for the advantages of assurance of a renewable supply and lower environmental effects. These sources cannot be counted on as an economic alternative to coal and nuclear power in the next three decades. They should be considered as possible alternatives to or competitors of breeder reactors, fusion, or coal later in the twenty-first century. It is important to recognize that society can depend on the availability of solar energy for the long run, at a cost it can afford, even though it would prefer to pay less.

GEOTHERMAL ENERGY

Geothermal energy is the natural heat contained in the earth's crust. The amount of energy stored is very large. The geothermal heat in the outer 10 kilometers of the earth's crust under the United States is some 6×10^{24} calories or about 1,000 times the heat contained in the total U.S. coal resources. Most of this vast store of energy is too diffuse to be economically useful but a substantial fraction is concentrated and probably recoverable with present-day technology.

Five types of geothermal energy resources are generally recognized:

- Hydrothermal convective: regions containing high temperature water at shallow depths. There are two kinds of formations: vapor-dominated areas, with temperatures between about 150° C and 250° C, producing generally superheated steam; and hot-water-dominated areas, with temperatures ranging from 90° C to 350° C, capable of producing hot liquid or a mixture of liquid and vapor.
- Hot dry rocks: not molten but very hot rock structures, essentially dry with temperatures up to 650° C.
- Geopressured resources: pressurized water reservoirs in sedimentary basins, capable of supplying heat, mechanical energy, and dissolved methane.
- Normal gradients: conduction-dominated areas produced by heat flows, radiogenic heat production, and thermal conductivity of rocks with temperatures ranging from 15° C at the surface to about 300° C within the first 10 kilometers of the crust.
- Magma: molten or partially molten rocks at temperatures greater than 650° C.

Estimates on how much energy is recoverable from each of these sources vary considerably. The following are given in the most recent publication of the U.S. Geological Survey (Geological Survey Circular 726): The combined potential of vapor- and liquid-dominated geothermal systems in the United States is about 7×10^{20} calories of recoverable geothermal energy in formations about 90° C. This energy, converted into electricity at an efficiency of about 5 percent, could supply present U.S. electrical consumption for twenty years. If the estimated 8×10^{22} calories of recoverable energy in hot dry rocks could be tapped to deliver electricity, it could meet U.S. electricity needs for 2,000 years at current consumption levels. The amount of energy in known geopressured sources and in molten rock (magma chambers) is comparable to or even larger than that in the hot dry rock resource, though the techniques for recovery are unclear and the costs might prove prohibitive. Most of the stored energy is in geological areas with only normal thermal gradient, the most difficult source to exploit.

These enormous geothermal energy sources would seem in principle capable of making a significant contribution to the U.S. and world energy needs in the relatively near future. However, one must examine carefully the problems associated with each source in order to assess its real potential.

The two hydrothermal convective systems occur in regions where the heat flux from the depths of the earth is several times the worldwide average. Within these regions are hot spots where water, almost entirely derived from surface precipitation, penetrates to a layer of permeable rock (characteristically at depths of about 3 kilometers) where it is heated by hot dry rocks below the permeable layer. The water can be tapped by drilling, or in some cases, such as the natural geysers in Yellowstone National Park, it makes its own way to the surface.

Electrical power is being produced commercially from hydrothermal sources on a small scale in several countries including Japan, New Zealand, Italy, the Soviet Union, the United States, Iceland, and Mexico. The total world electrical capacity at present is only about 1,200 MWe. There is also some minor utilization of geothermal power for nonelectric uses in the United States and considerably more abroad. The vapor-dominated hydrothermal sources are the oldest producers of commercial electricity. A plant at Larderello, Italy, demonstrated electrical production in 1904 and began continuous operation of a 250 kilowatt electric generator in 1913. The first commercial geothermal steam turbine in the United States was a 12.5 MWe unit that began operation in 1960 at The Geysers, California. This field, which now has a capacity of 502 MWe, making it the world's largest electrical generating geothermal site, is expected to grow to about 1,000 MWe within about five years. These vapor-dominated systems produce dry steam at temperatures of about 200° C which can be directly utilized by steam turbines, although because of the lower temperatures and pressures, they are only one-third as efficient as turbines in a conventional power plant and hence higher in capital cost per kilowatt hour.

Although hot-water-dominated geothermal sources are about twenty times more common than the vapor-dominated ones, they have not yet produced much commercial electricity. The major known hot water fields are in New Zealand, the Salton Sea area in California, and, of course, the Yellowstone geyser basin, which is presumably a permanent national park, never to be exploited for power. The principal reasons that hot water systems have not been used are the difficulties in separating the water from the steam and in handling the extremely corrosive, mineral-laden water. For geothermal fluids below 200° C, attention is now being given to the use of heat exchange systems in which the heat from the geothermal water is transferred to a low-boiling-point liquid, such as freon or isobutane, which is then allowed to expand in the turbine—a binary conversion system. The geothermal water is not allowed to boil and is reinjected into the ground. Above 200° C, fluids may be allowed to flash into steam; unflashed fluids plus condensate steam will have to be disposed of.

Most known hydrothermal reservoirs are in the western part of the United States. There are some 300 deposits with estimated subsurface temperatures above 150° C. Five deposits in California and one in New Mexico together contain more than 10^{19} calories. A very large region in Idaho is estimated to have 2.6×10^{20} calories. If one assumes that 25 percent of the stored energy is recoverable and that about 25 percent of this may be utilized, about 2×10^{19} calories are available. Assuming 10 percent efficiency in producing electricity, the Idaho hydrothermal deposit could supply a total of ten quads of electricity.

Current cost estimates of hydrothermal generating plants are uncertain. The original dry steam plants at The Geysers cost about $114/kwe. Production costs were estimated at about 7 mills/kwh of which 3.5 mills/kwh was the price of the purchased steam, including 0.5 mills/kwh for reinjection of water condensate. Recent estimates of installation costs range from $500 to $700/kwe for direct flash, and twice that for binary conversion systems. Operating costs range from 25-30 mill/kwh for direct flash to over 40 mills/kwh for binary conversion systems. Among the uncertainties in the total costs are those related to resource exploration and drilling and the operational lifetimes of production and disposal wells. Because of limited experience with these plants, it is difficult to compare their costs with those of more familiar forms of power generation.

The "national goal" for geothermal energy in 1985, as stated in the 1974 *Project Independence Blueprint, Final Task Force Report,* is 20,000-30,000 MWe. This projection is extremely optimistic and would require a great increase in resource exploration and commitment of funds both by government and by the private sector for higher-cost power than is obtainable from coal. A more recent estimate in the *First Annual Report of the Geothermal R&D Program* (January, 1977) is 3,000 to 4,000 MWe, including 2,000 MWe at The Geysers.

The technical problems in utilizing hydrothermal resources are for the most part mundane but serious. Drilling techniques and well completion methods developed in connection with oil and gas exploration are adequate for the shallower reservoirs, but some technical improvements will probably be required for economic exploitation of deep hot water sources.

There are also environmental considerations. Although geothermal energy is often viewed as environmentally attractive, waste waters from both steam and hot water wells are usually highly mineralized and may pose substantial environmental problems. A 1,000 MWe plant based on hot water with a salt concentration 10 percent that of sea water would produce 3 to 6 million tons of water per day containing some 20,000 tons of salt. Such water would have to be desalted or reinjected into deep wells. Reinjection may in fact be necessary to avoid land subsidence. For a thermal-electric power output from hot water equal to that available from a ton of oil, 1,500 tons of water must be removed from the earth. In fact, a single such 1,000 MWe power plant would produce as much water as falls as rain on 4,000 square kilometers of the earth's surface; the hot water flow through a hydrothermal plant is also about three times the ocean or river coolant flow in a once-through nuclear or coal-fired plant. Without a closed cycle involving reinjection, air pollution may also be a serious problem, since dissolved hydrogen sulfide is released from the hot water as it flashes into steam. For example, the hot water plant at Cerro Prieto, Mexico, may release more sulfur than would a conventional fossil fueled plant of the same output burning high sulfur fuel. Thermal releases of energy to the environment are similar in nature to those associated with other types of power plants, but three to five times as much by-product heat must be released per unit of power output, due to the low temperature of the source and the consequent inefficiency with which it may be used to generate electricity. While there are no insurmountable environmental problems associated with hydrogeothermal energy, it appears that the economic cost of controls might be appreciable.

If means can be developed to extract it, geothermal energy in hot dry rocks offers a promising source of energy. A hundred times more energy is stored in hot dry rock than in natural thermal convective sources. A reasonable value of the ambient thermal gradient in the United States is about 25° C per kilometer so that at depths of about 8 kilometers, temperatures reach 200° C. Oil wells are routinely drilled to depths of more than 8 kilometers using conventional rotary drilling equipment. To determine whether artifical reservoirs can be developed and operated economically by tapping the hot dry rock, the ERDA-supported Hot Dry Rock Geothermal Energy project is attempting to create natural circulation in a pressurized water loop in a hot granite formation at a site about 25 miles from Los Alamos, New Mexico, where the temperature at 3 kilometers is 200° C. The method involves hydrofracturing a significant volume of very low-permeability rock so that two bore holes can be connected below the surface by natural convective flow. The holes have been successfully connected, and experiments to demonstrate operation of a 20 MW (thermal) facility using air-cooled heat exchange are underway; a demonstration plant of about 80 MW (thermal) or 10 MW (electrical) is projected by 1980, and two to four commercial plants might be built before 1985.

Geopressured resources are found in sedimentary rocks along the Texas and Louisiana Gulf Coast. Although there is no energy production at present, such resources appear to be substantial; estimates by the U.S. Geological Survey project 30,000 to 115,000 MWe for thirty years. There is probably a similar amount of energy in entrapped methane. However, uncertainties are associated with the volume and permeability of the aquifers which store this high-pressure hot fluid. The depth and high pressures of the geopressured resources may require the development of less costly drilling technology and of high-pressure turbines before commercial exploitation is possible.

There are no good ideas for the extraction of heat from the normal temperature gradients encountered in the eastern and central portions of the United States, despite the size of the resource (8×10^{24} calories). Of course, a major difficulty is that as heat is withdrawn from a solid, an increasingly thick cool layer impedes the continued flow of heat.

Magma, which is molten or partially molten rock, generally lies beneath the crust starting at 20 to 50 kilometers. It is believed that in the United States as much as 2.5×10^{22} calories of energy are stored in magma chambers at depths of less than 10 kilometers. Serious technical obstacles must be overcome, however, before this huge source of energy can be tapped. To work in a hostile environment at high pressures and temperatures of 650°–1,200° C requires improvements in drilling techniques, methods to keep drill holes open during drilling and heat extraction, and techniques for extracting heat. There are no plans for exploitation in the foreseeable future.

Summary of Prospects for Geothermal Energy

Vapor-dominated hydrothermal resources are economic now for the production of electrical power and some are currently being exploited. These resources are very limited in magnitude and location, however, so that they can supply a total of only some 2,000 MWe. Hot dry rock resources are at the earliest stage of feasibility and exploration. The high capital costs for low-efficiency turbines, together with potential difficulties with mineralized water, will probably delay the advent of hot dry rock as a source of electrical power until coal and nuclear fuel become very expensive. Geothermal heat is more competitive with fossil fuel and nuclear power for low-quality heat for space conditioning and domestic hot water, but its use is probably limited to populations within a few kilometers of hydrothermal or hot rock resources. Thus, although there will be some small increase in the use of geothermal energy in the near term, substantial use will only develop with the advent of significantly increased competitive energy costs or through entirely new ideas for exploitation.

FUSION POWER

In fusion processes, light atoms react to produce a heavier element with the net release of energy. For example, at sufficiently high temperatures, deuterium and tritium, isotopes of hydrogen, react to produce helium and a very energetic neutron. Since the fuel for such reactions is essentially inexhaustible, fusion has long been regarded as a promising future source of energy. The sun, which is a giant fusion reactor held together by gravitational forces, amply demonstrates the potential of these reactions.

While there is no doubt that there is a net release of energy in the basic nuclear reactions (which are, of course, used in hydrogen bombs), the feasibility of this process as a practical energy source depends on additional factors. In order for a power plant to have a net energy gain (an output greater than the energy put in), large numbers of atoms must be made to react at a high rate. This rate must be great enough so that the total power output is larger than the power requirements of the various ancillary devices which are required to support operation. Because of the complexity of the process, demonstration of technical feasibility requires construction of a facility having a net energy output. Although the research and development toward that goal has now been proceeding for several decades, such a demonstration has yet to be made.

There are two basic approaches to a fusion reactor: equilibrium confinement and inertial confinement. In equilibrium confinement, reacting atoms are held together in approximate thermal equilibrium for a protracted period of time so that a reaction can proceed for a time sufficient to lead to useful power. In principle, this process can be made to work. The question is one of scale. As a fusion reactor becomes larger and larger, the losses of the reacting atoms across the surface of the reactor become proportionately smaller relative to the total volume in which the whole reaction takes place. For this reason, research, development, and demonstration of the confinement process is very slow and expensive. Experimentation on a small scale is frequently insufficient to give conclusive evidence as to the behavior of a large plant.

In inertial confinement, there is a rapid dynamic assembly of the hydrogen isotopes to produce a high density reaction for a short period of time. The classic example of this process is the hydrogen bomb. Power plant proposals in this category include small pellets containing deuterium and tritium which are to be compressed by a factor of 1,000 to 10,000. Compression might be achieved through the rapid heating effects of intense laser beams or by intense electron or ion beams from particle accelerators. The absorption of the incident radiant energy rapidly boils off an outer boundary layer; as this material recoils from the pellet, a shock wave is generated which compresses the pellet to a high density.

The use of the explosive power of hydrogen bombs to generate geothermal heat which can then be extracted has been repeatedly suggested. We do not consider such a proposal to be a useful energy alternative because it would require much of the same capital equipment required for an ordinary reactor, but would substitute a more expensive heat source for a compact reactor core. A 1,000 MWe plant would require about one 100-kiloton hydrogen explosion per day (or a number of smaller explosions with an equivalent yield) to heat water vapor to run the power plant.

The end product of the deuterium-tritium fusion reactions is a large number of high energy neutrons. These produce heat if absorbed in an enclosing "blanket." Liquid lithium appears to be the preferred substance for this use. The heat extracted can then be used to make steam which can be used to drive a turbine to make electricity, just as is done with fossil or nuclear fuels. At the same time, the lithium "breeds" tritium by neutron impact, thus replenishing the tritium used in the primary reaction.

Alternatively, it has been suggested that the fusion neutrons be injected into a "subcritical" fission reactor loaded with natural uranium or thorium, thus increasing the number of neutrons through fissioning of these materials. Ultimately, some of the larger number of neutrons would be captured in uranium or thorium, producing fissile plutonium-239 or uranium-233. Both heat and new fissile material would be produced in this manner. Such a "fusion-fission hybrid" might appear to be a potential competitor to the breeder reactor, since it would avoid the possible danger from criticality accidents, which are an inherent although remote possibility in the breeder. It would have, however, most of the other negative characteristics of the breeder in terms of safety and plutonium traffic and would involve all of the technical complexity of fusion reactors. Accordingly, there appears to be little reason to pursue this approach.

The fusion reaction used initially will almost certainly involve the reaction of deuterium and tritium at very high temperatures. In this reaction, an ionized deuterium atom and an ionized tritium atom with average energies of approximately 50 kilovolts produce a helium atom and neutron with an energy of 15 million electron volts (Mev)—an energy gain of about 300. In practice, of course, the energy gain will be seriously degraded below this theoretical amount. This energy gain is reduced by a factor of three by thermal inefficiency in converting heat sources into electrical energy. Other energy losses occur in particular fusion designs. For example, some of the energy is always shared by the electrons in the plasma, thus detracting from the energy available to the reacting nuclei.

To produce fusion reactions in equilibrium, it is necessary to create a "bottle" in which the reacting gases can be maintained at very high temperature for protracted periods of time. Since an energy of 50 kilovolts per particle corresponds to a temperature of some 150 million° C, it is clear that the walls of the bottle cannot be made of ordinary material. The only practical alternative appears to be confinement by a magnetic field. At the high temperatures involved, the deuterium and tritium will be in the form of a "plasma" in which electrons are stripped away from the parent atoms leaving them electrically charged. Magnetic fields can be designed to deflect these moving, charged particles in such a manner that they are confined within a volume, in effect creating a wall around the plasma. However, such confinement involves a number of problems. The resultant magnetic bottle tends to leak since there may be openings through which some plasma can escape at some rate and there may be instabilities in the motion of the plasma, which cause losses across a magnetic boundary. Moreover, the establishment of the magnetic field requires energy. To achieve a reasonable burning rate for deuterium-tritium, densities of the order of at least 10^{12} particles per cubic centimeter are required at a confinement time of 100 seconds or more. In turn, confinement of such densities requires pressures approaching one atmosphere, corresponding to a magnetic field strength of 6,000 gauss. A fusion reactor of the Tokomak type, which is currently the leading contender for a practical fusion reactor, might be 35 meters in diameter and 25 meters high. This corresponds to a magnetic field volume which is in excess of the largest built in the past. The energy contained in such a magnetic field is so large that the field would have to be maintained for a very long time rather than being replenished frequently; otherwise, the energy needed to restore the field would destroy the potential energy gain of the system. The coils required to produce such fields would generally have to be superconducting since the power required to excite a magnet by conventional copper or aluminum coils would use much, if not all, of the energy output of the device. The construction and maintenance of a superconducting coil under the condition of mechanical stress and neutron flux is a severe technical problem.

Since plasma is lost through reaction or escape at an appreciable rate, it has to be replenished. Substantial energy is required to create the new plasma which has to be introduced to make up for the losses. In addition, the introduction of new plasma in some approaches requires turning the magnetic field on and off. This again requires energy or a further capital investment for storage. (Resupply of fuel as frozen pellets or neutral beams is an alternative.) Thus, the success of any magnetic confinement machine (Tokomak, magnetic mirror, pinch systems, and so forth) depends on a reconciliation of the energy losses (involving establishment of a magnetic field, reintroduction of plasma, and requirements for other sources of energy) with the energy output of the reaction, its thermal conversion and electrical energy generation. Designers hope that a demonstration of a fusion reactor based on the Tokomak type which will yield a net positive energy output will be achieved in perhaps four to five years.

Although the deuterium-tritium reaction is the most promising initial candidate to fuel a fusion reactor, the deuterium-deuterium process also leads to a net power output, although at higher confinement temperatures.

Fusion reactions also exist in which the final products are charged particles rather than neutrons. If such reactions could be made self-sustaining, the excess energy could be extracted by slowing the particles in an external electric field thus directly converting the reaction energy to electric power. Such a "direct conversion" fusion reactor is, however, not a candidate for an initial practical device.

Although magnetic confinement appears to be the most promising approach to the fusion problem, nonequilibrium fusion reactions are also being pursued as a possible means to attaining the goal of a useful energy producer. This effort is currently focused on compressing solid pellets of deuterium and tritium to densities of 1,000 to 10,000 times normal. The efficiency of converting this explosive energy to electricity will certainly be well below the normal one-third conversion efficiency of thermal to electrical energy. For practical purposes, one ton of TNT appears to be a generous upper limit on the yield of the individual pellets in view of the delicate nature of the complex supporting equipment. (A 1,000 MWe plant would require at least one such explosion per second.) The production of competitive electric power by this method would require that the pro-rated cost of the entire process, including all capital costs of the plant, be held to about $1 per pellet—a cost much less than can now be envisaged. Thus, this approach does not appear to be a serious competitor with fission or fossil fuel for the foreseeable future even if a net energy output can be demonstrated. At this time, compression by use of ion beams may be a more promising approach since the explosive power from a single pellet can be considerably larger for such a device than can be sustained by the more sensitive optical elements needed to focus laser beams.

Prospects for Fusion Power

Among all the contenders for a fusion reactor at this time, the long-pulse magnetic confinement (Tokomak reactor) is most advanced. It is estimated that in about five years large scientific research machines in this country and abroad will generate neutron fluxes comparable to those required for a full-scale machine. Therefore, basic information from which a full-scale reactor can be designed and from which costs can be estimated might be at hand as early as

1985. However, at least two further generations of such designs leading to actual construction would be required before commercial use could be contemplated. Moreover, total research and development investment costs, necessary to reach this stage, have been estimated by ERDA to be in the $20 billion range, in constant dollars.

Although the fuel costs for the fusion reactor are small, capital costs are large and probably substantially greater than those of breeder reactors. This conclusion follows from the fact that the conventional components of the systems are comparable in cost to the breeder, while costs of shielding the fusion reactor are greater, and sensitive and costly elements of the plant must enclose this bulky shielding. Moreover, the basic fusion reactor is technologically a much more complex device than a breeder reactor core. Accordingly, development of central station power based on fusion reactors appears inherently more expensive than the breeder although associated social costs would appear to be less. It should not be concluded, however, that fusion reactors will avoid entirely the environmental and safety problems of fission reactors. Pure fusion systems will not be free from radiation. Indeed, more neutrons are produced per kwh in a fusion reactor than in a fission reactor, and these neutrons are of higher energy and therefore more difficult to shield. There will be no fission products or actinides, however, and the problem of long-term radiation is thus one hundred times smaller than for a fission reactor of comparable power. While the basic fuel (deuterium) is virtually inexhaustible, there may well be resource constraints on the special materials such as lithium, beryllium, and the materials required for superconducting coils.

As an overall assessment, we believe that even a technically successful fusion reactor is unlikely to be able to compete economically with the breeders as fixed station energy producers early in the next century. Nevertheless, fusion research and development should be pursued since it presents a major long-term energy option which is a hedge against problems involving coal and nuclear power. Fusion will avoid some of the social costs of coal and nuclear power, and its fuel supply is essentially inexhaustible. Moreover, new and more promising configurations for fusion reactors may emerge. In view of the large associated costs and the long time horizon for practical application, such research and development will remain a federal responsibility. Since the large size of research projects implies high costs, fusion presents unique problems in research and development strategy.

CONSERVATION

Conservation is one of the most effective means of making available additional energy to produce desired goods and services. Whatever the form of energy use, there will be ample opportunities for conservation. The higher the unit price of energy, the greater the benefit from a given energy saving. After many decades in which real energy costs have been falling, energy costs now seem destined to rise over the long term. This rapid reversal in the trend of energy costs has led to new interest in the techniques and economics of energy conservation.

Conservation encompasses a set of options to be evaluated for profitability (for business) or for income-equivalence (for individuals). For society, conservation actions free investment for alternative uses, at least until energy savings no longer repay further investments in conservation. Conservation also lengthens the time available to develop competitive means of energy generation.

The fourfold increase in the price of imported oil since 1973 justifies conservation investments today that would not previously have repaid their cost. The conservation business, however, has not yet expanded to the extent that its profitability would appear to warrant, and industry has not realized the range of opportunities to reduce its costs through more efficient energy use. In part, this reflects initial uncertainly about long-term trends in energy costs. After the 1973 increase in oil prices, there were predictions that such high prices could not be sustained. However, there is growing recognition that, in the long term, real energy costs will probably continue to rise slowly into the next century.

As higher energy costs become accepted as a fact of life, business will replace capital equipment, add insulation, buy diesel rather than spark-ignited vehicles, employ cogeneration, and adopt other conservation measures to the extent that they appear economically desirable. Government has a role in providing information, analyses, and in forecasting the results of such decisions. Government should also help support of research and development in conservation and, where necessary, promulgate energy performance standards. The critical government role, however, is to ensure that regulation does not impede the market in its adjustment to changing conditions. Regulations discouraging industrial production of electricity, differential freight rates on iron ore and scrap iron, and the pricing of energy (natural gas and electricity) below the marginal cost of production inhibit economically desirable conservation measures. For the market to function and to achieve the economically appropriate level of conservation, the value of a unit of energy saved should equal the cost of the last unit produced; that is, the marginal cost of energy supply.

Electricity is an example of energy pricing which does not directly reflect marginal costs. Generating capacity and capital investment are determined by peak load, while customers for the most part are charged a rate independent of system demand or time of day. Some system-demand-dependent rates exist for large users, and time-of-day pricing is spreading. In New York, where residential electricity prices average 7 to 8 cents per kilowatt hour, Consolidated Edison has initiated a time-of-day pricing experiment in which electrical energy used by a residential customer during peak hours will be charged at 14¢/kwh, while that used off-peak will be billed at 2¢/kwh. If such rates were widespread, there would be an incentive to reduce peak electrical consumption and the generating capacity to support it. The lower cost of electricity off-peak reflects the lower cost of generation in base-loaded nuclear or fossil-fueled plants and the low cost and low losses in transmission over facilities whose size is determined by peak demand. The much higher rate on-peak reflects the cost of additions to transmission and generation systems, and the high cost of fuel for standby low-efficiency gas turbine peaking plants.

In this study, the prediction of the effects of price-induced conservation is equivalent to an estimation of the price elasticity of energy demand. To determine aggregate national reductions in energy use at an unchanged aggregate output level necessitates a close look at many sectors of the economy. This could be analyzed in several ways: by type of energy used, by physical means employed for conservation, or by economic sector, e.g., commercial and residential, transportation, and industrial. We have adopted the third approach, identifying significant examples of energy use in various economic sectors and drawing upon recent reports to illustrate the potential for reductions in energy consumption and cost. The American economy's consumption of energy is determined by the designs, technologies, and modes of operation of its machines. In the past, these machines—lighting and heating homes and offices, transporting people, operating factories—were built to meet many objectives other than conserving fuel, which in relative terms, was cheap. The purchasers of homes and offices preferred to pay higher heating and cooling bills rather than to pay for better insulation. The

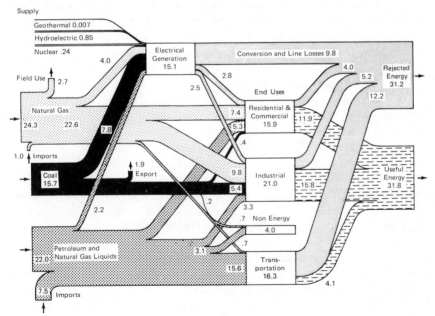

SOURCE: K.W. Ford (ed.) *The Efficient Use of Energy*, AIP Conference Proceedings #25, New York: American Institute of Physics, (1975).

*All values are in quads or units of 10^{15} BTU. Total production equals 71.6×10^{15} BTU.

Figure 4-1. U.S. Energy Flow Patterns, 1970*

American driver preferred size and comfort to fuel economy. Industry was content to rely on cheap power from public utilities for its needs.

These past practices should be carefully evaluated by energy users since many economies in energy use can be achieved at low cost. If the market mechanisms are not hindered by misguided regulations or by poor information, higher energy costs should make many conservation measures economically attractive. Indeed, the efficient use of energy and the potential gains from conservation completely overshadow the potential benefits from many of the second-order decisions about future energy technologies.

The chart of energy flow patterns (Figure 4-1) provides a map of conservation opportunities. On the input side, economic and technological choice may substitute a less expensive fuel for a more costly one, as in the use of coal or nuclear power plants instead of oil for electricity generation. On the output side, major economies are available by reducing the amount of energy thrown away (the "rejected energy" in Figure 4-1) using such techniques as combined-cycle electrical generation (a gas turbine followed by a steam turbine). In addition, industry can make more effective use of the high temperature achieved in fossil fuel combustion by cogeneration where electrical power is produced first and the "waste" heat is used for process steam or space heating, which was the basic industrial requirement. It is equally important to look for gains in the "useful energy" flow. In many cases this energy can be used more efficiently or its use can even be avoided. When useful energy needs are reduced, the corresponding rejected energy is automatically reduced as well.

Commercial and Residential Sector

The commercial and residential sector represented some 22 percent of U.S. energy consumption in 1970. The three largest items—residential and commercial space heating and residential water heating—account for 75 percent of the energy consumed in this sector and 17 percent of the total energy consumption nationally.

There is no difficulty in new construction in reducing heat loss (or gain, for air-conditioned spaces) through wall by a factor of three. If all existing homes were insulated to 1972 FHA standards (heat loss of 1,000 BTU/thousand cu ft/degree-day), 40 percent of the energy presently used for heating and cooling would be saved. A reduction to the technically feasible level of 700 BTU/thousand cu ft/degree-day would save 50 percent. Similarly, 30 to 40 percent of the energy used to heat (and maintain) hot water could be saved by simple add-on insulation without changing use habits.

Commercial buildings offer larger opportunities for conservation. Adapting heating and cooling levels and the rate of change of air supply to building and room occupancy, providing outside makeup air for combustion and fume hoods, moderating the temperature of chilled water when maximum cooling load is not required, providing individual switches for room lighting, and lighting suited to actual needs are all feasible and have short payback periods.

In new construction, limiting heat loss to 55,000 BTU/sq ft/yr has been achieved in many buildings, and the capital and installation cost of energy-efficient technology is in some cases less than that of conventional technology (although, of course, it need not be so in order to be economically justified). In new buildings, further savings can be made through increased efficiency of oil-fired boilers and of refrigeration systems, recovery of low-grade heat, use of heat pumps, and the use of thermal barriers to cover windows at night. Such barriers can reduce the heat loss through windows to less than that for a solid wall. Finally, total energy systems (cogeneration) can save fuel while saving money on electricity.

Transportation Sector

The transportation sector represents 23 percent of U.S. energy consumption, with the automobile consuming 52 percent of this amount: 32 percent in urban and 20 percent in intercity travel. Passenger aircraft account for 9 percent and freight trucks 22 percent of transportation energy use. Current laws require that average gasoline mileage of new cars double by 1985. This will be feasible without compromise of safety or serviceability. Many improvements contribute to this goal. Reduced vehicle weight, through redesign, can save 20 to 30 percent on fuel; radial tires, 5 to 10 percent; lowered air drag, 5 percent; better engine-transmission match, 10 to 15 percent. Adopting diesel engines would halve the fuel consumption of urban vehicles; it is available in foreign cars and is being introduced in domestic vehicles. Beyond 1985, further economies will be possible by such improvements as regenerative braking and dual-engine vehicles.

Economies for trucks will be found in reduced air drag (15 percent saving in fuel), removal of routing restrictions, and the like. But major savings in freight transport may well come from increased emphasis on barge and railroad transport, which consume about 700 BTU/ton-mile in contrast to about 2,800 BTU/ton-mile for trucks.

Industrial Sector

The industrial sector represents some 33 percent of total U.S. energy use. Of this, 85 percent is consumed in manufacturing and 7 percent each in farming and mining. The fifteen most energy-intensive U.S. industries consumed 45 percent of U.S. industrial energy use in 1967 while supplying 9 percent of the value added by all manufacturing. The ratio of energy consumed per dollar of value added was ten times as large for these industries as for all other industry.

Possible manufacturing energy economies are suggested by differences between the average energy consumed per unit of output in the United States and in other countries. In the production of hot-rolled steel, U.S. industry averages some 47 million BTU/ton while The Netherlands (basic oxygen production) uses 32 million. In the manufacture of a ton of polyethylene, United States, United Kingdom, and The Netherlands consume respectively 110, 55, and 41 million BTU. U.S. industry expends 240 million BTU per ton of aluminum; more efficient provision processes use as little as 180, while recycle of used aluminum consumed 9 million BTU per ton. Major savings are possible in aluminum production (accounting for 4 percent of U.S. industrial energy consumption) and steel (accounting for 12 percent). Other large industrial consumers can also save considerable energy including petroleum refining (9 percent of industrial energy consumption), industrial organic and inorganic chemicals (12 percent), and cement (4 percent); many of these users can also shift to lower cost fuel (coal). Overall, it seems possible to reduce present industrial unit energy consumption by 20 percent by 1985 and by 40 percent by 2000 with reduced overall costs.

Cogeneration

Roughly two-thirds of the energy content of nuclear and fossil fuels is lost as waste heat when it is converted to electrical energy. If some of the generating capacity were adjacent to industrial facilities or residential complexes, the waste heat could be utilized for industrial processes requiring steam, and for space heating. This would eliminate the need for many existing systems used for process or space heating alone.

This process, known as cogeneration, can combine generation of electricity and steam in an industrial, commercial, or multiple-residential complex. The concept is also known as total energy, district heating, or by-product power. Electrical power so produced can be used by the complex itself, saving transmission costs, and it may also be sold to the local public utility, which can provide a pool for cogenerating customers.

The overall energy efficiency of electrical generation, transmission, and distribution by a conventional utility system is 30 to 35 percent. The overall efficiency for a cogeneration system is 60 to 75 percent—that is, 60 to 75 percent of the heat of combusion of the fossil fuel is used either as steam or as electrical energy. A recent Dow Chemical Company study funded by the Federal Energy Administration (FEA) shows that, if industrial cogeneration were used where it would be profitable, it would save some 600,000 barrels of oil (equivalent) per day by 1985, assuming regulatory and institutional barriers could be overcome.

At present, West German industry cogenerates about 30 percent of that nation's total electrical power; the corresponding figure for the United States is 5 percent. Despite the apparent opportunities, the U.S. government has shown little interest in facilitating development of cogeneration.

The low fuel costs of the past and economies of scale in central generating plants, coupled with the low capital costs of process-heat boilers, created disincentives for cogeneration in the United States. These disincentives were magnified by regulatory policies which impose on cogenerating firms the same requirements as those imposed on public utilities. Utility reluctance to provide the necessary connections to utility power grids and reluctance to supply pooling services at a reasonable price has also inhibited the growth of cogeneration. Moreover, utility-owned small generating plants on user's premises may have seemed to require more management attention than they were worth. Industry, for its part, has preferred to avoid management costs associated with cogeneration. The economic realities are now vastly different; and the possibilities of cogeneration should be reassessed by industrial consumers, utilities, and their regulators. While the prospects for cogeneration are clouded by a number of uncertainties, it would appear that, if regulatory and institutional problems can be resolved, utilization of this wasted resource could result in savings equivalent to a million barrels of oil per day.

TRANSMISSION

Improved energy transmission technology can play an important role in the utilization of present and future energy sources. Electrical power plants now tend to be located near the large population concentrations they serve. This reflects not only the high costs and power losses associated with long-distance transmission of electricity, but also institutional factors such as the historical division of the market by utility districts.

If it were possible to transmit electricity more economically over long distances, coal and nuclear power plants could be sited more flexibly, reducing capital costs and associated social costs. In the case of nuclear plants, remote siting would be particularly significant since it could greatly reduce the consequences of an extremely serious accident—possibly by a factor of ten to one hundred. Alternative sources of energy, notably solar and geothermal, which are geographically limited and have high capital costs, could be utilized widely if their electrical output could be transmitted less expensively. If coal could be transported more economically to point of use, power from this energy source would be available at lower cost. Finally, using hydrogen produced from other energy resources would simplify energy transmission, storage, and utilization.

Long-distance transmission of electricity is relatively expensive. For example, transmission by a conventional power line from a single 1,000 MWe plant for a distance of 1,000 miles would add about 5 mills/kwh to the cost of electricity. This cost could be reduced by building transmission lines to handle very large blocks of power. Economical long distance transmission of large quantities of electrical energy could be achieved in the future with common-carrier power lines utilizing superconducting technology and direct current flow. Losses in such a line, due primarily to energy used to maintain it at low temperature, would be less than 1 percent. Transmission of large blocks of electricity (5,000 MWe or more) in this manner over 1,000 miles might add only 2 mills/kwh to the cost of electricity, with most of the cost being in the terminal equipment. While the economics of this situation is clear, major problems remain in establishing the technology commercially in manufacture, installation, operation, and maintenance.

An alternative to moving electric plants closer to coal sources is to improve the efficiency and economics of coal transport. Present costs for transport of coal and oil by various modes are shown in Table 4-1. For long distances, slurry pipelines would clearly provide for costs lower than those of railroads. Present barge transport figures are for small shipments, 1,000-3,000 ton capacity barges, on inland waters, and are not therefore directly comparable to tanker rates for oil. However, it should be possible to approach the oil transport rate for coal carried in supercolliers across the oceans. Ocean transport of coal might add about 4¢/MBTU per thousand miles or about 0.5 mill/kwh per thousand miles. Such rates could make coal competitve in international markets.

Hydrogen has often been suggested as a vehicle by which energy from intermittent or remote sources may be transported or stored for use elsewhere or at later times. Hydrogen may be separated from water by thermal or electrical means. For example, electricity from a reactor, operating at times when demand for electricity is low, could be used to make hydrogen. This conversion process is costly in terms of efficiency but may be justified by the high capital costs of generating facilities. Similar considerations apply to solar facilities where production is not only intermittent but is optimally located in areas of high insolation. Complex thermo-chemical processes to produce hydrogen through use of high temperature sources (900° C) are being investigated, although at a very low level of effort. Although conventional LWRs operate at too low a temperature to separate hydrogen from water, solar-thermal facilities, fossil fuels, or high temperature gas-cooled reactors could be used for this purpose. If these processes can operate at the estimated efficiencies and at reasonable capital and operational cost, they could be used to improve load factors, particularly for solar plants.

Hydrogen is an attractive intermediate fuel. It can be transported by pipeline at costs about 40 percent higher than natural gas, or for about $0.30/MBTU per 1,000 miles. Hydrogen can be substituted for many of the uses now met by natural gas, and new technologies, such as fuel cells, could extend its utility.

Table 4-1. Coal and Oil Transportation Costs

Fuel	Mode	Zero-Length Costs (Handling & Loading)	Additional Thousand Miles
		¢/MBTU	¢/MBTU
Coal	Rail	6	50
	Slurry	10	20
	Barge	4	14
Oil	Barge	1	8
	Pipeline	2	4
	Tanker	1	2

Note: Oil at 24,000 BTU/lb, western coal at 9,000 BTU/lb, eastern coal at 12,000 BTU/lb.

More rapid progress can and should be made in the transmission, transport, and storage of energy. Such progress would permit more effective and flexible use of existing energy resources at reduced social costs. Technological developments also improve the competitive economics of alternative sources of energy and hasten their introduction.

✳ *Part II*

Health, Environment, and Safety

✳ *Chapter 5*

Health Effects

The use of nuclear power to generate electricity inevitably results in risks to human health. The extent of these risks is uncertain and the subject of considerable controversy. To be meaningful in connection with public policy decisions, these risks cannot be considered in isolation but must be compared with the risks associated with coal-fired power plants, which are the principal alternative for electric power generation for the rest of this century.

In many industries, health risks have traditionally been measured in terms of occupational accident rates. In recent years there has been increasing awareness of possible health impacts of effluents on workers and the general public. This awareness has characterized assessments of nuclear power from its very beginning, since radioactivity was readily measurable and its effects understood to some extent, and since there was popular concern about radiation, heightened by the use and testing of nuclear weapons.

The assessment of nuclear power health risks involves analysis of occupational accident rates, radiation exposures for workers and populations under normal operating conditions, and the probabilities and consequences to public health of reactor accidents. Radioactivity is encountered at most stages of the nuclear fuel cycle—in mining and milling, in fuel fabrication and transportation, in reactor operation, and in waste management and disposal operations. Small quantities of radioactivity are normally released at each stage, affecting workers or spreading beyond facilities to reach the public locally or even globally. Worker exposures may range as high as fifty times normal environmental background radiation levels; population dose rates are usually extremely small fractions of normal background. The radioactive materials released, however, may be trapped in body tissues or remain in the environment for centuries, a continuing source of radiation exposure. Uranium mill tailings and other forms of radioactive waste may, depending upon the measures taken, continue to release radioactive materials into the environment for very long periods of time. Thus, while the rates of exposure may be small, the cumulative effects of contemporary power generation may be larger when summed over many generations.

There are many uncertainties in the assessment of the health effects of nuclear power. Some fuel cycle sources of radiation have not been determined precisely, and the many environmental and biological pathways to man are not well understood. Even if radiation dose commitments due to particular fuel cycle operations could be determined exactly, there is still considerable uncertainty about the relationship between radiation and biological effects, such as the incidence of cancer and genetic disease. However, these uncertainties are small compared to those associated with predicting the probabilities and consequences of accidents at reactors and other fuel cycle facilities. The accident problem is discussed in detail in Chapter 7. Resolution of these uncertainties will be a difficult and lengthy process. Thus, efforts to reduce risks through better treatment of radiation sources, such as mill tailings, improved siting policies, and other measures should not await resolution of existing uncertainties.

Coal-fired power plants impact health both through occupational hazards and emissions affecting the public. The latter effect dominates the adverse health impact of coal combustion, but there are also very large uncertainties in the quantitative relationships of the various emissions from coal combustion to human health. Present studies are not even in agreement as to whether there are thresholds for the health effects of particular pollutants. Despite these uncertainties, efforts should be made to reduce the effects of coal pollutants since they clearly have a major effect on human health.

This chapter examines the relationship of radiation exposures and health effects in man, the potential health impacts of the nuclear fuel cycle under normal and accident conditions, and the possible health effects of coal-fired power plants.

THE NATURE OF IONIZING RADIATION

The nuclei of most of the heavy elements and direct fission products encountered in the nuclear fuel cycle are unstable and emit radiation. The principal forms of radiation of biological significance are alpha (α) particles, beta (β) particles, gamma (γ) radiation, and neutrons.[a] When these particles penetrate biolog-

[a]The nuclei of the atoms of many of the heavier elements are inherently unstable. This instability usually results in the nuclei ejecting alpha (α) particles (the nuclei of the helium atoms consisting of two neutrons and two protons) or beta (β) particles (electrons or positrons) at very great speed. Since these particles are electrically charged, they perturb the electron clouds in atoms causing ionization of the atoms and rupturing chemical bonds, with the resulting breakdown of molecules. As β-particles are slowed by transferring their energy through such encounters, some energy may appear as ordinary X-rays. Gamma (γ) rays, which are electromagnetic radiations similar to but generally of greater energy than

ical or other matter, they give up their energy through a series of collisions with the atoms or nuclei of the molecules making up the biological material. As a result, many molecules are damaged by the breaking of chemical bonds and by the loss of electrons (ionization), which produces further chemical change. Energy is dissipated along the path of the radiation through these processes. A measure of the rate of linear energy transfer (LET) is the density of ionization events along this path. High rates of ionization are roughly equivalent to high intensities of local biological damage.

Different forms of radiation have different characteristic rates of linear energy loss, or ionization. Neutrons and α-particles, for example, have a high rate of linear energy transfer (high LET) and thus cause heavy damage locally. Their high rate of linear energy loss means that α-particles do not penetrate far; they are characteristically stopped by an inch of air or by the outside layer of human skin. Neutrons also have a relatively high LET, being effectively slowed by the light elements, such as hydrogen and carbon, in biological materials; when finally captured, intense γ-radiation is emitted. β-particles lose energy more slowly and may penetrate an inch or so into tissue. γ-radiation is very penetrating, requiring several millimeters of lead to reduce its intensity appreciably. β- and γ-radiation are therefore referred to as low-LET radiations.

The amount of radiation absorbed in tissue is expressed in rad units (one rad is the amount of radiation depositing 100 ergs of energy per gram of tissue). The biological damage caused by different types of radiation, however, is only approximately related to the energy deposited, since α-particles and neutrons, for example, generally cause more damage per rad than do β- or γ-radiation. To account for this difference another unit, the rem (radiation dose equivalent man), is used in setting radiological protection standards; while it lacks precision, it enables all radiation effects to be related to a single measure of dose. By definition, one rad of β- or γ-radiation has about one rem of biological effectiveness. A rad of α-particles is usually assigned a dose equivalence of 20 rem and a rad of neutrons an equivalence of 10 rem. Radiation exposures may occur over short periods of time or be accumulated slowly. The total dose is measured in rems, and the dose rate in rems per unit of time (for example, radiological protection standards often list permissible dose limits in terms of the number of rems per year which may be received). A radiation dose may be delivered externally to the whole body or to specific organs or tissues through ingestion, inhalation, or

---

X-rays, are directly emitted by certain radioisotopes. When positive β-particles (positrons) are stopped, they are annihilated with electrons, emitting energetic γ-rays. Neutrons are emitted in fission, that is, in the splitting of the nuclei of certain heavy elements into two fragments. Because a neutron is electrically neutral, it undergoes many collisions with nuclei in the process of slowing down and is ultimately absorbed in a nucleus with the production of γ-radiation. This event produces a higher isotope that is itself often radioactive. But the principal damage comes from the speed, and hence ionization capacity, imparted to nuclei with which the neutron collides since these nuclei create dense ionization over a short distance, much like an α-particle.

other routes. A dose to specific tissues may be translated into a comparable whole-body dose.

The average whole-body dose rates from natural and man-made radiation are shown in Table 5–1. Natural sources account for about 0.130 rem or 130 millirem (mrem) per year. About 25 mrem of this come from ingested potassium-40 and other radioisotopes in food. The man-made radiation exposures were estimated in 1972 to amount to about 80 mrem of which about 73 mrem came from medical sources. The medical and occupational exposures are actually averages of larger exposures received by a small fraction of the population. Federal standards require that average population exposures from all man-made nonmedical sources be kept below 170 mrem per year and that individuals in the general population not receive more than 500 mrem per year. Occupational limits are set at 5 rem per year. To date, nuclear power has contributed only a very small amount to the total man-made background. The 1970 average individual exposure from nuclear power, 0.003 mrem per year, increased to about 0.023 mrem per year in 1975.

The current dose rate, however, is not a comprehensive measure of the eventual health effects of current nuclear power activity. Releases of radioactive materials in nuclear fuel cycle activities may continue for years after power generation occurs and gradually build up in the environment or in human tissue. The tailings from uranium mining and milling activities, for example, continue to emit radioactive materials for many thousands of years at a nearly constant rate. Moreover, radioactive materials may remain in both human bodies and in the environment, delivering radiation doses to individuals over a period of a few years, over a lifetime, or to subsequent generations. The fuel cycle activities associated with the generation of one reactor-year of power may thus be responsible for radiation exposures for many years. This commitment to future radiation exposures due to present activity is referred to as the dose commitment; its estimation for each radioisotope and fuel cycle step requires analysis of environmental and biological pathways and residence times in the environment and in the body. Since some radioisotopes are extremely long-lived or may be emitted over long periods of time, it is necessary to prescribe a period of time over which their effects are to be included in the dose commitment. A period of thirty or fifty years, corresponding to the time over which an individual might receive radiation from current nuclear power activities, is conventional. For certain sustained emissions (from tailings, for example), or for radioisotopes with relatively long residence times, it is necessary to assess potential health impacts over longer periods of time and to decide what weight they should be given in current decisions.

Biological Effects of Radiation

While radiation may damage any of the molecules present in cells, the DNA macromolecules carrying the genetic information necessary to cell development,

Table 5-1. Estimates of Annual Whole-Body Dose Rates in the U.S., 1970[a]

Source of Radiation		Average Dose Rates (mrem/year)
Natural		
Environmental		
Cosmic Radiation		45 (30–130)[b]
Terrestrial Radiation		60 (30–115)[c]
Internal Radioactive Isotopes		25
	Subtotal	130
Man-Made		
Environmental		
Global Fallout		4
Nuclear Power		0.003[d]
Medical		
Diagnostic		72
Radiopharmaceuticals		1
Occupational		0.8
Miscellaneous		2
	Subtotal	80
	TOTAL	210

[a]From National Academy of Sciences, "The Effects on Populations of Exposure to Low Levels of Ionizing Radiation," Report of the Advisory Committee on the Biological Effects of Ionizing, BEIR, 1972; and Klement *et al.,* "Estimates of Ionizing Radiation Doses in the United States, 1960–2000," U.S. Environmental Protection Agency, 1972.

[b]Values in parentheses indicate range over which average levels for different states vary with elevation.

[c]Range of variation (shown in parentheses) attributable largely to geographic differences in the content of potassium-40, radium, thorium, and uranium in the earth's crust.

[d]This rose to about 0.023 millirem per year in 1975; see the fuel cycle discussion below for magnitude of dose commitment from current capacity. The dose rate is the annual amount of radiation due to all preceding nuclear power generation activities; the dose commitment is the total amount of radiation eventually delivered, over an assigned period of time, due to a given amount of electricity generated by nuclear means.

maintenance, and division are the most critical biological targets. Radiation can break one or both strands of the DNA, at one or many sites, destroying or altering some of the information contained. Much of this damage is reparable, but in a small proportion of cells the DNA is permanently damaged. The consequences of this damage may be death or altered function of the affected cell, transformation of the cell to a form which may ultimately be manifest as cancer in the organism, or, in the case of germ cells, genetic disease in later generations. The amount of permanent damage induced by radiation depends on the type of radiation and on its intensity and duration.

Very high doses of any radiation result in acute bodily injury: vital tissues, such as bone marrow, may be damaged to the point where death ensues. Whole-body doses of 500 rem or more are generally fatal within weeks. Even at 100 rem, many cells are destroyed, individuals may contract radiation sickness, and fetuses may be killed or their development impaired.

At lower doses and at low dose rates, radiation effects are more subtle and difficult to predict on the basis of experience with high doses. At high doses, sensitive regions of a given cell may be subjected to many thousands of traversals by radiation. If this occurs in a short period of time, cell death is likely. At lower doses, particularly if spread out in time, the biological effects depend on how the damage occurs and on the opportunities for repair. The exposures of interest in assessing the risks of nuclear power range from fractions of a millirem, well below background in the case of normal emission, to a few rem in the case of occupational exposures, and to exposures of a few millirem to hundreds of rem in the event of accidents. These doses are generally spread out in time, over periods of years in many cases. At background levels, the sensitive region of a given cell may be traversed by radiation only about once a year; at occupational exposure levels, traversals may occur once a month; and under accident conditions, as many as several times a day.

The relationship between biological responses and the doses and dose rates of interest must, at present, be estimated from experience at higher doses and dose rates and from experimental data on nonhuman systems. The results are clearly tentative. There is considerable evidence for a few general distinctions, however. With high-LET radiation (neutrons and α-particles), it is believed that traversal of the DNA by a single particle has a high probability of causing irreparable damage. These conditions are more rarely achieved with single particle with low-LET radiation, with which it is believed that traversal by two or more particles, in close spatial and temporal proximity, are usually required to cause irreparable damage. The consequence of this is that high-LET radiation has much greater potential for biological damage at low doses and dose rates than does low-LET radiation. Since our data on biological effects are derived largely from observations at high doses, low-LET radiation may be less than proportionately harmful at lower intensities, where on traversal of a sensitive region of a cell is not likely to be irreparably altered before repair can take place. The effects of high-LET radiation, on the other hand, may remain proportionately severe at low doses and low dose rates. Thus a linear extrapolation of the dose-response curve from the high dose region, where it may be studied, to the low dose region may be more justifiable in the case of high-LET radiation than for low-LET radiation. Since present background radiation and much of the normal emission from nuclear plants is low-LET radiation, actual biological effects may be smaller than a simple linear extrapolation would suggest. Linear extrapolation may not be unduly conservative, however, in the case of alpha-emitters, such as plutonium. There is even experimental and epidemiological evidence that the risks per rad of high-LET radiation may, in certain instances, actually increase with decreasing dose and dose rate, as would be expected if carcinogenic or mutagenic damage were prevented by killing of the affected cells at higher doses.

The effects of radiation depend strongly on the tissue involved and on the biological pathways by which radioactive materials enter and concentrate in the body. Tissues with rapidly dividing cells, especially embryos, are particularly sensitive to radiation from external or internal sources. Some organs are vulnerable because particular radioisotopes concentrate in them; the concentration of radioactive iodine in the thyroid and strontium-90 in bone are examples. Local radiation doses to certain tissues may thus be higher. Finally, radioisotopes have very different residence times in the body. Tritium has a half-life residence of only eight days in the body, while strontium-90, which is chemically similar to calcium, is incorporated in the bone structure, where it remains on the average for many years. The route of entry into the body is also important. Plutonium, for example, is extremely radiotoxic if inhaled as small particles or if it enters the blood stream; it is much less dangerous if ingested, since it is not easily absorbed through the gut wall. Since the biological effects of radiation depend on a large number of variables, the effects can only be partially characterized by simplifying models, such as the linear dose-response relationship, and then only as long as important variations and uncertainties are recognized.

Radiation-Induced Cancer

The principal nonhereditary delayed effect of radiation is cancer. There is usually a delay of years or even decades between irradiation and the appearance of cancers in excess of normal frequencies. This long latency usually makes it impossible to identify a particular cancer as due to radiation. But there are substantial data on humans testifying to an increased incidence of certain types of cancer as a result of irradiation.

The largest single source of data is the study of the Japanese atomic bomb survivors; among about 24,000 people who were exposed to an average dose of roughly 50 rem, more than a hundred additional cancer deaths have occurred to date. In Britain, among 15,000 people who were treated with X-ray for arthritis of the spine, corresponding to an average whole-body dose of 250–500 rem, more than a hundred additional cancer deaths have occurred thus far. Other groups who have developed excess cancer include Marshall Islanders who were subjected to bomb-test fallout in 1954, uranium miners exposed to radon in the air of mines, pioneer radiation workers, and various groups of radiotherapy patients. There is also substantial evidence from animals consistent with the observations on humans.,

The interpretation of these and related data has been the work of a series of expert committees: The Advisory Committee on Biological Effects of Ionizing Radiations (BEIR, 1972),[b] the United Nations Scientific Committee on the Effects of Atomic Radiation (UNSCEAR, 1972),[c] the National Council on

---

[b]National Academy of Sciences, "The Effects on Populations of Exposure to Low Levels of Ionizing Radiation," Report of the Advisory Committee on the Biological Effects of Ionizing, BEIR, 1972.
[c]United Nations, "Ionizing Radiation: Levels and Effects, A Report of the United Nations Scientific Committee on the Effects of Atomic Radiation," 1972.

Radiation Protection and Measurements (NCRP), and the International Commission on Radiological Protection (ICRP). A statistical relation between dose and cancer deaths has been established, but with considerable uncertainty attached. Several complications deserve attention.

One complication is that the excess of cancer deaths may persist for the entire lifetime of an exposed population, whereas existing data are based on only the first twenty to forty years after irradiation. Estimation of the total cumulative lifetime excess in a population must thus involve assumptions about the minimal and maximal latent periods required for development of the induced cancers, as possibly influenced by age.

A second complication is that it is not clear from existing data whether the excess risk of cancer resulting from a given dose of radiation causes a constant number of additional cases in all adult age groups (i.e., a constant "absolute" risk per rem) or a constant percentage increase above the natural risk in all adult age groups (i.e., a constant "relative" risk per rem). Since the natural risk of cancer increases sharply with age during adult life (e.g., the chance of dying from cancer at age 61 is 100 times higher than at age 26), projection of a constant relative increase in risk throughout adult life would predict a higher cumulative lifetime cancer excess in those exposed at young ages than would projection of a constant absolute increase. Evidence on some kinds of cancer favors the absolute risk assumption and others the relative risk assumption. In the BEIR Report, ranges of estimates based on both approaches are presented, along with approximate averages of the two assumptions.

A third complication is extrapolating the dose-incidence relation from the assessed dose region of 20–400 rem down to doses a thousand or more times smaller. The usual procedure has been to fit the data with a straight line passing through zero—a linear nonthreshold relation. This procedure will tend to overestimate the effects of low doses if deaths rise more sharply than linearly with increasing dosage. The effects of prolonged low-level doses are also likely to be less than the same dose delivered at high dose rates over short times. Both of these conjectures may be valid for low-LET radiation; but it is likely that they are less valid for high-LET radiation, which may be expected to cause more nearly the same amount of damage per rad even at low doses and low dose rates.

Assessments of nuclear power require projection of the risk presented by possible increases in the continual low-level background of radiation and the risk resulting from accidents, where the exposure might be larger and delivered over a shorter time. The BEIR analysis considers the effects of a continuous ongoing radiation (whole-body) dose of 0.1 rem per year to the (1967) U.S. population and estimates that cancer deaths could be increased by 1,700 to 9,000 per year as a result of such an addition to the background level. The wide range is due to uncertainty about whether the risk is absolute or relative, and whether the duration of risk (the plateau period) extends for thirty years or for duration of life.[d]

[d]BEIR (1972). The absolute risk model gives 1,726 annual cancer deaths for a thirty-year plateau and 2,001 for a lifetime plateau; the relative risk model gives 3,174 for a thirty-year, and 9,078 for a lifetime plateau.

These figures can be translated into a lifetime risk estimate for average persons given a one-time dose of radiation; i.e., in a million people a 1 rem exposure would, according to the above estimates, result ultimately in 90 to 470 excess cancer deaths, depending on the assumption about risk model and plateau. The figure 180 deaths per million man-rem is frequently cited as an average. This expected affect of a 1 rem exposure might be compared with the roughly 200,000 persons out of the million who will naturally die of cancer from other causes.

The BEIR estimates are based on the linear hypothesis: that a number of small doses given to many individuals or spread out in time have the same effect as a smaller number of larger doses, equal in the aggregate to the sum of the smaller ones, given to fewer individuals at one time. As explained above, there is some eivdence that this procedure overestimates the risk from low doses or low dose rates in the case of low-LET radiation. Some groups have recommended that dose-effectiveness factors, reducing the risk by factors of 1 to 5, be applied to low doses or low dose rates of low-LET radiation which characterize most background radiation and most of the routine emissions of the nuclear fuel cycle, except for the α-emitters in the emission from mine tailings. High-LET radiation, however, appears to be more effective in causing irreparable damage at low doses and low dose rates, which is important in the case of α-emitters like plutonium.

The Reactor Safety Study (WASH-1400) published by the Nuclear Regulatory Commission in 1975 applied dose-effectiveness factors to all low-level radiation exposures in deriving its central estimate for carcinogenic effects. For an upper bound, it adhered closely to the original BEIR estimates; for a lower bound, estimates were based on the threshold hypothesis that below 1 to 25 rem there is no risk of radiation-induced cancer. The NCRP Scientific Committee 40 is now considering smaller dose effectiveness factors than those used in WASH-1400 and only for low-LET radiation. The NCRP risk estimates thus fall between those of WASH-1400 and those of the BEIR Committee. Table 5–2 shows a tabulation of the annual cancer deaths resulting from a continuous exposure of the U.S. population to 0.1 rem per year, according to the three estimates.

On the basis of existing evidence and the associated uncertainties, we believe that for our purposes the original BEIR estimates are not yet in need of substantial revision. On completion of the present NCRP study and of its review by the scientific community, some reappraisal of the BEIR conclusions may be warranted but is not likely to result in drastic reductions.

Genetic Effects

The fact that radiation could induce mutations and hence cause genetic effects was discovered fifty years ago. While this response has been studied extensively in other species, the dose-response relationship in humans is not known owing to the rarity of mutations at any given locus and hence the need to study very large numbers of individuals to obtain statistically significant information.

Table 5-2. Comparative Estimates of the Number of Cancer Deaths Per Year in the U.S. Population Attributable to Continuous Exposure at Rate of 100 Millirem Per Year

| Age at Time of Irradiation | Duration of Plateau Period | Estimated Excess Cancer Deaths Resulting from 0.1 rem Per Year[a] | | | | | | | No. of Cancer Deaths Occurring Naturally[e] |
| | | BEIR Estimates[b] | | WASH-1400 Estimates[c] | | | NCRP Estimates[d] | | |
		Absolute Risk Model	Relative Risk Model	Upper Bound	Central Estimates	Lower Bound	Absolute Risk Model	Relative Risk Model	
In Utero	10	150	110	90	20	0	150	110	–
0–9 years	25–30	240	800	240	50	0	120	400	3,000
	life	290	6,000	–	–	–	140	3,000	–
10+ years	25–30	1,300	2,200	1,300	500	0	650	1,100	308,000
	life	1,600	3,000	–	–	–	800	1,500	–
Total	10–30	1,700	3,200	1,650	600	0	850	1,600	311,000
	life	2,000	9,000	–	–	–	1,000	4,500	–
Average[f]		3,500		650			2,000		

[a]Values in table have been rounded.

[b]From BEIR, 1972, p. 169, based on vital statistics for U.S. population, 1967.

[c]Values are rough calculations, based on application of (WASH-1400) risk models to U.S. population, 1967.

[d]Values are rough calculations, based on application of NCRP (1976) risk model to U.S. population, 1967.

[e]Values denote rounded numbers of deaths in age groups indicated, from vital statistics for U.S. population, 1967.

[f]3,500 is the average of the range 3,000–4,000 suggested by BEIR (1972); 2,000 is the arithmetic average of the four totals.

Studies of the genetic effects of radiation are complicated by the variety of distinct types of mutation and by the great range of manifestations of genetic disease. Radiation may alter a small region of the DNA molecule, such as a single gene, that carries instructions for making a single protein; or it may alter genes at two or more sites; or it may result in chromosomal aberrations, which are rearrangements (including loss) of materials in chromosomes or changes in the number of chromosomes. Single gene mutations are subclassified into dominant, recessive, and sex-linked types. Dominant mutations, such as congenital cataract, show up rapidly; recessive mutations, such as cystic fibrosis, and forms of muscular dystrophy are not as immediately manifest in subsequent generations; sex-linked disorders, which include color-blindness and hemophilia, appear quickly in subsequent generations, much like dominant mutants. Chromosomal aberrations, which usually result in spontaneous abortion but which include such common disorders as Down's syndrome (or mongolism), can now be studied microscopically. The radiation dose-response relationship may be studied directly in this class of disorder. Finally, there are those classes of complexly inherited diseases, called diseases of multifactorial origin, for which an interaction of genes and environmental influences appears necessary. These diseases include birth defects, such as cleft palate and other skeletal anomalies, and diseases that develop later in life, such as heart disorders, epilepsy, schizophrenia, asthma, and diabetes, as well as other degenerative diseases. These complexly determined diseases constitute the bulk of human genetic disorder, but it is extremely hard to determine the magnitude of the genetic contribution to these diseases, and how much the genetic component will be influenced by an increase in mutation rate.

The results of these genetic effects range from invisible to conspicuous, trivial to lethal; some occur primarily in the first generation, others show up later, and some persist for tens of generations. Unlike cancer, where deaths provide a rough measure of the impact of the disease, genetic defects cannot be measured by a single index. Many hundreds of different diseases, all of which are significant but of varying severity, are involved. Moreover, genetic effects may include health disorders whose origin would be difficult to detect: one estimate (BEIR, 1972) is that as much as 20 percent of ill health, in addition to specifically identifiable genetic disease, is due to genetic effects.

Estimates of the genetic effects produced by radiation must be made by an indirect approach. For a given type of naturally occuring mutation, such as mutation of a single recessive gene, one estimates the amount of radiation necessary to double its incidence, the doubling dose. To do this, experiments are done on animal germ cells. The rate of spontaneous human single gene mutations is not well known; a commonly accepted range is $(0.5 - 5.0) \times 10^{-6}$ per gene per generation. To obtain an estimate of single-gene mutability by radiation, irradiation-induced mutations in mouse germ cells have been studied. An early estimate (BEIR, 1972) was that 1 rem of radiation, administered continuously over a long period, induced an average rate of 0.25×10^{-7} mutations per generation.[e] The human doubling dose would then be taken as $(0.5 - 5.0 \times 10^{-6})/(0.25 \times 10^{-7}/\text{rem})$ or 20–200 rem.

The situation, however, may be more complicated for multifactorial disorders than that described above. As mentioned earlier, the bulk of genetic diseases (more than 80 percent) result from complexly inherited traits. The BEIR Committee suggested that the incidence of these disorders were less susceptible to increase by change in the mutation rate and that it would require some two to twenty times as much radiation as do simple gene mutations to cause a doubling in their incidence. This issue is under considerable debate since one school of geneticists believes that these disorders are maintained in the population exclusively by selective factors, which means that the frequencies of these diseases would be independent of changes in the mutation rate and yet be dependent on changes in the environment such as diet and pollution. In our estimates described below, we have employed the range used by the BEIR Committee although we recognize that this range may underestimate the increased disease burden from radiation.

The doubling dose can then be used to estimate the number of genetic diseases in first and subsequent generations due to exposure of a population to a given amount of radiation. For example, the current incidence of dominant single gene disorders is estimated to be 10,000 per million live births. It is further estimated that about 20 percent of these, or 2,000, are new mutants (a comparable number being selected out from the previous generation). It is this number that would be doubled by exposing a single generation to 20–200 rem. From this it can be deduced that 5 rem exposure of those persons responsible for the next million live births would result in 50–500 dominant mutants (2.5 to 25 percent of 2,000). If subsequent generations were exposed to similar radiation levels, the incidence of this type of mutation would rise from its present value of 10,000 to somewhere in the range 10,250–14,500 per million live births. Five rem is the maximum public exposure, exclusive of natural background and medical radiation, now allowed for a generation (170 mrem per year for thirty years). It should be noted that this is about 1,000 times the average annual exposure from the nuclear fuel cycle at present and 100 times the level that might be expected by the end of the century.

Since information about the dose-response factors for radiation-induced genetic disorders has been in a state of flux, estimates of the potential impacts of radiation have varied. Table 5-3 shows the BEIR estimates of 1972, the Reactor Safety Study (WASH-1400) estimates of 1975, and estimates based on more recent studies which employed a 100 rem doubling dose as the basis for their estimates, available since the WASH-1400 study was completed. These more

[e]A later analysis has suggested that the actual value may be four times higher. See, S. Abrahamson and S. Wolff, "Reanalysis of Radiation-Induced Specific Locus Mutations in the Mouse," *Nature* 264 (1976):715.

Table 5-3. Comparison of Three Estimates of First Generation Genetic Effects of 5 Rem Per Generation Per Million Live Births

Type of Disorder	BEIR		WASH-1400		Possible Revisions	
	Current Incidence	Disease Cases	Current Incidence	Disease Cases	Current Incidence	Disease Cases
Single gene disorders						
Autosomal dominance	10,000	50–500	10,000	100	3,000	60–600
X-chromosome link	400	0–15	400	3	400	7–70
Autosomal recessives	1,500	0	1,500	0	1,500	0
Multifactorial disorders	40,000	5–500	40,000	10–100	85,000	42–4,200
Effects of chromosome Aberrations						
Unbalanced rearrangements	1,000	60	1,000	60	500	100–200
Aneuploidy	4,000	5	4,000	5	4,000	5–130
Total		120–1,080		178–268		214–5,200
For 3 × 10^6 live births (U.S. population)		360–3,240		534–804		642–15,600
Risk estimates (1st generation defects per man-rem)		12–100 × 10^{-6}		18–27 × 10^{-6}		21–5,200 × 10^{-6}

recent studies[f] have shown that the incidence of genetic diseases, primarily multifactorial diseases, are twice as frequent as indicated in previous studies carried out twenty years earlier, and that the indicated mutation rate may be as much as four times larger than previously believed, leading to a corresponding decrease in the doubling dose. These two factors together might lead to as much as an eightfold increase in the number of multifactorial diseases induced by a given amount of radiation as compared to the estimates in the original BEIR Committee report. Uncertainty ranges are clearly large, extending now to include consequences considerably higher than those indicated by the WASH-1400 analysis. If the present legal limit for man-made nonmedical radiation (170 mrem per year or 5 rem in thirty years) were to be reached through use of nuclear power, there could be an increase of 0.2 to 5.5 percent in genetic effects in the first generation. If radiation at this level continued during subsequent generations, genetic disease might ultimately occur at a rate 2 to 50 percent higher than at present. As noted above, to reach such levels of radiation would involve annual exposures from the nuclear fuel cycle at least 100 times greater than are expected to result from use of nuclear power during this century.

The aggregate genetic effects of radiation may be compared with the cancer induction rates shown in Table 5-2. The incidence of genetic disease in the U.S. annual total of three million live births, attributable to continuous radiation of 170 millirem/year to one generation, is variously estimated (see Table 5-3) from 360-3,240 (BEIR), 534-804 (WASH-1400), and 642-15,600 (Possible Revisions). The yearly increase in cancer deaths from the same continuous radiation dose would be 1,500 to 8,000 (according to the NCRP estimates of Table 5-2 adjusted from 100 to 170 mrem per year). Thus, for the same exposures, the genetic defects arising in the first generation due to radiation exposure may be roughly comparable to the cancer death incidence in terms of numbers of persons affected. While some of the genetic disorders considered here have only small impact on life, others may have as serious an impact on life as the ten or so years of average reduced life expectancy due to radiation-induced cancer. Earlier assessments of the potential genetic impact of nuclear power have underestimated the range of uncertainty, and perhaps the actual magnitude, of this impact. Moreover, if dominant genetic disorders are selected out at a rate of about 20 percent per generation, the long-term burden will be five times that of

[f]B.K. Trimble and J.H. Doughty, "The Amount of Hereditary Disease in Human Populations," *Ann. Human Genetics* (London) 38 (1974). This study of 750,000 newborns in British Columbia suggests that multifactorial disorders have been underestimated and dominant mutations overestimated in the past. This reassessment is reflected in the third column of Table 5-3. The mutational factor in multifactorial diseases is still in dispute. The BEIR estimates for unbalanced chromosome aberrations have been reduced (see P.A. Jacobs, M. Melville, and S. Ratcliffe, "A Cytogenetic Survey of 11,680 Newborn Infants," *Ann. Human Genetics* [London] 37 [1974] :359), and the doubling dose for induced translocations decreased (see J.G. Brewen, R.J. Preston, and N. Gengozian, "Analysis of X-ray Induced Chromosomal Translocation in Human and Marimoset Spermatogonial Stem Cells," *Nature* 253 [1975] :468).

those displayed in the first generation. Uncertainty about actual rates of selecting out prevents making a confident quantitative estimate.

HEALTH RISKS OF NUCLEAR POWER

Health risks, potentially involving deaths, injuries, and illness, arise at all stages of the nuclear fuel cycle, from uranium mining to plant decommissioning. The risks also depend on the specific fuel cycle employed. Assessments of health impacts are conventionally divided into occupational and public categories and assigned as attributable to a year's operation of a 1,000 MWe power plant, assumed to operate at 70 percent of capacity. Such a power plant would provide, at present consumption rates, for the direct and indirect electricity use of about one-half million people.

Assessments of health effects from nuclear power are complicated by the fact that there has been relatively little operational experience, and data accumulated thus far have been derived from practices that are changing. Where there are improvements, as in uranium mining, future impacts will be lower; where care becomes lax, they may be higher. In many areas, data are lacking and assessments must be derived from projections based on design and performance standards. To the extent that these idealized conditions are not met, health impacts will be larger. The uncertainties about radiation effects on man, discussed in the preceding section, and about environmental pathways to man also affect assessments. Finally, assessments must include not only the health effects from normal operations but also those resulting from accidents.

This section reviews health impacts from normal operations and accidents at various stages of the nuclear fuel cycle. For each stage, occupational and public health impact are treated separately and accidents not involving radiation are distinguished from radiation exposure. To give a rough measure of the effect of radiation doses received, we have used the linear dose-response relation, with a risk factor of 180 cancer deaths per million man-rem (whole-body equivalent), discussed in the preceding section. Unless otherwise indicated, radiation exposures are expressed as fifty-year dose commitments per reactor year, received by people over a period of fifty years, due to the generation of power by a reactor for one year.

Mining

Mining is a primary locus of occupational hazard. Accidents in underground mining result in about 15 deaths per 10,000 miners[g] or about 0.2 deaths per

---

[g]*Comparative Risk-Cost-Benefit Study of Alternative Sources of Electrical Energy*, WASH-1224, U.S. Atomic Energy Commission; December 1974. We have used a uranium ore concentration of 0.1 percent U_3O_8 and assumed that half of mining production is from open pit mines. Lower concentrations and greater dependence on underground sources would likely increase health impacts.

reactor-year. Estimates for open pit mines show smaller numbers. Another estimate[h] puts deaths at about 0.15 per reactor-year. Nonfatal accidents are equivalent to about 1,100 days of incapacity per reactor-year.[i]

In addition to accidents, miners are exposed to external radiation in mines and to inhaled dust on which radon daughter decay products are adsorbed. Radon daughters inhaled in this way may become trapped in tissue in the lower respiratory tract, where they can be carcinogenic. One estimate of external whole-body dose in mines is 100 man-rem per reactor year.[j] Lung dose commitments are reported as being about 960 man-rem per reactor year.[k] Using the average dose response figures described in the preceding section, the whole-body exposure would result in about 0.02 delayed cancer fatalities and the lung exposure about 0.04 delayed fatalities (at 40 deaths per 10^6 man-rem) per reactor year.[l] Actual rates will depend upon particular mining practices.

Radon emissions during mining may also have a public health impact, though estimates of its magnitude vary greatly. Earlier studies, noting the remote location of uranium mines, generally concluded that population exposures were negligible. The recent GESMO study comes to a different conclusion, however. In addition to its direct decay, radon-222 results in the formation of daughter decay products and several of these, notably polonium-210 and lead-210, appear to enter the food chain by deposition on crops. Modeling the environmental fate of radon-222 releases, about 5,000 curies of which are released per reactor-year of ore mined, the GESMO concludes that the population dose commitments are about 440 man-rem per reactor year (628 man-rem per gigawatt-electric year at 70 percent capacity factor). The dominant pathway is through the diet and not through direct lung dose. Using the simple average linear dose-response relation above, one would obtain about 0.08 delayed deaths per reactor year. This makes mine emissions the source of one of the largest contributions to radiation exposure in the LWR uranium fuel cycle. The calculational model used appears to be somewhat conservative and may thus overestimate exposure. Realistic, as opposed to conservative, assessments are important in indicating remedial actions and in making comparative assessment with the alternative fuel cycle involving reprocessing and recycle of plutonium.

[h]L.A. Sagan, "Human Costs of Nuclear Power," *Science* 177 (1972):487.
[i]Ibid.
[j]E.E. Pochin, *Estimated Population Exposure from Nuclear Power Production and Other Radiation Sources*, Nuclear Energy Agency, Organization for Economic Co-operation and Development, Paris, January 1976.
[k]*"Final Generic Environmental Statement on the Use of Recycle Plutonium in Mixed Oxide Fuel in Light Water Cooled Reactors,"* (GESMO), NUREG-0002, Vol. 4, U.S. Nuclear Regulatory Commission, Washington, D.C., 1976.
[l]We have used the BEIR risk estimate, assuming a thirty-year duration of risk. The GESMO uses a lung risk factor of 22.2 deaths per 10^6 man-rem and a latency period of fifteen years, arriving at lower consequence estimates. In subsequent computations, where lung dose is not dominant, we have simply used the GESMO whole-body dose commitment, multiplying by our representative risk factor of 180 delayed deaths per 10^6 man-rem.

Milling

Occupational accidents in the milling of uranium ore are poorly documented but appear to be relatively low (of order 0.001 deaths per reactor year). In addition, the GESMO estimates a lung dose commitment for workers of about 700 man-rem. Using our average dose-response relations, this exposure would be expected to result in a total of about 0.03 delayed cancer fatalities per reactor year.

Radon emissions from milling operations were cited by the AEC[m] as being about 75 curies per reactor year. The GESMO projects emissions of about 900 curies, but apparently includes those from active tailings piles, that is processed ore residues which have not been stabilized as described below. The GESMO estimates a dose to the general population of about 100 man-rem per reactor year (whole-body dose equivalent); this is also very likely a conservative estimate, with actual exposures being somewhat lower. Latent cancer deaths per reactor year would amount to about 0.02 for the GESMO estimate.

Milling operations result in the creation of huge tailings piles near the mills. These piles continue to emit radon-222 for many thousands of years and are thus a prospective source of long-term exposure. The GESMO estimates annual whole-body population dose-commitments due to the tailings from one reactor year, of about 10 man-rem. This assumes that the piles are dried and stabilized, with two feet of earth placed over the tailings material. On an annual basis, this rate of exposure is not large; however, it accumulates to a commitment of about 1,000 man-rem per century, resulting, according to average dose-response estimates, in about 0.2 latent cancer deaths per century for each reactor-year of operation. Another estimate by Ellett and Richardson[n] is slightly higher. This hazard could be greatly reduced by preventing escape of radon until it can decay (the daughter products are then trapped in the tailings material) by covering piles with an impermeable asphalt or other barrier or by reburial of the tailings. It should be noted that simple surface stabilization of tailings, through earthen or other covering, is very unlikely to provide assurance against release over the many centuries the tailings remain a hazard.

Transportation, Conversion, and Fabrication

Relatively small volumes of material are involved in nuclear power generation after the initial mining and milling operations. Moreover, the levels of radioactivity present at these stages are small, at least for the uranium fuel cycle. While accident rates for these operations do not appear to have been specifically analyzed, they are probably comparable to, or lower than, those for similar industrial activities. For transportation, Rose, et al.,[o] project 0.01 deaths per reactor

[m]Atomic Energy Commission, *Environmental Survey of the Nuclear Fuel Cycle*, WASH-1237, November 1972, B-4.

[n]H.W. Ellett and A. Richardson, "Estimate of the Cancer Risk due to Nuclear-Electric Power Generation," U.S. Environmental Protection Agency, 1976.

[o]D. Rose, P. Walsh, and L.L. Leskovjan, "Nuclear Power—Compared to What?", *American Scientist* 64 (1976):291.

year. Radiation exposures are projected by GESMO to be negligible. Ellett and Richardson estimate about 0.01 delayed fatalities from both occupational and population exposures. The risks may vary with fuel cycle option: the presence of plutonium in the fuel cycle would increase hazards in transportation and fabrication. The GESMO projects occupational lung dose commitment from mixed oxide fuel fabrication of about 1 man-rem per reactor year. Population dose commitment is projected to be about 3 man-rem to the bone per reactor year. If exposures are actually constrained to be this small, health effects will be insignificant. However, this would require advances over past containment practices.

Reactor Facility Construction

Many studies have neglected fatal accidents which accompany the construction of reactor facilities. Construction typically involves fatalities of 2 to 8 deaths per 10,000 man-years,[P] with the higher figure typical of the heavy construction required by reactor plants. On this basis, Pochin (1976) estimates four fatalities per plant. Sagan (1972) reports two fatalities. Prorated over a twenty- to thirty-year plant lifetime, this leads to an occupational loss rate of about 0.07 to 0.2 deaths per reactor year.

Normal Reactor Operation

During operation, radioactive gases and volatile radioisotopes migrate out of fuel, through small cladding defects, into the coolant. These effluent gases, which include iodine, xenon, and krypton isotopes, are held up and treated. At present, about 500 curies of krypton-85 are ultimately released annually from each reactor. Krypton-85 is a chemically inert gas emitting β- and γ-rays, with a half-life of 10.7 years. Release of krypton-85, which diffuses through the atmosphere worldwide, appears to have very limited health effects, primarily through radiation exposure to the skin. Its global health risk is on the order of 7×10^{-5} estimated deaths per reactor year (Ellett and Richardson, 1976). If necessary, krypton-85 can be captured and stored until its hazard is reduced to any desired degree by radioactive decay. If reprocessing occurs, much larger quantities of krypton would be released from spent fuel—a prospect discussed below.

In the past few years, it has been recognized that some carbon-14 is produced in commercial reactors by neutron-induced reactions with oxygen-17 and nitrogen-14 in the cooling water and oxide fuel. Although quantities are uncertain, it appears that 20–30 curies are formed per reactor year. The fraction of this released during reactor operation is also uncertain; the remainder would be released only during reprocessing. The GESMO (1976) projects the release of about one-third of the carbon-14 during reactor operation (as carbon dioxide gas) and projects a resulting population whole-body commitment of about 50 man-rem per reactor year. This would result in about 0.01 latent cancer fatalities per re-

[P]*Accident Facts*, National Safety Council (Chicago, 1972).

actor year, using our average dose-response relation. Ellett and Richardson, using a 100-year environmental dose commitment, project 0.09 delayed fatalities per reactor-year. The radioactivity of carbon-14 is potentially of greater concern than this, however, since it is very long-lived (with a half-life of 5,730 years) and is incorporated in biological material, a property which makes naturally occurring carbon-14 useful in dating fossils and artifacts. The magnitude of the long-term hazard to man depends on the environmental path of released carbon-14, a highly uncertain matter. But the amount of reactor-generated carbon-14 is minute compared to the 220 million curies of carbon-14 in the atmosphere due to cosmic rays and the 6 million curies due to atmospheric weapons tests. Carbon-14 release from reactors can be greatly reduced using existing technology, if the hazard is deemed excessive.

While the public radiation dose-commitment from normal operations is relatively small (of order 100 man-rem per reactor year according to GESMO), and much smaller if carbon-14 and krypton-85 emissions are kept low, the occupational dose is generally higher. Workers appear to be exposed primarily during refueling and repair operations. Data for existing plants indicate that occupational dose commitments are about 800 man-rem per reactor year, with variations of up to a factor of 2. Some new data[q] suggest that exposures go up with plant age, presumably because of equipment activation and accumulated contamination. To these exposures should be added those eventually resulting from decommissioning; the magnitude of this is uncertain, however, and we are thus unable to include it. If occupational dose-commitments from normal reactor operations remain at 800 man-rem, the associated latent cancer incidence will be about 0.15 per reactor year. Future trends in exposure levels are clearly deserving of attention.

Impact of Core Melt Accidents on Health Risk Assessments

Public concern with nuclear safety has not focused on the elements of the fuel cycle treated thus far but rather on the potential consequences of reactor accidents, involving core melt and breach of containment. Although reactor accidents and their analysis in WASH-1400 are examined in considerable detail in Chapter 7, a few points relevant to the health effects deserve emphasis here.

The types of radiation effects resulting from release of some of the radioactive inventory from a large reactor vary widely. The dose within the immediate environs may be high enough to cause rapidly lethal radiation sickness, while the dose at greater distances diminishes ultimately to background levels as the radio-

[q]Pochin, *Estimate Population Exposure*; J.C. Golden, "Effect of a Change in Occupational Radiation Exposure Limits on Manpower and Total Dose at a Nuclear Power Station," *Health Physics*, in press (1977); J.C. Golden and R.A. Pavlick, "A Review of Effluents, General Population Doses and Occupational Doses Resulting from Commonwealth Edison's Use of Nuclear Power," *Symposium on Population Exposures,* CONF-741018, pp. 199–266, National Technical Information Services, Springfield, Virginia 22161 (1973).

activity is gradually diluted by dispersion, fallout, and decay. The consequences to an exposed population, in turn, will depend on the kinds and amounts of fission products released, prevailing weather conditions, the population density at various distances within the path of the radioactive cloud, and factors influencing the degree to which potentially exposed individuals are protected against irradiation by early evacuation, shelter, shielding, and other measures minimizing the inhalation, ingestion, or uptake of radioactive material.

The risks of early radiation sickness and death are estimated in WASH-1400 to be confined to the area within 10–15 miles of the reactor but depend heavily on plume temperature, weather conditions, population density, and speed of evacuation of the area. Effective evacuation speeds exceeding 5–7 miles per hour were estimated to reduce the risk of early mortality toward zero in the absence of heavy local contamination due to rain. Under the most pessimistic assumptions, however, involving ineffective evacuation and high population density, the number of individuals developing radiation sickness was estimated at 45,000, with 3,300 prompt fatalities (Table 7-1).

In addition to estimating the early effects on exposed populations, WASH-1400 also projected the later occurring carcinogenic and genetic effects, with the use of various dose-effect models (Tables 5-2 and 5-3), those used for arriving at upper bound estimates being derived from the BEIR report. The dose-effect models, applied to a hypothetical population of ten million people residing under the radiation plume within 500 miles of the most serious type of reactor accident, led to estimates of 45,000 fatal cancers, 240,000 thyroid nodules, and 30,000 genetic effects as long-term consequences of exposure (Table 7-1). The cancers, which were projected to be distributed over a thirty-year period and to represent roughly a 9 percent increase above the natural incidence in the exposed population as a whole, resulted in large measure from irradiation of the lung through inhalation of the passing cloud. Because a substantial portion of the doses to the lung were calculated to be in excess of 25 rem, the reduced dose-effectiveness factors mentioned earlier in this chapter were not considered to be fully applicable; hence the estimate was intermediate between the corresponding value for the WASH-1400 upper bound estimate and that for the central estimate (Table 5-2).

The thyroid nodules, which also were projected to develop over a thirty-year period, were considered to be nonfatal in 90 percent of cases. It was suggested, moreover, that the induction of these growths could in principle be largely prevented by the administration of stable iodine to the population at the time of exposure, in this way blocking the uptake of radioactive iodine and greatly reducing the radiation dose to the thyroid.

The genetic effects were projected to become manifest in the descendants of the exposed individuals, distributed over the first five generations, amounting to roughly 200 cases per year among the 140,000 offspring born annually, as compared with a natural incidence of 8,000 cases per year in the same population.

It will be noted that this estimate corresponds to the lower end of the range of estimates presented by the BEIR Committee and updated herein (Table 5-3).

In addition to its assessment of the most serious type of accident, discussed above, the probability of which was estimated to be about one in 200 million years of reactor operation, WASH-1400 assessed the expected impact (early fatalities, latent cancer deaths, genetic effects, property damage, etc.) of a spectrum of lesser accidents, ranging from those with relatively high probabilities and minor consequences through an intermediate range of events. Integration over the entire range of probabilities and consequences analyzed in WASH-1400 leads to the conclusion that on an average rate-of-loss basis, there would be about 0.023 fatalities per reactor year (0.002 early fatalities and 0.021 latent cancer deaths per year). Considered strictly from a health perspective, these numbers are very uncertain and do not represent a conservative assessment. The possibility exists that the expected number of cancers could be several times higher, depending on the assumed dose-response model used in deriving the risk estimates (Table 5-2). These figures also do not include deaths attributable to genetic effects, which could be substantially higher than those estimated in WASH-1400. Despite the uncertainties, the WASH-1400 assessment indicates that reactor accidents contribute only a small percentage of the total average rate of loss attributable to the fuel cycle.

The matter does not rest here, however. As the discussion in Chapter 7 indicates, the crucially important probability estimate in the WASH-1400 report, the probability of a core melt with breach of containment per reactor year of 5×10^{-6}, is of low reliability because of the large uncertainties involved in its calculation. Although we have not attempted to make an independent estimate of this average value, the analysis in Chapter 7 indicates that the WASH-1400 probability estimate could be low, under extremely pessimistic assumptions, by a factor of as much as 500. On the other hand, it could even be on the high side as well. In the most pessimistic case, where the estimates are low by a factor of 500, the average (over the first 100 reactors) rate-of-loss would be about 10 fatalities per reactor year for a 1,000 MWe nuclear power plant. It is significant that even under such extremely pessimistic assumptions, the fatalities are less than the high end of the range of estimated deaths associated with coal-fired power plants, discussed below. As the discussion in Chapter 7 stresses, this is not a prediction but a limit to which no probability is attached. It is used solely to give an upper bound on the range of estimates that are possible.

A further important consideration arises if the estimate of health risk is to be used not as an index of present performance of nuclear power plants but as a guide to making a future choice between nuclear and coal or other energy sources for electricity generation. The central average rate-of-loss estimate in WASH-1400 of 0.023 fatalities per reactor year derives largely from about 10 percent of the 100 reactors surveyed. Indeed, more than half its value is probably contributed by only a few reactors whose location with respect to dense populations is such that certain weather conditions at the time of accident could expose very large populations and thereby lead to unusually large numbers of presumptive fatalities. Thus, to the extent that reactors could be located at less potentially risky sites, the average rate-of-loss risk for a particular new reactor could be lowered by a factor of 10 to 100. Therefore, even the higher risk probabilities that could possibly enlarge the WASH-1400 values upward toward an average rate-of-loss of 10 latent cancer deaths per reactor year could be reduced by prudent site selection to average values that are low relative to other contributions of the nuclear fuel cycle.

Waste Management and Disposal

Public and occupational doses resulting from waste management and disposal are often regarded as insignificant (Rose et al., 1976; GESMO, 1976). However, this may be in part due to a lack of experience with commercial waste management and disposal beyond the care of spent fuel in reactor cooling ponds. Data concerning exposures from commercial burial sites are only now being acquired; however, it will be of limited applicability to future practices if necessary improvements over past practices are achieved. Successful disposal of spent fuel (or reprocessed waste), as considered in Chapter 8, appears to involve little long-term public risk. However, it is at present difficult to estimate the health costs of constructing and filling geological repositories or the actual effects of repository failures. These factors should be assessed as plans for specific repositories go forward. It is possible that mill tailings, as presently treated, will be found to have larger health effects than other wastes disposed of in geological repositories. The health risks of waste management, before disposal, are likely to depend on whether reprocessing and recycle occur.

Reprocessing and Recycle

If the present uranium fuel cycle, as described above, is extended to include reprocessing and recycle, new potential health risks are introduced. Reprocessing would involve the dissolution of spent fuel, with a consequent liberation of gaseous and volatile radioisotopes, and the separation of uranium and plutonium from the other radioactive waste materials. Recycle of plutonium, either in LWRs or breeders, would introduce plutonium into most of the fuel cycle, including transportation and fuel fabrication.

The occupational hazards of reprocessing and recycle are almost entirely radiological and arise from handling of radioactive waste materials in the reprocessing operations, and in subsequent steps to treat wastes for disposal, and from the variety of chemical and mechanical operations involving plutonium. Experience with reprocessing plants has not been entirely satisfactory, though inadequate records make assessment difficult. Pochin (1976) reports British experience of about 1,400 man-rem per reactor year, or 2 man-rem per MWe-year, which is higher than estimates of exposures at the Nuclear Fuel Services plant, where incomplete records hamper analysis.[r] Earlier plants, however, did not attempt to resolidify liquid waste materials or convert plutonium nitrate solution to solid plutonium oxide, measures now thought to reduce subsequent radiation risks, particularly in transportation. Complete plants built to earlier standards would thus have had even higher exposures. The most difficult question in projecting future occupational exposures is the extent to which better technology and higher standards will reduce exposures. The GESMO projects an occupational dose commitment equivalent to about 20 man-rem per reactor year, with exposures about evenly divided between separations, plutonium nitrate to oxide conversion, and waste facilities. For the separations facility, the GESMO projection is one to two orders of magnitude below experience. If the 20 man-rem figure is achieved, occupational latent cancer fatalities would be at the low level of about 0.004 deaths per reactor year.

Population exposures as a result of reprocessing appear to be due predominantly to carbon-14, krypton-85, and other radioactive materials emitted in normal operation under present regulations. Carbon-14 not released in reactor operation would probably be liberated at the reprocessing plant as carbon dioxide gas. According to the GESMO, release results in a population dose commitment of about 100 man-rem for each reactor year of experience,[s] or less than 0.02 latent fatalities per reactor year. Ellett and Richardson (1976) cite a figure ten times higher, using a 100-year environmental dose commitment for carbon-14.

About 300,000 curies of krypton-85 would be released from reprocessing for each reactor year. According to the GESMO, this results in a U.S. population dose commitment about 20 percent of that due to carbon-14. Tritium releases account, according to the GESMO, for about 200 man-rem population dose commitment. A lower estimate of 15–30 man-rem is reported by Pochin (1976). Collectively, worldwide population dose commitment is projected as being about 450 man-rem per reactor year (GESMO, Table IV, E-9); using our average dose-response conversion, this would result in about 0.08 latent cancer fatalities worldwide. Virtually none of the dose commitment in the GESMO analysis comes from plutonium or other actinides. The population dose commitment is more than a factor of 20 larger than the occupational dose, according to the GESMO. This impact could be reduced, perhaps by a factor of 10, by effluent control technology.

The GESMO results and other projections for future reprocessing plant opera-

---

[r]Data presented in "Comparative Risk-Cost-Benefit Study of Alternative Sources of Electrical Energy," WASH-1224, U.S. Atomic Energy Commission, December 1974, for NFS estimate an average occupational exposure equivalent to about 250 man-rem per reactor year for the period 1968–1970.

[s]This is a sum of direct exposures from the plume of gas from the reprocessing plant and the worldwide dose forty years after release. The GESMO assumes an environmental half-life of six years for carbon-14. It is not clear whether long-term exposures would be correctly included by this procedure.

tions should, however, be considered as tentative. Exposure estimates are based on design specifications. For the separations facility, projected exposures are much smaller than experience; for waste solidification and treatment and for plutonium conversion, comparable facilities have never been built. Thus, exposures will depend upon whether engineering specifications are met in actual plants. Small accidents, spills, and gradually increasing contamination, experienced in past plants, would increase occupational exposures particularly. Estimates of occupational exposures also do not include interim clean-up and ultimate decommissioning. Since reprocessing plants involve large contaminated areas (much larger, say, than reactors), decommissioning could be costly, in both dollars and man-rem.

Reprocessing and recycle reduce the need for uranium mining and milling and thus potentially reduce occupational and public health consequences of these operations. For the remainder of this century, a reduction of about 20 percent is possible if all plutonium is recycled. This reduction in uranium mining and milling might thus reduce occupational fatalities (accidental and cancer) in these operations by about 0.05 per reactor year and population delayed fatalities by about 0.02 per reactor year. The net impact of reprocessing and recycle, however, is to increase the health consequences of the nuclear fuel cycle. This is a relatively large increase (more than 50 percent) in population exposures. If the projection in the GESMO for occupational exposures is correct there would be a small net decrease (10 percent) in occupational health effects. However, this projection is highly uncertain and perhaps overly optimistic if historical experience is any guide.

According to the GESMO analysis, plutonium plays essentially no role in occupational or public health effects from reprocessing and recycle. This conclusion is due to the assumption that it is controlled to design requirements throughout the fuel cycle. It is impossible to assess the correctness of this without an independent detailed engineering analysis and projections of probabilities for accidental releases. Such an effort has not yet been made. It is possible, however, to form preliminary judgments about the health effects of plutonium should it enter the environment.

Health Risks of Plutonium

The health risks of plutonium depend on the chances for its release to the environment and its subsequent behavior in environmental and food chain pathways to man, its mode of uptake and translocation to various organs, and its particular radiotoxic effects. All of these are uncertain to some extent, though not to so great a degree as controversy suggests.

The health hazards of plutonium are special in that very long-lived energetic α-emitters are involved.[t] With the present uranium fuel cycle, most radiation is

[t]Plutonium from spent reactor fuel is a mixture of isotopes, and its hazard depends on this fact. Plutonium-238, -239, and -240 are α-emitters. Plutonium-241 emits low-energy β-

low-LET. The only appreciable source of high-LET radiation is via the radon from mines, milling, and tailings, a problem which can be greatly reduced. Reprocessing and recycle would thus introduce a new source of high-LET radiation into the active fuel cycle. This does not necessarily present large risks. The fact that α-radiation cannot penetrate skin means that plutonium is only dangerous if it enters the body where it can come in close contact with living cells. If this happens, it is highly carcinogenic and presumably mutagenic. Measures to prevent this can reduce risks very considerably. The plutonium oxide powder which would be used in recycle fuels is chemically very inert and can be isolated by mechanical barriers. Plutonium does not appear to concentrate in most food chains, with the exception of some marine plant and invertebrate species where concentration (over ambient levels in seawater) may be a factor of 1,000 or more. However, the evidence for low food chain concentration is based upon periods of observation short compared to the duration of the hazard. Indeed, there is some evidence that the rate of uptake in plants may increase in time, perhaps as plutonium slowly forms new chemical compounds in the soil. It should also be noted that americium, associated with reactor plutonium through decay of plutonium-241, has higher biological concentration factors.

The largest risk from plutonium is through inhalation of small particles which become lodged in the lower respiratory tract. Plutonium present in the lung can induce cancer; it can also translocate with the same effect to other tissues through absorption and transport in the lymphatics and blood. The health consequences of this translocation are expected to be largest in liver and bone, in that order. Plutonium also concentrates, to a more limited degree, in human gonads. The extent of the genetic effects of low concentrations of plutonium have not been quantitatively assessed. Ingested plutonium presents less risk since absorption in the gastrointestinal tract is small. There is evidence that americium is absorbed more readily in both lung and gastrointestinal tract. Most experiments and most experience in the weapons program involves relatively pure plutonium-239. Commercial reprocessing and recycle would involve use of a mixture of plutonium isotopes with α-activity about eight times higher per gram than that of plutonium-239. While the higher activity puts more rigorous constraints on mixed oxide fuel cycle facilities in preventing exposures, it is not

particles. The most prevalent isotope is plutonium-239 (60 percent by weight); however, its half-life is 24,000 years and its level of radioactivity is consequently relatively small (7 percent of the α-radioactivity present five years after removal from the reactor). Plutonium-238 is only 2 percent by weight, but, with a half-life of eighty-six years, accounts for 80 percent of the α-activity in reactor plutonium. Plutonium-240 is 24 percent by weight, but 13 percent of the α-activity. Plutonium-241 decays rapidly, with a half-life of thirteen years, emitting large numbers of relatively low-energy β-particles. However, it decays to americium-241, an intense α-emitter. The amount of americium in separated plutonium depends on when spent fuel is reprocessed. If spent fuel is allowed to age for several half-lives of plutonium-241, most of the americium-241 produced by its decay goes out in the waste stream in reprocessing. If plutonium is reprocessed immediately and stored, the stored product gradually becomes contaminated with americium-241.

clear whether the biological effects of reactor plutonium once released (per curie) are different from those of plutonium-239. Experiments with particles of plutonium-238 in animal lung tissues imply that their high level of α-activity may result in enhanced translocation to other organs.

More than twenty years ago, the International Commission on Radiological Protection and the National Council on Radiation Protection and Measurements set standards for the maximum permissible body burden of plutonium at 40 nanocuries of α-activity for plutonium workers and at 4 nanocuries[u] for the general public. Very rough calculations suggest that 40 nanocuries may increase the risk of delayed fatal cancer by about 0.2 percent, though there is considerable uncertainty about the precise value. The limit of 40 nanocuries was determined in part by referring to much more extensive work on radium-226, for which a limit of 100 nanocuries had long been set, and to experiments with dogs. There is some uncertainty attached to both these procedures. Since the radiation from plutonium and radium is almost identical, they would have equal carcinogenic effect if identically distributed throughout the body. However, the concentration patterns differ in ways that make plutonium a greater hazard. Whether the radium-plutonium comparison, based primarily on bone studies, and the extrapolation from dogs to humans are sufficient to lower the standard by only a factor of 2.5 has been frequently disputed, but reviews have thus far left the maximum permissible body burden unchanged.

One of the objections raised to present standards has been the "hot particle" hypothesis. According to this hypothesis, small particles lodged in lung tissue emitting α-radiation to a small surrounding volume are far more likely to induce cancer than the same amount of radiation distributed uniformly throughout the tissue. It was asserted by supporters of this hypothesis that the maximum permissible lung burden (which is 40 percent that of the body burden or 16 nanocuries) should be lowered by a factor of 2,000. Most studies discredit the "hot particle" hypothesis although it cannot be said to have been rigorously disproven. In principle, one would not expect high-LET radiation to be more carcinogenic per rad if concentrated in small regions containing relatively small numbers of cells, since under such conditions the vast majority of cells would be killed and much of the dose thereby dissipated without inducing the kinds of changes which might lead to cancer. The principal radiation protection boards in the United States and in Britain have agreed that the nonhomogeneous dose distribution from inhaled plutonium should not be expected to result in demonstrably greater risk than that assumed for a uniform dose distribution.

Finally, as noted above, there are several studies which appear to lead to the sobering conclusion that cancer induction may be greater, not smaller, if a specified total dose of α-radiation is given at a low level over a long period of time

---

[u]A nanocurie is one-billionth of a curie; one nanocurie of reactor plutonium weighs about two-billionths of a gram.

than if it is given rapidly. As one example, a 1972 report[v] found that rats exposed to varying doses of plutonium-238 showed a sharp rise in the incidence per rad of lung tumors with an initial aveolar burden of 5 nanocuries, the effect of leveling off at higher burdens. One explanation is that cells which might otherwise have become cancerous are killed at higher doses and dose rates. This is an area in which further experimentation is warranted as a guide to assessing health impacts and setting standards.

There is no evidence that present standards for exposures and body burdens of plutonium are grossly in error; however, this conclusion rests primarily on animal experiments[w] and incomplete follow-up studies of several groups of radiation workers with elevated body burdens of plutonium. More than 5,000 such workers have been identified, of whom only thirty have died and been autopsied to date. Evaluation of the data from these autopsies, on individuals with body burdens up to 1.5 times the maximum permissible body burden, has revealed no evidence of plutonium toxicity. Although eleven of the thirty autopsied cases died with some form of malignancy (double the normal incidence), the cancers were of a wide variety of types and arose in various sites, without any discernible relation to the levels of radioactivity in the affected organs. Moreover, the relatively high frequency of cancers among the thirty cases, as compared with the incidence in the general population, is consistent with the elevated frequency of cancer characteristic for most autopsied groups, which represent selected populations. While these data are of limited statistical significance, surveys of the larger group of living plutonium workers reveal no significant excess of nonfatal cancer. If present standards were a factor of one hundred too high, many tumors should have been found by now. Smaller uncertainties in risks and standards, however, cannot be resolved on the basis of available statistical evidence in humans; at best we can conclude that present assessments of risks are not generally wrong by more than roughly an order of magnitude.

Summary Assessment

Health impacts of the present nuclear fuel cycle include occupational accidents and radiation-induced disease in workers and the public due both to routine emissions and to accidents. Under normal operating conditions, each reactor year of power production is estimated to involve somewhere between 0.2 and 0.5 accidental worker deaths, with roughly half of this the result of mining activity and half the result of reactor construction. Occupational dose commitments (whole-body equivalent) are estimated to be between 1,000 and 1,500 man-rem per reactor year, with about two-thirds of this coming from reactor

[v]C.L. Sanders, "Carcinogenicity of Inhaled Plutonium-238 in Rat," *Radiation Research* 56 (1972):540.

[w]W.J. Bair and R.C. Thompson, "Plutonium: Biomedical Research," *Science* 183 (1974):715.

operations and repair and about one-third from mining and milling. Most of the reactor exposure is low-LET radiation while most of that in mining and milling is high-LET. If an average dose-response conversion of 180 latent fatalities per million man-rem is assumed, occupational exposures account for about 0.2 to 0.3 delayed fatalities per reactor year. As noted in the preceding section, this assumed dose-response relationship is very uncertain, perhaps overestimating the effects of low-LET radiation[x] and underestimating those of high-LET radiation. Total occupational fatalities may thus be in the range of 0.4 to 0.8 deaths per reactor year.

Assessment of public health consequences is a much more uncertain endeavor. Population dose commitments, for normal operations without reprocessing and in the near term, appear to be dominated by radon emissions in mining and milling and by routine effluent emissions, notably carbon-14, tritium, and krypton-85. If the GESMO calculations are correct, it is dietary intake of the decay daughters of radon-222 that dominates population exposures. The total population dose commitment is below 1,000 man-rem per reactor year according to most estimates. However, continuing emissions of radon from mill tailings piles could, over a period of centuries, greatly increase this number. While the total population dose commitment is the sum of a great number of small doses delivered at low dose rates, the fact that the dose commitment is dominated by α-emitters suggests that application of the usual linear dose-response relation may not be grossly in error. A population dose commitment of 1,000 man-rem would then correspond to about 0.2 latent fatalities per reactor year. This would bring the total health risks for workers and public to 0.6 to 1.0 expected deaths per reactor year.

Reprocessing and plutonium recycle, if added to the present uranium fuel cycle, would change the distribution and magnitude of health risks, causing a reduction in occupational deaths in mining and milling (largely a reduction in accidents) and a small decrease in population exposure from radon and its daughters, but all at the expense of new, and very uncertain, impacts. Using the GESMO figures for reprocessing and recycle the net effect would be a reduction in occupational fatalities of about 0.04 per reactor year and an increase of 0.07 delayed deaths in the general population. However, it is possible that the health effects of reprocessing and recycle would be greater than projected if new facilities do not achieve the desired great improvements over past containment practices.

The average public health consequences of reactor accidents are even more uncertain. Probabilities cannot be predicted with certainty, and past efforts are flawed by methodological problems; consequence calculations depend heavily on modeling and are subject to the dose-response uncertainties described above. The Reactor Safety Study (WASH-1400) projected average accident consequences of about 0.02 latent fatalities per reactor year. As discussed in Chapter 7, however,

---

[x]It should be noted, however, that occupational exposures are sometimes accumulated at relatively higher dose rates during brief periods.

this probably underestimates the average risk; as an extreme upper limit, if all uncertainties are viewed pessimistically the risk might be as high as ten deaths per reactor year (with a comparable number of genetic effects) for the first 100 reactors as considered in WASH-1400.

HEALTH RISKS OF COAL-FUELED POWER GENERATION

Coal is now the leading source[y] of power generation as well as the source most competitive with nuclear reactors for the expansion of capacity in the next few decades. The state of knowledge of the health impact of coal-fueled generating plants is similar to that of nuclear power in that both are characterized by large uncertainties which could dominate the total average rate-of-risk.

Mining

Generating plants tend to be supplied from large underground mines and from open pit mines where more accidents occur with only half the frequency as in small mines. On this basis, it is estimated that there are about 0.5 fatal mining accidents per 1,000 MWe plant-year.[z]

Traditionally, miners suffered severely from pneumoconiosis, also known as black lung disease, a progressive inflammatory disease of the lungs due to the permanent deposition of inhaled coal dust. Although incidence rates appear to have been very high in the past, with welfare and death benefits for this disease amounting to more than $1 billion annually, it is anticipated that mining activity meeting the conditions of the Federal Coal Mine Health and Safety Act of 1969 will involve little further occupational health impact from this disease. If this is the case, deaths from pneumoconiosis associated with future power plant experience should be negligible compared to deaths from mining accidents.

Transportation and Other Steps

About 10 percent of freight car haulage is due to coal transport for power plants. With 2,300 persons killed by trains each year, mostly in automobiles at grade-crossings by freight trains, one could apportion 230 deaths to the 180 equivalent 1,000 MWe plants and arrive at 1.3 deaths per plant year (Sagan, 1974). An alternative calculation (WASH-1224, 1974) projects 0.55 deaths (90

[y]The electric power produced in the United States in 1974 can be expressed in terms of equivalent 1,000 megawatt plants (using a 60 percent capacity factor) as follows: coal—177; oil—63; gas—67; nuclear—24; hydro—63. A December 1976 announcement from the Federal Energy Administration requires that 142 proposed fossil-fueled power plants all burn coal and that 74 existing plants convert to coal from oil. If this is carried out, coal demand would be increased by 236 million tons per year or 38 percent.

[z]L.A. Sagan, "Health Costs Associated with Mining, Transport and Combustion of Coal," *Nature*, 250, 197 (1974); The 1975 National Safety Council Injury Rate Survey finds 0.70 deaths plus total permanent disabilities for the 1,200,000 man-hours of mining estimated to be required to fuel a 1,000 MWe power plant for one year.

percent in the public) per plant year on the same basis. Since this effect is proportional to the length of haulage, it would increase if western coal were hauled to generating plants in the East, unless more protective measures were taken at crossings. Alternatively, mine mouth generation and transmission of power to population centers could reduce impacts. While one might question the validity of including automobile fatalities in such a comparison, the fact remains that the nuclear cycle does not involve comparable railroad transportation.

Construction of generating plants presumably involves an average occupational fatality rate for this kind of heavy construction of about 0.05 deaths per reactor year, somewhat less than in the case of nuclear power plants, which involve more extensive construction.

Although there would appear to be some health impact from waste disposal at various stages of the coal cycle, detailed estimates of these effects do not appear to have been made.

Electricity Generation

Air pollution has been recognized as a cause of illness and death as well as discomfort for centuries. An ordinance of 1273 prohibited the use of coal in London as "prejudicial to health." Only within the last two decades, however, has progress been made in identifying the causative agents and in realizing the magnitude of the effects not only on health, but on plants, erosion of stone due to acid rain, and other environmental damage. Major air pollution emergencies have created considerable concern. In 1948, at Donora, Pennsylvania, a prolonged period of inversion produced respiratory complaints in almost half the town's inhabitants. A similar situation accompanied by a severe fog in London in December 1952 led to 3,500 to 4,000 deaths. Levels of particulates and sulfur dioxide (SO_2) rose sharply during this period. Studies in Europe, the United States, and Japan confirmed that an abrupt rise in smoke and sulfur dioxide is associated with excess mortality. Those most affected were the aged and those with chronic obstructive pulmonary disease. Programs were mounted to reduce the most visible pollutant, large particulates, in numerous cities where coal burning was extensive, and considerable success has been achieved.

As a consequence of actions taken since the Clean Air Amendments were enacted in 1970, the levels of major pollutants, for which ambient air quality standards have been set, except for nitrogen oxides, have been slowly decreasing. Particulates from stationary fuel combustion diminished from 8.3 million tons in 1970 to 5.9 million tons in 1974. Sulfur dioxide (SO_2) from the same sources peaked at 27.0 million tons in 1970 and fell to 24.3 million tons in 1974 largely through the use of low-sulfur coal. Half of the SO_2 came from coal-fired power plants.

The principal health effects of coal-fired plants are thought to result from sulfur related pollutants and particulates. Numerous studies have attempted to elucidate the manner in which SO_2 and particulate emissions lead to subsequent health effects. A critical examination of this problem was prepared by the National Academy of Sciences in 1975.[aa]

The causal relation of severe air pollution episodes to increased mortality and morbidity (aggravation of heart and lung diseases, asthma and bronchitis) is firmly established. What is not clear is the identity of the pollutants responsible, or the manner in which the effects fall off as pollution decreases. Numerous studies show a correlation of health effects with SO_2 concentrations. On this basis, air quality standards were set for sulfur dioxide concentration—80 micrograms per cubic meter ($\mu g/m^3$) annual average. However, the correlation of health effects with concentrations of suspended sulfates, which are oxidation products of SO_2, appears stronger than that with sulfur dioxide. The suspended sulfate level corresponding to the sulfur dioxide limit depends on the rate of conversion of SO_2 to sulfates as influenced by various pollutants and weather. However, measured suspended sulfate levels for average conditions are around 12 $\mu g/m^3$ in the eastern United States and in many urban areas in this region levels are in the range of 16-19 $\mu g/m^3$.

The relationship between suspended sulfates and health is a relatively recent discovery, and thus earlier studies of sulfur-related air pollution and respiratory disease mortality use SO_2 concentrations. In London in the 1950s, SO_2 above 120 $\mu g/m^3$ was accompanied by a marked immediate rise in mortality if smoke levels were also high. Many other British studies led to similar conclusions: bronchitis was often listed as the cause of death. Good correlations between bronchitis deaths by area and mean SO_2 concentrations in these areas suggested that continuous exposure to lower concentrations had similar effects. A number of studies of individual American cities were carried out in the 1960s. In Buffalo, it was found that respiratory disease mortality in white males 50-69 years old doubled when the SO_2 concentration rose from below 80 to above 135 $\mu g/m^3$. Several New York City studies found major increases in deaths associated with days when SO_2 exceeded 800 $\mu g/m^3$; increases of 2 percent in mortality occurred when concentrations were above 500 $\mu g/m^3$.

As studies multiplied, particularly of New York City, it became increasingly evident that there was a strong correlation between mortality and morbidity and suspended sulfate concentration. However, the quantitative relationships varied widely suggesting that other variables were involved. As an example, data from New York, London, and Oslo, shown in Figure 5-1, indicate wide variations in excess mortality. Figure 5-1 shows that the percentage of expected excess mortality owing to acid sulfates in the air is estimated from suspended particulates and sulfur dioxide levels in three cities. The "mathematical best fit" line is currently considered to be a pessimistic assumption.

Informed policy guidance requires an estimate of deaths and disease on a countrywide basis due to the added burden of pollution from additional power

[aa]National Academy of Sciences, "Air Quality and Stationary Source Emission Controls," March 1975. Prepared for the Senate Committee on Public Works, U.S. Government Printing Office, Washington, D.C.

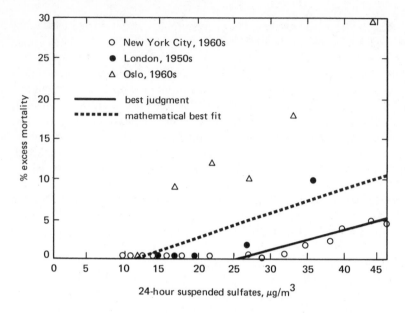

Figure 5-1. Excess Mortality from Suspended Sulfates

Source: National Academy of Sciences, 1975. "Air Quality and Stationary Source Emission Control." Prepared for the Senate Committee on Public Works, March 1975.

plants. This is an enormously more complex problem than the correlation studies made within cities. Only a start has been made in this direction.

The general pattern of SO_2-related health effects is thought to be as follows. Local high concentrations of SO_2 at groundlevel are prevented by tall stacks. Ejected at altitudes exceeding 500 feet, the sulfur dioxide can be carried by air currents for several days during which its concentration is reduced by dispersal, deposition, adsorption on particulates, washout by rain, and oxidation through intermediate compounds which lead to sulfates. The oxidation products end up principally as components of aerosols or particulates which are often quite small (< 1 μm). Such particles can lodge in the lung and the bronchial tubes; their presence interfers with respiration, aggravates heart and lung diseases, and in extreme cases, leads to death. The groups most at risk are infants, aged, asthmatics, and individuals with cardiovascular and pulmonary diseases.

Given various sulfate concentrations, superimposed on various population distributions, dose-response relations for each kind of impairment are required to deduce effects on health. As in the case of nuclear radiation, the question arises whether or not there is a threshold dose below which there is no effect and whether or not the dose-response relation is linear.

The National Academy of Sciences (1975) survey did not accept the existence of a threshold response, insisting that there is no convincing evidence that repeated low-level exposures will not lead to eventual respiratory impairment.

However, experts continue to disagree as to whether a threshold actually exists. Although the Academy survey concluded that the variable nature of mortality data does not permit a determination of whether life span is shortened primarily for relatively unhealthy people whose life expectation is short, some experts hold that this is the case. It may be that simple dose-response relations cannot represent the effects of sulfur-related pollution. Nevertheless, since decisions must be made on allowable levels of pollutants, best estimates of dose-response relationships must be used, with full recognition of the uncertainties, to calculate the incremental effects of incremental sources of pollutants.

The dose-response relations most used are those deduced by Finklea, who has drawn on various sets of data to deduce "best judgment" linear dose-response relations for five different health effects: (1) respiratory disease deaths; (2) aggravation of heart and lung disease in the aged; (3) aggravation of asthmas; (4) excess acute lower respiratory disease in children; and (5) excess risk for chronic bronchitis.[bb] It should be noted, however, that there is wide disparity in the data and considerable controversy over the interpretation of the results.[cc]

The relationship developed by Finklea for respiratory disease deaths is shown as the "best judgment" line in Figure 5-1. The other adverse health effects have lower thresholds and higher slopes. These lower thresholds are consistent with the tentative conclusions drawn by the U.S. Environmental Protection Agency from its studies as quoted in the National Academy of Sciences Report (1975, p. xvii):

> there appears to be an association between a level of sulfate to 8 to 12 $\mu g/m^3$ in the ambient air and adverse health effects in elderly persons with heart and lung disease and persons with asthmas. Somewhat higher levels of sulfate (13 to 15 $\mu g/m^3$) appear to be associated with increased prevalence of chronic bronchitis in adults, increased acute lower respiratory disease and decreased lung functions in children, and increased acute respiratory disease.

The calculations based on the assumption of a threshold yield results that are highly dependent on existing pollution patterns in two ways. Wherever ambient sulfate concentrations are near the threshold or above it, the added sulfur-related pollution from a single power plant may have a large effect whereas in a much less polluted environment it may have only a small effect. In addition, existing

---

[bb]J.F. Finklea, et al., "Health Effects of Increasing Sulfur Emissions," Draft Report, National Environmental Research Center, U.S. Environmental Protection Agency, Research Triangle Park, North Carolina (1975). See also p. 604 of the National Academy of Sciences (1975).

[cc]The Environmental Protection Agency's Research Program with Primary Emphasis on the Community Health and Environmental Surveillance System (CHESS): An Investigative Report, Report Prepared for the Subcommittee on Special Studies, Investigations and Oversight and the Subcommittee on the Environment and the Atmosphere of the Committee on Science and Technology, U.S. House of Representatives, 94th Congress, Second Session, Serial SS, U.S. Government Printing Office, Washington, D.C., November 1976.

pockets of pollution (usually in urban areas) may contain catalysts that can greatly accelerate the conversion of SO_2 to sulfates.

Health effects would be very sensitive to daily fluctuations in sulfate levels that may carry ambient concentrations above the threshold. In fact, the majority of adverse health effects may arise from such fluctuations, in which case the distribution of fluctuations about the mean would be crucially important to the outcome.

Most coal-fired power plants are located in the northeastern quadrant of the United States, where ambient levels of pollutants are highest. Because of prevailing wind patterns, effluents from plants in this region will exacerbate pollution problems in highly populated areas. As an example, one set of calculations[dd] has shown that a power plant in Chicago emitting 800 tons of sulfur dioxide per day will generate particulate sulfate concentrations of 1 $\mu g/m^3$ in parts of Illinois, Indiana, Michigan, and Ohio and of 0.1 $\mu g/m^3$ over all of the northeastern United States and eastern Canada. These calculations are consistent with the distribution of SO_2 and suspended sulfate found throughout the United States in 1974. In view of the high levels of sulfates already present in this region, additional emissions could be expected to produce adverse health effects. It should be noted that the level of emissions from the power plant used in these calculations would be high by present standards for new sources; a new 1,000 MWe coal plant meeting EPA standards will produce only 100–200 tons of sulfur dioxide per day.

Due to prevailing wind patterns, the construction of coal-fired power plants in the West may add to the burden already borne by midwestern and eastern states. Conversely, new plants located in less populous regions of the eastern coast (north of Massachusetts and south of Virginia) would probably contribute much less to the national health risk.

There have been very few attempts to develop models of the sulfur-related pollution problem to permit actual estimates of health effects from specific plants. The bulk of work to date is that described by D.W. North and M.W. Merkhofer in the National Academy of Sciences Report (1975). One of their studies estimated the additional adverse health effects produced by increasing the concentration of suspended sulfates in the New York Metropolitan area, using the dose-response relations postulated by Finklea. The population of 11.5 million was divided into three age groups of different susceptibilities. The annual average suspended sulfate concentrations were taken as 16 $\mu g/m^3$ with a standard deviation of 5.6 $\mu g/m^3$. The questions asked are, (1) what are the health effects of a 1 $\mu g/m^3$ increase in the annual average level? and (2) what are the health effects of increasing the ambient concentration for the top one percent peak days by 10 $\mu g/m^3$? The results of this study are summarized in Table 5–4. According to these calculations, the increase by 1 $\mu g/m^3$ of suspended sulfates has a signifi-

[dd]L.D. Hamilton, "Health Effects of Air Pollution, Informal Report," Brookhaven National Laboratory, Upton, New York, 1975.

Table 5–4. Estimated Annual Adverse Health Effects Attributable to Increased Suspended Sulfates in New York Metropolitan Area[a]

Health Effect	Average Number of Cases Per Year	Additional Cases Attributable Annually to Increased Suspended Sulfate Concentration of:	
		$1 \mu g/m^3$ Increased Every Day	$10 \mu g/m^3$ Increase on Top One Percent Peak Concentration Days
Premature Deaths	118,000	22.7	7.7
Aggravation of Heart and Lung Disease (man-days)	24×10^6	406,000	8,950
Asthmatic Attacks	2.5×10^6	84,100	439
Lower Respiratory Disease in Children	127,000	9,900	–
Chronic Respiratory Disease Symptoms	366,000	40,600	–

[a]National Academy of Sciences (1975), pp. 607, 609.

cant health impact. However, these effects probably could not be detected against the background of high normal incidence of the health effects in question. The effects of an increase of 10 $\mu g/m^3$ on peak days (only a few per year) is less than the increase of 1 $\mu g/m^3$ throughout the year. Despite the uncertainties that abound, the excess deaths and health impairments, resulting from one additional power plant in the New York area, are likely to be significant.

In presenting the above estimates of health effects, the authors emphasized the large uncertainties in the data. In a sensitivity analysis of the estimates, they considered a range of "reasonably extreme values" ranging from 10 percent to 200 percent of the nominal values on which the study was based. The basis for this judgment on the range of uncertainties was attributed to the representatives of the Assembly of Life Sciences Committee that prepared the National Academy of Sciences Report (1975).

An alternative calculation in the same study compared the estimated health effects of a remotely located plant and an urban plant, taking into account the broad areas affected by each. The remotely located plant was 230 miles upwind from New York City. The plant was representative of those now operating, a 620 MWe (capacity) plant burning 3 percent sulfur coal without flue gas desulfurization and emitting 213 tons of sulfur per day. The principal health effects are listed in Table 5-5. The more remote location reduces the estimated number of deaths attributable to this representative coal plant from forty-two to fourteen per year. If this estimate is normalized for a 1,000 MWe plant, operating at a 60 percent capacity factor typical of nuclear plants rather than the assumed 80 percent capacity factor, the range of deaths resulting from variations in loca-

Table 5-5.   Estimated Annual Adverse Health Effects From Sulfur-Related Pollutants From a 620 MWe Plant

	Remote Location	Urban Location
Cases of Chronic Respiratory Disease	25,600	75,000
Man-days of Aggravated Heart-Lung Disease Symptoms	265,000	755,000
Asthma Attacks	53,000	156,000
Cases of Children's Respiratory Disease	6,200	18,400
Premature Deaths	14	42

Source: National Academy of Sciences Study, 1975, Chapter 13. Illustrative calculations based on distributive models, postulated conversions of SO_2 to SO_4, and EPA epidemiological data for representative power plants in the northeast emitting 96.5×10^6 pounds of sulfur per year—equivalent to a 630 MWe plant.

tion would be eighteen to fifty. When the range of extreme uncertainties of health effects is applied to this estimate, the number of premature deaths from a coal-fired plant would range from about two to one hundred per year per 1,000 MWe plant.

These results apply to alternative siting in a highly populated region that is already highly polluted. Were the ambient sulfate concentration lowered by 25 percent, the projected number of premature deaths could be cut in half for the urban plant. The SO_2 emissions are more easily controlled than are location or background ambient levels of sulfates; and in any event, these emissions would have to be reduced in new plants to meet standards for new sources.

To meet current standards for new plants the 3 percent sulfur coal used in the example could be replaced by 0.6-0.8 percent sulfur coal, thus lowering SO_2 emissions fourfold to fivefold. This would reduce the estimated range of premature deaths to four to twenty-five per year per 1,000 MWe plant. Alternatively, if lime scrubbers were used with 3 percent sulfur coal, 90 percent of the SO_2 could be removed. This would result in an estimated range of premature deaths of from 0.2 to 10 per year per 1,000 MWe plant.

Two other, less well-documented, studies yield estimates of fatalities lower and higher than the above. A study at the International Institute for Applied Systems Analysis (IIASA) concludes that "SO_2 emissions from a single coal-fired power plant (1,000 MWe) can amount to thousands of days of human illness, and some premature fatalities."[ee] However, these results can scarcely be translated into the conditions used for the studies just described. The IIASA Study

[ee]W.A. Buehring, et al., "Evaluation of Health Effects from Sulfur Dioxide Emission for a Reference Coal-Fired Power Plant." Research Memorandum RM-76-23, International Institute for Applied Systems Analysis, Laxenburg, Austria (1976).

limits effects to areas to within fifty miles of the power plant, the population density is about one-tenth that of the New York Metropolitan area, no account is taken of the increased rate of suspended sulfate conversion in urban areas, and the standard deviation about the mean year-long concentration is relatively small. But, even taking these factors into account, the results appear low. Using data for a plant emitting 70,000 tons of SO_2 per year into an environment of mean SO_2 concentration of 100 $\mu g/m^3$ (20 percent above EPA standards), the expected deaths are only 0.03 and 0.004 per year for an urban and a rural site, respectively. Other effects are 25–1000 times lower than in the North-Merkhofer study. Given these boundary conditions, this analysis does not seem to address the problem faced in typical power plants in the eastern United States. Extensive calculations would be required, however, to see whether this and the National Academy of Sciences study are otherwise self-consistent.

By contrast, a study cited by Hamilton[ff] leads to higher estimates for the average adverse health effects of a 1,000 MWe plant per year. Using 3 percent sulfur coal, this study estimates fatalities in the range of 15–150 per plant per year. Other estimates fall within the broad limits set by the IIASA and Hamilton studies.[gg]

The effects of numerous other pollutants (nitrogen oxides, carcinogenic hydrocarbons, radioactive particulates, and heavy metals) remain to be assessed. The uncertainties here are even larger than in the case of sulfur-related pollutants. Although the effects appear to be much less than for sulfur-related emissions, they undoubtedly add to health impairment and fatalities.

Overall Health Impact

The average rate-of-risk for a 1,000 MWe coal-fired plant is estimated to be about two deaths per year when the effects of effluents from combustion are not considered. When effluents are considered the rate-of-risk is substantially increased but is subject to much greater uncertainty. The North-Merkhofer model used in the National Academy of Sciences Report (1975), indicates an extreme range of two to one hundred deaths per year for a 1,000 MWe plant burning 3 percent sulfur coal. This wide range reflects the effect of variations in plant location and uncertainties in dose-response relations for health effects of sulfur-related pollutants. In the case of new plants meeting current new source standards for sulfur emission by burning low-sulfur coal, the estimated range would be 0.4 to 25 deaths per year for a 1,000 MWe plant. With lime scrubbers, the sulfur emissions could be reduced by a factor of 10 resulting in a range of 0.04 to 10 deaths per year depending on whether low-sulfur coal or 3 percent

[ff]L.D. Hamilton, "Health Effects of Air Pollution, an Informal Report," Brookhaven National Laboratory, Upton, New York (1975); and "Proceedings of the Conference on Computer Support of Environmental Science and Analysis," (CONF-750706) U.S. Environmental Protection Agency, Albuquerque, New Mexico, July 1975.

[gg]C.L. Comar, and L.A. Sagan, "Health Effects of Energy Production and Conversion," *Annual Review of Energy*, 1, 581–600 (1976).

sulfur coal was used. As sulfur dioxide emissions are decreased the relative importance of other pollutants, such as nitrogen oxides, which are presently not well understood, would increase and at very low sulfur dioxide levels these other pollutants would probably become the dominant health effect.

For new coal-fired plants meeting new source standards, this analysis indicates a range of premature deaths from occupational and public effects of the coal fuel cycle and the effluents of coal combustion in the range of two to twenty-five per year for a 1,000 MWe plant. These numbers could be reduced by the use of lime scrubbers alone or in conjunction with low-sulfur coal or by the use of other new technology, such as fluidized bed combustion.

Despite these large uncertainties, the general conclusion is that on the average new coal-fueled power plants meeting new source standards will probably exact a considerably higher cost in life and health than new nuclear plants. However, both coal and nuclear power plants built in the rest of this century could have much reduced health risks relative to existing plants. This can be accomplished in the case of coal plants by limiting sulfur dioxide and other emissions in conformity with present or improved air quality standards and by prudent siting; and, in the case of nuclear power plants, by improved siting and safety controls. In the immediate future, a major effort should be made to improve the assessments of the health effects of the pollutants from coal combustion. The most pressing demand, however, would appear to lie in upgrading the research and development directed at the reduction of the adverse health effects associated with coal-fueled power plants.

✳ *Chapter 6*

Environmental Effects

The generation of electricity by nuclear power or fossil fuels has environmental effects on air, land, and water, and on global climate even when all performance standards are met. These effects are additional to the direct effects on human health discussed in Chapter 5. An assessment of nuclear power requires a comparison of its environmental impacts with those of fossil fuels, particularly coal, which is the main alternative for the generation of electricity in this century. In this chapter, the principal environmental impacts of the nuclear fuel cycle are compared with those of the coal fuel cycle. Direct effects of the two fuel cycles on human health, the consequences of reactor accidents, and problems of radioactive waste management are considered separately in Chapters 5, 7, and 8.

The most serious potential impacts on the environment from greatly increased power generation are changes in global climate. The nature of such changes, however, is not well understood and any estimates are extremely uncertain. Both fossil fuel and nuclear power heat the atmosphere directly in producing electricity. A more serious heating effect, however, results from the carbon dioxide (CO_2) produced in the combustion of fossil fuel since it can change atmospheric properties enough to perturb the present balance between solar radiation incident on the earth's surface and radiant heat flow back from it.

Fossil fuels have already released enough CO_2 to raise the average global temperature slightly, possibly about $0.3°$ C, above what it would otherwise have been. A substantial increase in the use of fossil fuels could have important consequences for the global climate since we do not yet know a practical way of containing the CO_2 emissions. Emitted oxides of sulfur and nitrogen and particulate matter may also play significant roles in perturbing future climate. Nuclear plants do not emit any of these pollutants. Radioactive gases released by present modes of nuclear power generation would appear to have negligible climatic impact and could be more completely contained if necessary.

The amounts of thermal, chemical, and radioactive wastes that would be introduced into the atmosphere by various kinds of power plants can be estimated reasonably well. However, even if the expected alteration of our present climate from these pollutants could be calculated with great precision, the unperturbed natural climate expected in several decades cannot now be predicted. Thus, even if it were known how much future electric power generation might raise the average temperature, it would not be possible to know what the consequences would be, especially in any particular region of the earth. The climate might still be cooler than our present temperature, in which case some additional warming could be desirable in many areas, or it might be considerably warmer so that the additional heating could be much less acceptable or even possibly catastrophic in some regions. The considerations of these long-term global consequences of greatly increased power generation are characterized by great uncertainties. Unfortunately, the timescale for resolving them or for monitoring the climatic effects of power production is likely to be decades since trends in climatic variables are generally not evident over shorter periods.

Impacts on the local environment from nearby power generation depend on the details of the fuel cycle. Power generation by fossil fuels is a relatively straightforward process involving extraction, burning, and disposal of waste products. The nuclear fuel cycle is more complex. Extraction and milling of the ore is followed by conversion to uranium hexafluoride, enrichment, fuel fabrication, transportation, reactor operation, and subsequent disposal of nuclear wastes. The total process is even more complicated if reprocessing and recycle of plutonium are added to the fuel cycle.

Both coal mining and uranium mining and milling use and scar large land areas, but at present considerably more land is disturbed by coal extraction to support the same power output. The amount of land disturbed by uranium mining will increase, however, as lower concentration uranium ore is exploited. Coal and nuclear power have similar adverse local thermal effects on the air and water, particularly on small rivers or lakes.

Local harmful environmental impacts unique to nuclear power include the possibilities of land contamination from runoff from mining and milling and high local heat concentrations if many reactors are collocated for security or economic reasons. There are also environmental effects unique to the coal fuel cycle. Coal mining produces acid runoff, which makes land reclamation difficult and damages water resources. Coal burning, in addition to the atmospheric pollution which affects humans directly (Chapter 5), produces effluents which increase the acidity of regional rainfall with deleterious effects on plants, animals, and materials.

The magnitude and significance of these global and local environmental impacts are considered in the following sections.

GLOBAL CLIMATE EFFECTS FROM ATMOSPHERIC POLLUTION

Past and Present Climate

The impact of man-made power on the world's future climate depends upon what the natural climate would have been in the absence of man's intervention. Regional and global climates have varied greatly in the past and could change significantly even in the next few decades entirely from natural causes.

While a great deal is known about the history of the earth's changing climate on both long and short time scales, very little is known about how to predict future trends. A billion years ago, the earth was about 10° C warmer than it is today with a much higher ocean level and almost no polar ice caps. More recently, ice ages have appeared on the average every hundred thousand years. The past ten-thousand-year period has been comparatively warm but important temperature fluctuations have occurred; during several millenia in this interval, glaciers grew rather than contracted. The most recent period of pronounced cooling was the so-called Little Ice Age, which lasted from 1550 to 1850. Subsequently, there was a gradual warming that peaked in the early 1940s. During the relatively high temperatures of 1930–1950, the crop-growing season in the northern regions of the Western Hemisphere was about two weeks longer than it was during the cooler last decades of the nineteenth century. Between the 1940s and the late 1960s, the average Northern Hemispheric temperature dropped about 0.3° C. Since then there have been some indications that this global cooling may have stopped, but given short-term fluctuations it will be some time before a new trend can be determined with confidence. Regional climates have shifted even more dramatically. The droughts of the past few years in Africa provide a tragic example of the effects on a specific region. Local climate patterns are often fragile and may be susceptible to changes from what appear to be negligible sources.

It is unclear from recent climate history whether the temperature reduction of the past several decades is the beginning of a more sustained reduction or was a temporary anomaly in a continued warming trend. Prediction is hampered by the lack of observed regularity in past behavior and by the absence of a widely accepted theory of the origin of climate variations. Global and regional temperatures may fluctuate because of complicated interactions among terrestrial environmental components or because of subtle changes in solar input or both.[a] A change in average temperature of several tenths of a degree centigrade,

[a]For example, sunspot activity was absent during the first century of the Little Ice Age, but any causal relationships are still obscure. The eleven-year cyclic variation in sunspot activity appears to correlate with rainfall and crop yields in some places, including the U.S. upper plains. There is independent, significant support for correlations between certain

even within the next decade, would not be inconsistent with past experience. Regional temperature variations of a full degree or two could accompany average global changes of a few tenths of a degree.

Any attempt to forecast the effect on climate of greatly expanded energy use is complicated by our present inability to understand natural climate fluctuations well enough to predict them, and by the possible effects of other pollutants from human activities. Thus, even if we could predict the influence on global climate of greatly increased electric power generation, our ignorance of natural variations would still prevent us from determining the future climate. We cannot yet unambiguously identify the effect of the spectacular growth in thermal and other pollution on the global temperature record of the past century, since, except for some local effects, at present natural climate variables appear to dominate man-made perturbations. An example of these local effects is seen in Manhattan where the annual average input of energy from human activities is some 700 watts per square meter (watts/m^2), but the average absorbed solar radiation at this latitude is less than 100 watts/m^2. Thus, the heat from industry, power generation, heating, air-conditioning, transportation, and other energy uses dominates the natural flux and causes significant perturbations in climate over this small region; there are more clouds, precipitation, and higher temperatures here than in similar geographic areas with less energy-intensive activity.

When man-made heat inputs are compared with the solar input that drives the global weather system or when CO_2 from fossil fuel burning is compared with that from photosynthesis, these artificial perturbations are small and their effects may be masked by normal fluctuations. However, although unrecognized, these effects may be significant and the environmental damage irreversible.

Global Heat Pollution and the Radiation Balance

When coal or uranium is used to produce electricity, essentially all the energy released eventually reaches the atmosphere. Even the part converted to electric power will be dissipated into the atmosphere as the power is used. Because it is less efficient in converting heat, a nuclear plant eventually rejects to the environment about 20 percent more heat in the generation of a given amount of electricity than does a coal plant.

To understand the magnitude of thermal pollution, and to compare it with the climate effects of other emissions, we first consider the overall thermal energy generation by man's actions relative to the thermal effects of solar radiation. The amount of solar energy intercepted by the entire earth's surface and

---

meteorological measurements in the Alaskan Gulf and changes in direction at the earth of the solar magnetic field sweeping over the earth. Both of these effects suggest that there are some solar determinants of climate which do vary enough to have a significant meteorological effect. On the scale of tens to hundreds of thousands of years, the earth's inclination and distance from the sun seem to be crucial features in establishing terrestrial ice ages.

returned to the atmosphere averages 341 watts/m^2. This energy is sent back into the atmosphere at the same rate it is received. On the average, 30 percent of the incident solar flux is reflected back, 45 percent is absorbed and immediately reradiated (within minutes to days) as infrared radiation, and 25 percent is used in evaporating water; 0.2 percent drives winds, waves, and currents; 0.02 percent is used in photosynthesis. Ultimately all of the 25 percent of the solar flux involved in powering the earth's meteorological phenomena (as well as that used in photosynthesis) is converted to heat and sent back as infrared radiation. This energy, which is neither reflected nor immediately reradiated, is called the radiation balance; it varies with latitude and averages about 100 watts/m^2. Man's activities directly affect the radiation balance because power plants, other industrial processes, and indeed any energy use, produce a source of heat additional to that from the sun.

The man-made contribution to the radiation balance in 1975 was about 9 X 10^{12} thermal watts. This amount corresponds to an average global input of 0.02 watts/m^2 (0.05 watts/m^2 over the continents). Since this input is only of the order of 0.02 percent of the radiation balance, it has probably not affected climate on a large scale. However, as discussed above, some local areas do show effects. If energy production should grow at an average annual rate of 4.5 percent, then in about fifty years the average global artificial heat input will be about 0.2 watts/m^2 or of the order of 0.2 percent of the radiation balance; in eighty years, the heat input would be about 0.6 percent of the radiation balance. An increase of a few tenths of 1 percent of the global radiation balance could over several decades cause melting of polar sea ice. This man-made heat input may have dramatic effects on the earth's climate over a relatively short time span.

Gaseous Pollution and the Greenhouse Effect

Atmospheric components, such as CO_2 and water vapor, are transparent to visible light and therefore do not prevent incident solar energy flux from reaching the earth's surface. They do, however, absorb infrared radiation given off by the land, oceans, and clouds at temperatures much lower than that of solar radiation. They reradiate back to the surface a portion of the absorbed energy so that some fraction of the heat that otherwise would be lost to space remains to warm the atmosphere, sea, and land. This effect is commonly known as the "greenhouse effect."

Although CO_2 constitutes only one three-thousandth of the atmosphere, it greatly influences climate. An increase in the atmospheric CO_2 content by 10 percent with other parameters fixed, leads to a warming of the entire lower atmosphere by about 0.3° C. Measurements indicate that CO_2 concentration has increased by some 10 percent since the beginning of the Industrial Revolution, and thus the atmospheric temperature is probably about 0.3° C higher today than it would have been without increased industrial growth. Temperature increases caused by additional CO_2 input may, however, be even more sensitive to atmospheric CO_2 injection from fossil fuel burning because of important feedback loops. For example, as the temperature rises, CO_2 dissolved in the oceans tends to escape into the atmosphere; and this additional atmospheric CO_2 will further increase warming. Increased temperature can reduce the albedo (the ratio of solar radiation reflected from the earth to that incident upon it) by melting snow and ice. Since less energy is reflected, more is available to heat the surface. Higher temperatures also cause the atmospheric water vapor content to increase, which in turn causes more absorption of infrared radiation and thus adds to the greenhouse effect warming. However, higher temperatures also increase cloudiness at higher altitudes, which increases the albedo and causes a cooling effect. Thus the net effect of feedback mechanisms is complex and uncertain.

There might be little future change in the CO_2 content of the atmosphere if it were not for human activity since the amount of CO_2 consumed in photosynthesis is almost exactly in balance with the amount released to the atmosphere through natural processes such as the oxidation of dead organic material. However, worldwide fossil fuel combustion already emits about 15 billion tons of CO_2 per year (and there is no appreciable return of carbon to the fossil reservoir) compared with 110 billion tons of carbon dioxide used annually in photosynthesis. In the time span of a few decades, fossil fuel burning is threatening to upset a natural balance between production and take-up which has developed over geologic time.

Today the atmospheric mass of CO_2 is about 2.4 trillion tons. Of the total amount of CO_2 emitted annually during combustion, about one-third, or 5 billion tons, stays in the atmosphere and the remaining two-thirds are absorbed by the oceans and land masses. Thus, even if the rate of fossil fuel burning did not increase, the equivalent of 0.2 percent of the total atmospheric CO_2 mass would be emitted each year, or 2 percent per decade. If the rate of fossil fuel use were to double every fifteen years, the CO_2 concentration in the atmosphere would increase to around 30 percent above pre-Industrial Revolution levels by the year 2000. Such an increase would cause a $1°$ C increase in atmospheric temperature and probably have major impact on climate. The actual effects would depend on the magnitude of the concurrent natural fluctuations. For the past several decades, world temperature has been decreasing about $0.1°$ C per decade, presumably from natural causes. Should this trend be reversed, an increase in temperature from carbon dioxide emission could have particularly significant consequences. Some areas of the world would suffer but some would benefit. A $1°$ to $2°$ C rise in temperature would almost certainly reduce the amount of ice and raise sea levels around the world, but on a scale of centuries. The impact on local climate would be most uncertain but would likely be profound, particularly in the polar regions where shifts in temperature during warming and cooling periods are accentuated. If, on the contrary, the cooling trend were to continue, or to accelerate, the carbon dioxide could perhaps reduce the severity of planetary cooling.

Nitrous oxide (N_2O) is also emitted to the atmosphere during coal burning and can affect surface temperature directly by greenhouse heating in the same way that CO_2 does and also, indirectly, after being converted to nitric oxide (NO), by catalytically destroying ozone. The ozone effect is complicated since ozone both reduces surface temperature by absorption of solar ultraviolet radiation and contributes to the greenhouse effect by infrared absorption. At present only 0.1 percent of the atmospheric burden of nitrous oxide is estimated to have come from combustion of fossil fuels, but major increases in coal burning could add significantly to this. Some greenhouse effects are also expected from methane liberated by fossil fuel burning, but analyses of the effects and consequences of nitrous oxide and methane emissions have only recently begun.

Particle Pollution

Particle pollution, whether from fuel burning, industrial aerosols, or agricultural or volcanic dust, can affect the thermal balance of the atmosphere in several ways. Small particles in the atmosphere scatter incoming solar radiation but only the small part that is backscattered increases the earth's albedo (reflectivity). The decrease in solar radiation reaching the surface will lead to lower temperatures. Early calculations suggest that a decrease in solar transparency of 3 to 4 percent could lead to a surface temperature reduction of 0.4° C. Some small particles, however, can absorb incoming solar radiation and also outgoing long-wave radiation. This could decrease the earth's albedo and heat the lower atmosphere. The net effect on the total radiation balance depends on the abundance, location, size, composition, and altitude range of the small particles. Since the normal albedo of the earth varies in different regions, the total direct effect of atmospheric particles in increasing or decreasing natural reflectivity may also depend on geography.

Temperatures can be affected indirectly when dust particles provide condensation nuclei that form clouds composed of small water droplets. An increased steady-state concentration of dust particles could lead to increased cloudiness. Calculations indicate that a 2.4 percent increase in cloud cover could lead to a 2° C drop in temperature because of the reflection of solar light before it reaches the ground. Local increases of fog and low cloud cover in urban regions are well documented, but global trends require further analysis.

Some 200–400 million tons of man-made particles, mostly sulfates from combustion, are injected into the atmosphere each year. These particles are introduced into the lower atmosphere and remain there for about ten days before raining out. Volcanic eruptions introduce an estimated 800 million to 2 billion tons of particles per year. These particles can be carried to higher altitudes and thus remain suspended much longer. Occasional large volcanic eruptions can introduce much more material. The Tambora eruption of 1815 was estimated to have introduced some 50 billion tons into the atmosphere, much of which remained for many months. Worldwide temperature drops the following year are attributed to this eruption: alpine temperatures decreased an average 1° C, and 1816 was widely known as "the year without a summer." The Krakatoa eruption in 1883 depressed worldwide temperatures an average of 0.5° C and the smaller Mt. Agung eruption of 1963 had about half that effect beginning about six months after the eruption. Presumably SO_2 injected into the stratosphere, which ultimately contributed to sulfate particle formation, more than the dust injected into the atmosphere caused the delayed climatic effects. Evidence suggests that emissions from normal volcanic activity are more important than combustion-related particles, but this relation may be reversed if fuel combustion increases by a factor between two and ten. However, even a factor of ten would give very much less atmospheric particulate material than the largest historic volcanic eruption. The comparisons between the two are not intended to suggest that the man-made effects should be ignored, but only to use the empirical results from volcanic eruptions to set a scale for the effects of particulates.

In sum, the overall effects of particle pollution on climate are unknown. Increases in purely reflecting particles should depress average temperature, but the magnitude of man-made changes in either transparency or cloud cover are not known. Increases in absorbing particles should increase average surface temperatures. The mix of reflecting and absorbing particles ultimately put into the atmosphere by fossil fuel burning is not yet well understood. Technology to reduce man-made particulate materials injected into the atmosphere is considered later in this chapter.

Radioactive Pollution

The nuclear fuel cycle inevitably releases some radioactive effluents to the environment. The atmospheric ionization caused by this radioactivity could conceivably have some climatic effects in the future. The electrical properties of the atmosphere contribute to cloud formation and precipitation, and thunderstorm development, but no specific climatic consequences of increased atmospheric ionization have been identified.[b] In present operations the isotopes krypton-85, xenon, tritium, and carbon-14 are released to the environment. Of these, krypton-85 contributes the most to atmospheric ionization.

Present reactors are reported to emit between zero and 100,000 curies of krypton-85 per 1,000 MWe per year depending on design; with reprocessing, this amount could increase to 500,000 curies. If, in the next thirty years, world-

[b]Solar magnetic activity affects the earth through the modulation of ionizing cosmic radiation above the polar regions and through solar cosmic rays, which are ejected in solar flares. If this is related to the claimed variations of certain meteorological variables with the solar sunspot cycle, then such ionization may be affecting climate and weather in subtle ways and to a larger extent than has been assumed. If this proves to be the case, the significance of krypton-85 ion pair production may also be increased.

wide nuclear power generation increased to 1000 gigawatts, and the krypton-85 produced were freely released, this could increase the present atmospheric ionization levels by the order of 10 percent over the oceans and of the order of a few percent over the land.

Since krypton-85 has been regarded as innocuous, its removal during reprocessing is not presently planned. However, a variety of methods for removing most of the krypton-85 in gaseous waste streams seems technologically feasible. Although not inexpensive, removal of krypton-85 would not add significantly to the costs of nuclear power.

LOCAL AND REGIONAL EFFECTS FROM THERMAL POLLUTION

Local Thermal Pollution

Nuclear and coal power plants both cause thermal pollution in natural bodies of water that are used for cooling purposes. Nuclear power plants, however, can have a somewhat more serious effect than comparable coal plants.

New coal power plants convert 36 to 40 percent of fuel energy to electricity, while nuclear power plants convert only 30 to 33 percent. Moreover, while a coal plant discharges part of its waste heat directly into the atmosphere, a nuclear plant discharges all of its heat into cooling water. As a consequence, a nuclear power plant can result in 50 percent or more local heating of a body of water than a comparable coal plant.

Two different systems are used to provide cooling water and to dissipate heat. The "once-through" method circulates water from a river or lake through the reactor and returns it to its source. In a typical 1,000 MWe nuclear plant, this method usually heats a thousand cubic feet of water some $10°$ C each second. The second method uses a recirculating evaporative system, which transfers the heat of the water to the air by means of large cooling towers. The water can be recycled through the reactor or discharged into the original body of water. This method reduces the thermal effect on water but introduces new problems such as salt drift (where ocean water is used for cooling), water vapor, noise, and evaporation of about 2 percent of the amount of water as contrasted with temporary diversion of water.

Thermal pollution affects water-based ecosystems: It changes feeding habits and reproduction rates of fish; it increases nutrient levels, photosynthesis, decomposition rates, and eutrophication. Oxygen levels can decrease, and new circulation currents can be induced. When a power plant shuts down, fish that have become accustomed to warm water go into shock and die as they are subjected suddenly to cold water. With the exception of this cold kill, most of the effects of thermal pollution are quite subtle and very site-specific.

From an environmental point of view, estuaries are extremely poor sites for power plants. Estuaries are highly productive and vulnerable parts of the oceanic ecosystem. Since estuaries are shallow, fish cannot escape to cooler depths when the water is heated. Organisms that normally live on the bottom are also destroyed by the increased temperature.

Provided a lake is deep and large enough, thermal heating need not harm it environmentally. The bottom layers will not be heated if the cooling water is taken from the colder depths and released on the surface. In small lakes, however, excess heating can initiate aging by altering the oxygen distribution. Once this damage is done, it is irreversible; the lake can no longer support its normal ecological system. Since a nuclear plant has a greater thermal discharge than a comparable coal plant, it would cause a somewhat faster aging process in small cooling lakes.

Rivers are potentially good sites for plants using once-through cooling because the rapid flow of the river causes the heat to dissipate more quickly than it would in a lake or estuary, and the area of higher temperature is thus smaller. Rivers have no salinity gradients, and there is no stratification to be disturbed. In quickly flowing rivers, plant biomass is low even when nutrient levels are high. The major effect of heating river water is on the fish. If the heated plume extends across the width of the river, it can prevent fish from running upstream. If fish become acclimated to warmer temperatures, then cold kills can result. A major problem with river flow is that it is variable. If it is assumed that all the organisms taken in the cooling water are killed, then it is important that the amount of water withdrawn never be more than about 10 percent of the lowest flow.

Regional Thermal Pollution from Nuclear Parks

The collocation of 10 to 40 nuclear 1,000 MWe generating units, together with the possible collocation of fuel cycle centers, into a so-called nuclear park has been suggested as a means of reducing costs and hazards from sabotage, fuel diversion, and other societal problems of dispersed nuclear sites. By so doing, serious environmental problems could arise. For example, although the total energy released from a nuclear park would be minute compared with the global radiation balance, the great quantities of heat could cause major atmospheric perturbations on a local or regional basis.

A 40,000 MWe nuclear park would emit heat at a rate about four times that of Manhattan and one-tenth the rate of latent heat released in a typical thunderstorm. In Manhattan, the released heat causes more cloudiness, increased frequency of fogs and precipitation, and higher temperatures than would otherwise be the case. If nuclear parks generate heat at a rate comparable with that released in thunderstorms, the potential for severe storm activity near those parks may increase.

Nuclear parks will require substantial river flow capacity since each 1,000 MWe plant requires a flow of about 1,000 cubic feet per second (cfs) if once-through cooling is used and about 30 cfs if an evaporative system is used. If the maximum withdrawal permitted is 10 percent of the river flow, then a forty-unit facility would require a river capacity of at least 400,000 cubic feet per second at low flow for once-through cooling and 12,000 cubic feet per second for an evaporative system. The average flow rates of some of the larger U.S. rivers are only: Arkansas (45,100 cfs), St. Lawrence (243,000 cfs), Columbia (262,000 cfs), and Mississippi (640,000 cfs). Thus, nuclear parks would likely require evaporative systems, but when used on such a large scale might significantly affect local precipitation patterns.

LAND REQUIREMENTS

The principal impact on land usage by the nuclear fuel cycle results from the mining and milling of uranium ore. The amount of land used for mining and milling depends on the concentration of uranium in the ore. It has been estimated that at the present ore concentration of around 0.2 percent some twenty to fifty acres of land must be mined annually to produce enough uranium to support a 1,000 MWe nuclear power plant. In addition, the amount of land required for milling operations and for storage of tailings, which comprise essentially all of the material mined, may run to about thirty to seventy acres per 1,000 MWe plant per year for 0.2 percent uranium. The land used for tailings storage is of limited use because of radiation hazards; some of the uranium and other associated radioactive elements, particularly radium, remain in the tailings. The radium is easily leached by water; and radon (a chemically inert, radioactive gas formed by the natural decay of radium) can escape into the air at a rate of about 56 curies per 1,000 MWe per year. Left undisturbed, most of the radon, which has a half-life of only about four days, would have decayed before it reached the surface. The tailings, along with the liquid effluent from the leaching process, are placed in shallow holding ponds to evaporate, thereby allowing the radon to escape. The Nuclear Regulatory Commission (NRC) recommends that all tailings waste be placed in tailings ponds that are off limits to the public. After the mills are shut down, the ponds are to be covered with earth and planted, but the area will still have to be restricted because of excessive radiation levels. This practice, although an improvement over the negligence in the 1950s, still presents radiation leaching hazards.

The land required for the coal cycle to generate 1,000 MWe per year consists of that used in mining (about 100-400 acres per 1,000 MWe plant per year, depending on the thickness of the seam), power plant operations, and disposal of ash (from burning), and sludge (from pollution control equipment). Because so much of what is mined in the coal cycle is burned, the overburden usually can be used to fill the mines from which the coal was removed and can thus be reclaimed at costs of about $925-$2,750 an acre or 0.02-0.2 mills per kwh.

As the richer and shallower uranium ores are depleted, greater quantities of overburden must be removed to secure the same amount of uranium, which today is found in narrow thin strips, a few inches to a few feet thick, in beds which lie from the surface to several hundred feet below. At concentrations of 0.03– 0.07 percent uranium, the land requirements for the nuclear fuel cycle become comparable to those for the coal cycle, i.e., about 100–400 acres per 1,000 MWe per year.

The bulk of U.S. uranium mining occurs in arid and semiarid regions characterized by harsh environments and delicate ecosystems where rainfall is sporadic and averages only about 10 inches per year. Reclamation is difficult if not impossible in some of these areas. In its undisturbed state, the land used for uranium mining at present is probably of less economic value than that used by the coal fuel cycle.

REGIONAL POLLUTION

Uranium Mining Runoff

Water in the open pits of uranium mining sites can become contaminated by uranium, thorium, and radium. If pumped from the mine, this water can then contaminate local streams, and incidents of such contamination have been reported. The mismanagement of radioactive tailings has also resulted in several serious incidents. In the 1950s, the liquid tailings from uranium milling were discharged directly into the Colorado River system. In the Animas River several miles downstream from an AEC-licensed uranium mill at Durango, Colorado, the river fauna were seriously affected by the radioactive effluents. A repetition of the situation was avoided by ponding the effluents, but the radiation level did not return to normal until after about ten years. Such problems probably can be avoided by better management, but remain a potential risk.

Acid Runoff from Coal Mining

Today, acid mine drainage and runoff associated with coal mining appear to be more serious than the runoff associated with the uranium mining. Acid mine drainage occurs when sulfur-bearing minerals carried by water seepage and percolation from coal mines or refuse piles react with oxygen to form sulfuric acid. Acid mine drainage is more prevalent in the East, where the sulfur content of the coal is higher and the annual rainfall is greater than in the West. Runoff, or surface flow, carries heavy metals, acid and alkaline compounds, suspended and dissolved solids, and toxic substances from the mines and spoil piles that, prior to mining, were unexposed to air and water and therefore presumed to be innocuous. The effects of acid mine drainage and contaminated runoff are evident in destroyed plant and animal life in some 10,000 miles of streams and lakes and polluted water unfit for industrial, recreational, or residential use.

These problems have been recognized for some time, and research on specific causes, effects, and means for control is ongoing, but correction appears costly. The situation is perhaps even more serious with abandoned mines than with active mines because, even after operations cease, water continues to flow for decades before the acidity is removed, and there is no incentive to assume the high costs of corrective action.

Oxides of Sulfur and Nitrogen: Acid Rain

Sulfur dioxide is released when coal and other fossil fuels are burned and is further oxidized in the atmosphere to form sulfuric acid, which increases the acidity of rain. A similar series of steps converts nitrogen oxides to nitric acid.

It appears that in recent years the rainfall over some large areas in the northeastern United States is about twenty times more acidic than natural rainfall. (Natural rainfall has a pH of about 5.5 to 5.7, while rainfall over these areas has dropped to about 4.1.) There is also evidence that rainfall in general has become more acidic since 1950. Acid precipitation can affect the environment by direct action on plants and by changing the acidity of the soil and water bodies. When acid rain falls on vegetation, part of the acid is neutralized in the leaves. This causes a loss of such nutrients as potassium, magnesium, and calcium. Acid rain that falls on the soil is neutralized by forming soluable sulfate salts of nutrients such as calcium and magnesium. Continued exposure of the soil to acid precipitation can lead to depletion of nutrients and acidification of the soil.

Fish can be affected by acid water in lakes and streams. Young fish are particularly vulnerable to acid waste water and, since the spawning time coincides with the period of greatest acid concentration (when melting winter snowfall accumulates in lakes and streams), many fish can die prematurely.

Natural emissions of sulfur compounds, which result from decaying vegetation, constitute a background atmospheric level of about 0.5 to 4 micrograms of sulfates per cubic meter. This is estimated to be one-and-one-half times the amount currently resulting from man-made sources. However, in industrialized regions, the concentrated emissions from fossil fuel combustion are much greater than those from natural sources. In the eastern industrial belt of the United States, sulfur oxides have a concentration greater than 30 micrograms per cubic meter, and in some urban areas concentrations reach 80 to 120 micrograms per cubic meter. The geographical variation is great, with eastern urban sites having concentrations averaging three times those of western sites, and with the northeast industrial belt showing the highest concentrations. Over half of the man-made sulfur oxide emissions in the United States have been estimated to come from electric power plants, and the contribution from this source increased by almost 100 percent from 1960 to 1970. Present regulations place limits on sulfur oxide emissions from power plants so the rate of buildup of this pollutant should be slowed or reversed in the future.

Analysis of rainfall indicates that about 25 percent of the acidity results from nitric acid and most of the rest from sulfuric acid. The rate of increase of the nitric acid component appears to have been more rapid than that of sulfuric acid. The ratio of nitric to sulfuric acid is important since the technology for removing the oxides of nitrogen is not well developed compared to that for removing sulfur dioxide.

Current technology permits removal of sulfur oxides from the flue gas by various processes of desulfurization or scrubbing. Although numerous problems have been encountered with early installation of scrubbers, it is generally believed that these difficulties will be solved. One problem concerns the disposal of the large quantity of waste by-products, or sludge, from lime and limestone scrubbers. These by-products are temporarily stored in lined ponds near the power plant, and thus the availability of space near the power plant is an important consideration in a decision to use scrubbers. It is likely that eventually this sludge can be put through a fixation process that would make it suitable for road beds and landfill. Another way to reduce sulfur oxide emissions is to remove part of the sulfur before burning. Washed coal has about 40 percent less sulfur than fresh eastern coals, but this reduction is not sufficient to meet the Environmental Protection Agency's (EPA) New Source Performance Standards. Alternatively, low-sulfur western coal can be burned, but this raises questions of land reclamation in the dry western states and of high transportation costs to eastern markets. Use of tall stacks can achieve acceptable sulfur dioxide air quality standards at relatively low cost in the vicinity of the power plant; however, this does not solve the real problem since the same amount of sulfur dioxide is dispersed over a wide area. The sulfur dioxide is subsequently converted to sulfates that produce acid rain and the health effects which are the main problem with coal combustion (see Chapter 5). Fluidized bed combustion (see Chapter 2) can remove sulfur from coal by reaction with limestone during combustion yielding a dry product less difficult to handle than sludge from scrubbers.

CONCLUSIONS

The possible impact on global climate appears to be the most serious environmental consequence of greatly increased electric power generation. The thermal output of both coal and nuclear power contribute directly to the long-term heating of the atmosphere. However, a much more serious threat appears to be posed by the carbon dioxide (CO_2) produced in fossil fuel combustion. Carbon dioxide, control for which does not appear practical, heats the atmosphere by the greenhouse effect. The heating problem is complicated by the uncertain effects of particulates and other pollutants such as sulfur dioxide and nitrous oxide. The heating of the earth's surface as a result of all of these effects is very uncertain. Even more significantly, the current unpredictability of natural climate variations and the untested effects of a global temperature rise on cloud formation, rainfall, and regional weather patterns preclude any clear assessment of the actual impact of increased use of fossil fuels. Since short-term fluctuations in climatic variables would generally mask any trend for many years, these questions are not likely to be resolved in the near future. However, major increases in the use of fossil fuels could have a significant effect on climate. Whether this will combine with natural changes in climate to the net disadvantage of mankind or to particular regions such as the United States cannot be judged at this time.

With regard to local and more immediate environmental impacts, the situation is somewhat clearer. Except for local thermal pollution where nuclear power has a somewhat more serious effect, the coal cycle generally has more harmful environmental impacts than the nuclear power cycle. Coal mining has more extensive effects on land use than uranium mining and milling although this difference will diminish as lower grades of uranium ore are mined. Coal mining results in acid drainage, and coal combusion in acid rain that adversely affect the general environment and are only partly controllable.

On balance, the local environmental consequences of the nuclear power cycle in normal operation are not as serious as those from fossil fuel power generation. These local effects, however, are less critical in an overall evaluation of potential environmental impacts than the effects of increased greenhouse heating on global climate. While it is not possible at this time to judge the nature of the potential impact on civilization, it may develop on the basis of greater knowledge that this global climatic effect could be overriding in a comparison of coal and nuclear power. This argues against putting complete reliance on coal power at this time.

※ *Chapter 7*

Reactor Safety

The safety of nuclear power plants is a central issue in the debate on the future of nuclear energy. Thus far the safety record has been excellent. The 200 reactor-years of operation of U.S. commercial light-water reactors[a] have had no demonstrable adverse effects on public health. The lack of serious accidents in the past, however, is of only limited value in predicting a future in which U.S. commercial nuclear plants may accumulate a total of some 5,000 reactor years by the year 2000.

The consequences of a reactor accident could be very serious. Moreover, the psychological impact of a major accident might exceed its physical consequences. Accordingly, safety regulation for nuclear power is justifiably more stringent than in competing energy technologies. A by-product of such regulation is a profusion of data on "abnormal occurrences," varying from trivial malfunctions to serious events. This flood of information has increased public concern over reactor safety. Recent events, such as the serious fire at the Brown's Ferry reactor in March 1975 uncovered significant defects in design, procedures, and enforcement. One would expect improvements in safety design as a consequence of these events as in any technological endeavor.

Critics of nuclear power cite these "abnormal occurrences" as demonstrations that reactors are unsafe. Advocates cite the absence of adverse effects on the public as demonstrations that the "defense-in-depth" safety design philosophy is successful. However, neither the occurrence of these events nor the absence

[a]The number of reactor years is sometimes given as 2,000. This figure includes U.S. naval reactors, which have some ten times the operating history of commercial reactors with an excellent safety record. However, the Navy's experience is only partially applicable to commercial reactors since naval reactors are of much lower power, are subject to different quality control and operating procedures, and are of different design.

213

of adverse public consequences provides statistically meaningful predictions about the future impact of nuclear power on public safety.

Since nuclear safety policy cannot depend only on experience, the evaluation of risks must also be based on analysis and judgment. Moreover, determining the significance of nuclear risks depends on a comparison with the risks of competing energy sources, primarily coal. The comparison is difficult because the safety problems of each technology are so different. The mining, transportation, and combustion of coal have exacted a steady toll of human lives and have affected the health both of hundreds of thousands of industrial workers and of the general public; in contrast, experience with nuclear energy has been relatively benign. However, while even a major coal mine disaster produces relatively few casualties, a single major nuclear accident could be very large with "extremely serious accidents" resulting in as many as several thousand immediate fatalities and several tens of thousands of latent cases of cancer that will be fatal within thirty years.

Because of the dissimilarity of risks, some analysts have avoided comparing competing technologies in searching for standards of "acceptability" of nuclear risks. Instead, they have drawn comparisons with natural disasters, such as hurricanes or earthquakes, and man-caused disasters such as dam failures or air crashes. Neither comparisons with the safety record of competing technologies nor comparisons with accidents unrelated to energy are fully satisfactory. Comparisons with competing technologies enable one to examine the effect on health of alternate technologies, but suffer from differences in the technical character of the competing cases. Disasters unrelated to energy provide experience of impact and recovery from events producing casualties in the thousands. Such events are unsatisfactory comparisons for nuclear reactor accidents since the context of the comparison is so dissimilar.

With extensive accident experience an expected rate-of-loss can be computed. This expected rate is the sum for all possible accidents of the probability of each type of accident multiplied by the consequences of that accident. Since we do not have sufficient experience in reactor safety, both the probability of occurrence and the consequences of an accident must be estimated. Estimates of probabilities of designated accident sequences are necessarily very uncertain, especially when probabilities of very small magnitude are estimated by formal methods. Estimates of the consequences of a specified accident can be made to a reasonable degree of approximation.

Calculation of an expected rate of loss leaves a great deal of room for controversy whether the risk is "acceptable." The calculated rate-of-loss, although small, is very uncertain. Moreover, the consequences of an extremely serious accident may be considered unacceptable by some no matter how low its predicted probability. Consider the range of reactions to a comparison of some hypothetical situation in which, say, a thousand individuals die each year with another situation for which it is predicted that once in a thousand years a single accident will kill one million people. Some would argue that these situations are comparable since the expected average rate-of-loss is a thousand lives per year in both cases. Others would argue that any situation in which a million people might die as a result of a single peacetime event is unacceptable. Still others would assert that, since it is very unlikely that the large accident would occur in this century, the consequences should be substantially discounted, like any other event in the distant future. Moreover, in the course of time, corrective action might be taken and a technology would be changed for the better so that the event would not occur at all.

Clearly, reactor safety is an area where the judgments of reasonable people can differ. In our examination of reactor safety, we have attempted to reflect a range of concerns by considering whether each of the following general conditions is met:

- On a predicted average rate-of-loss basis, nuclear power compares favorably with competing technologies.
- The health and property consequences of a single extremely serious accident would not be out of line with other peacetime catastrophes that our society has been able to handle.
- Despite large uncertainties, a reasonable upper limit or ceiling, that is not in itself unacceptable, can be placed on the probability of the class of the extremely serious accidents.

Within this framework, we will seek conclusions about the safety of nuclear power after presenting an overview of the physical nature of possible accident sequences.

THE NATURE OF A REACTOR ACCIDENT

Almost all the commercial reactors now operating or under construction in the United States and most of the reactors elsewhere in the world are light-water moderated reactors (LWRs), either in the pressurized (PWR) or the boiling water (BWR) version. For the next generation of reactors, the U.S. nuclear R&D program is largely directed toward the development of the liquid metal fast breeder reactor (LMFBR). The LWR and the LMFBR are therefore the focus of the current debate on nuclear reactor safety. A general understanding of the nature of the accidents possible in both types of reactors is necessary to appreciate the safety problem.

The LWR Accident Sequence

An LWR accident of sufficient severity to have public health consequences must rupture the outer containment structure enclosing the reactor and disperse a significant fraction of the radioactive material accumulated in the fuel elements. This can only happen if the reactor core melts and there is a succession of events leading to failure of the containment. The core can only melt down after the water coolant that removes heat from the core has been lost. In the LWR, the low enriched fuel cannot sustain a chain reaction in the absence of the water coolant, which also serves to slow down neutrons to an energy at which they are more efficient in producing fission. Therefore, an LWR cannot explode like a nuclear weapon since, once the coolant is removed, the chain reaction stops. However, a core meltdown can occur in the absence of water coolant after the chain reaction has been terminated, since substantial heat continues to be produced by the radioactive decay of the accumulated fission products.[b]

A sequence of steps leading to a core meltdown could be initiated by any of the following events leading to a loss of coolant:

- Failure of the primary cooling loop, comprising the reactor vessel and primary piping.
- Failure of the secondary cooling loop intended to transfer heat generated by the reactor to a turbine generator.
- A "transient" or upward excursion in the reactivity of the reactor that has not been controlled by the reactor control rods.

If one of these "initiating events" occurs, safety provisions come into play to prevent core meltdown. If these fail and the core melts, rupture or leakage of the containment will occur only in certain circumstances. Finally, the severity of the effects of dispersal of radioactive material will depend on weather conditions in relation to the distribution of population near the reactor site.

All of the accident sequences involve loss of coolant to the reactor core at some point. Therefore, as part of the regulatory process, the vendor must demonstrate to the satisfaction of the Nuclear Regulatory Commission (NRC) that specified "design basis accidents" will not lead to serious consequences. Since there is no full-scale experience with these accidents, the vendor has to demonstrate by calculations that the safety system will work as intended.

The most controversial of the required safety features is the emergency core cooling system (ECCS), which supplies additional cooling water if the regular cooling system fails. If emergency cooling operates effectively after a major pipe break, core meltdown will be prevented. Some radioactivity might be released into the containment structure, but the pressures and temperatures should be sufficiently low that the containment would not be breached. More-

[b]Initially, this heat amounts to about 7 percent of the thermal power of an operating reactor or some 240 megawatts of thermal power (MWt) in the case of a standard 1,000 megawatt electric (MWe) power plant. Thereafter, the heat generation decreases rapidly, declining after ten minutes to about 2 percent, after one day to 0.5 percent, and after one year to 0.02 percent of that before shutdown.

over, special devices such as sprays and filters would reduce the temperature and remove some of the radioactive material.

There has been a major public debate about the reliability of the ECCS. The problem is complicated by the fact that the ECCS has never been tested in a full-scale reactor accident and even realistic ECCS experimentation is conceptually difficult. Therefore, prediction of ECCS performance rests on elaborate computer calculational codes which, because of the complex nature of the problem, depend on simplifying assumptions. Although the designers of such codes generally consider the underlying assumptions to be conservative, it is difficult to have confidence that the ECCS will operate as predicted under particularly stringent circumstances.

If the ECCS should fail, a core meltdown would occur. Once the core melts, the containment could fail for any of the following reasons:

- The pressure might rise to the failure point of the containment if the cooling systems were not operating or if noncondensible gases (CO_2 or H_2) were released in sufficient amounts.
- The molten core might penetrate the floor of the reactor permitting some radioactive material to escape to the air through the ground.
- A steam explosion resulting from the reaction of the molten core with water might rupture the containment.

The problem of steam explosions in which a hot fluid mixes with a cooler fluid of low boiling point has recently received increased attention. Such explosions are hazards in other industries and can result when volcanic lava flows into the sea. Three early accidents in experimental reactors involved steam explosions. In a commercial LWR the explosive power that would be available if the heat of the entire molten core was instantaneously transferred to the surrounding water is equivalent to about ten tons of TNT. The actual energy released would probably be much less, since the steam and core material cannot mix intimately during an interval of only a few milliseconds. Steam explosions will therefore be smaller than the energetic maximum and may not endanger the integrity of the containment. However, neither theory nor experiment provides data easily applicable to the core meltdown problem, the solution to which clearly depends on further research.

If containment fails, the specific sequence of events that leads to it will determine the amount and composition of the radioactive material that escapes. Unless a steam explosion takes place, containment failure would probably be delayed by several hours, which would reduce the amount of short-lived radionuclides that would be released. Once containment has been breached, the radioactivity would be carried upward in a thermal plume and dispersed in the direction of the wind.

The LMFBR Accident Sequence

The safety assessment of the liquid metal fast breeder reactor (LMFBR) is still in an early stage. While all of the operating LWRs have gone through the licensing process, the safety assessment of a commercial LMFBR will not be completed before 1986.

In the LWR, the chain reaction stops once the water coolant has been lost; and after meltdown the nuclear fuel in a compact form (even under water) cannot sustain a chain reaction. In contrast, the LMFBR is a fast reactor in which the absence of the sodium coolant increases rather than decreases the reactivity. Voids caused by the boiling of the sodium coolant near the center of the reactor core thus can increase reactivity and temperature. Changes in the geometry of the reactor core to a more compact form can also increase reactivity. On the other hand, an increase in reactor temperature decreases reactivity since the broadening of the neutron absorption resonances by the Doppler effect removes neutrons that would otherwise contribute to the fission process. The net effect of all the factors on reactivity must be analyzed in detail before the behavior of the reactor core under abnormal conditions can be understood. If reactivity becomes excessive, a control system must act to shut down the reactor. In both the LMFBR and LWR, shutdown is achieved by the insertion of neutron-absorbing control rods; in addition, the LWR has provisions for the injection of neutron-absorbing "poisons" into the cooling system. To achieve redundancy more than one control rod system is provided, and the control systems are designed to be as independent of one another as possible.

Should the reactivity of an LMFBR increase more rapidly than can be handled by the control rod systems,[c] the fission reaction will eventually be stopped by mechanical separation of the core components. This mechanical disassembly will give rise to both heat and mechanical forces that are transmitted to the reactor vessel. It is also conceivable that after core disruption or melting, the fissionable material will settle in the bottom of the vessel in such a manner as to lead to "recriticality" of the fissionable material and restart of the fission process. As a consequence, the material will again disassemble and be dispersed mechanically. In other words, in contrast to the LWR, the LMFBR could possibly sustain a low-grade nuclear explosion directly driven by the fission chain reaction. However, such an explosion would be fundamentally different

---

[c]In all cases, control rods can control a reactor only because of the existence of the so-called delayed neutrons. In contrast to prompt neutrons, which replace themselves within milliseconds in an LWR or microseconds in an LMFBR, these delayed neutrons, which constitute about 1 percent of neutrons emitted, take about a second to emerge after the fission reaction. In neither the LWR nor the LMFBR can the control rods be inserted rapidly enough to act on the time scale of the "prompt" neutrons, moderated or not. Thus, the delayed neutrons provide the necessary margin for control of reactivity. A sudden increase in the number of neutrons by more than about a percent thus cannot be controlled by the mechanical motion of the control rods. Under these circumstances, the fuel temperature would rise, and the sequence of events would be determined by the reactivity of the reactor as affected by the changing temperature or by changes in the mechanical configuration.

from that produced in a nuclear explosive device, in which the fissionable material is precompressed (imploded) at high velocities before the fission chain is initiated. In an LMFBR criticality accident, fissioning would occur in material of below normal density, and the reaction would cease as soon as disassembly began. As a result, the accident would probably not produce explosive power equivalent to more than a fraction of a ton of TNT, in contrast to the kiloton or megaton TNT-equivalent power of nuclear weapons. Such an explosion in itself would not necessarily breach the containment.

In some areas the comparative safety advantages of the LMFBR and the LWR are not as yet clear. Clean liquid sodium is less corrosive than water in its action on the steel in reactor vessels and piping and corrosion embrittlement is not a problem. At this time, not all of the effects caused by extended high-temperature exposure of the steel used in LMFBR vessels have been fully evaluated. These effects can be controlled by reduction in operating temperature, but at the expense of operating efficiency.

The high flux of "fast" neutrons in an LMFBR creates a materials problem not encountered in an LWR.[d] Such fast neutrons can produce swelling of materials, embrittlement, and fracture occurring at high temperatures. Accelerators, and an experimental fast reactor (the Fast Flux Test Facility), to test these phenomena are now being designed or built. However, it will be five years or longer before definitive radiation-damage data are available and analyzed.

Another set of safety considerations inherent in the LMFBR stems from the large quantities of liquid sodium used for cooling. Since liquid sodium interacts chemically with water or air, an inert gas atmosphere must be provided for the liquid sodium pools in which the reactor components are immersed. While sodium leaks can present a serious fire hazard, fire fighting methods are well established. Accident sequences can be envisaged in which sodium dispersal contributes to radioactivity spread. On the other hand, the release of radioactivity might be reduced by possible "scrubbing" or absorbing action by the sodium coolant. Finally, in assessing the potential consequences of an LMFBR accident, one must take into account the fact that an LMFBR may have an inventory of several tons of plutonium, in contrast to the maximum of several hundred kilograms in an LWR.

There are some areas where the LMFBR has safety advantages over the LWR. In contrast to the LWR, where cooling water is under high pressure, the LMFBR is cooled by low pressure sodium, which can absorb large amounts of heat without boiling. Heat transfer in the reactor-core-to-coolant interface is more efficient in the LMFBR than in the LWR. Although larger temperature excursions can occur without undue consequences in the LMFBR than in the LWR, such ex-

---

[d]Fast neutron fluxes as high as 10^{16} n/cm^2 sec are generated in an LMFBR, in contrast to fluxes of 10^{13} n/cm^2 sec in an LWR. Serious mechanical effects can occur above an integrated exposure of 10^{23} n/cm^2, which may occur in a year for an LMFBR but not in the thirty-year lifetime of an LWR.

cursions could occur more rapidly in the LMFBR. Finally, the consequences of a major pipe break in the cooling loop between the reactor core and the steam generator are much less severe in the LMFBR. For instance, should an LMFBR suffer a "guillotine break" in the cold leg of its primary cooling loop, a serious accident would be much less likely to occur. In an LWR, such an accident could lead to loss-of-coolant and possibly to a serious accident. Had the U.S. LMFBR program adopted the pool-type design, used in the U.S. EBR-2 research reactor and the French Phenix reactor, in which the entire primary cooling loop is immersed in liquid sodium, a break would have even less significance.

In summary, the most serious threat to safety in the LMFBR will arise from an excursion in reactivity leading to a situation that cannot be controlled by the control rods or that is accompanied by control rod failure. If, for instance, the coolant pumps stop at the same time the control rods fail, the resulting boiling of the sodium coolant would create voids that could increase reactivity faster than it was decreased by the Doppler effect. Under these extreme circumstances, mechanical core disruption could occur and "recriticality" might follow. It is difficult to compute whether the energies released in these sequences would result in failure of the primary reactor vessel. If the vessel should break, it is possible to envisage a series of steps that will eventually lead to failure of the outer containment with release of radioactivity.

As in the case of the LWR, an LMFBR accident involves breaching a series of safety barriers designed to absorb excess mechanical energy and to limit consequences. The detailed engineering of the LMFBR has not proceeded far enough to make a safety evaluation of a commercial system. There do not, however, appear to be any fundamental physical barriers to engineering an LMFBR so that the risk of a major accident would be as low as for the LWR. A necessary condition for such a result is the continuation of intensive safety research.

CONSEQUENCES OF A MAJOR LWR ACCIDENT

Damage Mechanisms

The estimate of the consequences to public health and property of a particular nuclear accident starts with the estimated quantity and composition of the radioactive materials (radionuclides) that escape containment. The dispersion of the radionuclides is then calculated in terms of its concentration in the air and on the ground over time after the accident and distance from the reactor. Physical phenomena included in this estimate are: height of radioactive plume rise, decay of released radionuclides, atmospheric dispersion, diurnal and seasonal weather variation, and rate of deposition of radionuclides. Radiation doses to humans are calculated from the following types of exposure: external irradiation from the passing cloud and radionuclides deposited on the ground, internal irradiation from inhaled radionuclides, and radionuclides that enter the food chain from contaminated crops, milk, and water.

Except for core material melting through the bottom of the containment vessel, all LWR accident sequences involve failure of the aboveground portion of the outer containment. A plume of hot gases containing radionuclides from the reactor core will rise to some height (from hundreds to thousands of feet). The quantity of radioactive release depends on the nature of the accident, and the mixture of radionuclides will depend on the time since the last refueling.

Dispersal of the plume depends on the weather, and the consequences of contamination depend on the distribution of population and property in its path. As a highly simplified illustrative model, one can imagine a steady wind pattern that would carry the plume downwind at a constant altitude, spreading laterally in the form of a wedge. Since deposition and decay of the important radionuclides are relatively slow, the decrease of the radiation intensity and rate of disposition are primarily controlled by the lateral spread of the cloud. Thus, while the intensity of radiation decreases through this lateral spread, the number of people exposed to the radiation will generally increase. In an area of nearly uniform population density, to a fair approximation these two effects cancel except for the loss due to deposition. Therefore, the product of individual radiation dose received times the number of people exposed at a given distance from the accident is approximately the same for each incremental increase in distance. This illustrates why significant latent effects can occur at a considerable distance from the reactor. In reality, populations are not distributed uniformly. Major urban areas can have a hundred times the average population density. To the extent such an area is crossed by the cloud, there will be a proportional increase in the total exposure. Thus, the consequences of accidents are very sensitive to the density of population and weather patterns in the vicinity of the accident.

Radionuclides released by a reactor accident would produce both prompt and latent health consequences. These effects are discussed in the chapter on health. Prompt consequences are illness or death within weeks or months of the accident as a result of acute radiation exposure. About half the people receiving some 400 rem[e] will die in a short time. However, such high exposures would normally only occur quite close to the accident. Latent effects include certain types of cancer and other illnesses that become evident over a period of several decades as well as genetic defects that become evident over future generations. As a first approximation, it can be assumed that these latent effects are proportional to the total dose received by the entire exposed population (expressed in man-rems, i.e., the product of the dose received by each individual times the number exposed) since the probability that any individual will contract these

---

[e]rem = roentgen equivalent man; the measure of radiation exposure to man.

illnesses varies with the level of his exposure. A population exposure of ten thousand man-rems will produce very approximately one latent cancer case during the next thirty years and one genetic defect in later generations.

Two messages that emerge from this simple model of weather and population density are:

- The risk to the public, both prompt and latent fatalities, is very site-dependent; population distribution and meteorological conditions are both important factors.
- The latent consequences can affect population hundreds of miles from the accident.

The WASH-1400 Calculations

A comprehensive assessment of the consequences of reactor accidents is contained in the Reactor Safety Study (WASH-1400, commonly referred to as the Rasmussen Report), that was published by the NRC in October 1975. In general, the WASH-1400 calculations constitute an elaborate version of the above illustrative model. The detailed calculations in WASH-1400, however, contain a number of debatable assumptions. Some of these are examined below to identify the uncertainties and to suggest possible improvements in future calculations.

Population and Meteorology. The first one hundred commercial LWRs will be at sixty-eight different sites in the United States. For its calculations, WASH-1400 groups these plants in six "composite" sites, each of which represents a distinctly different portion of the United States. Population distributions from U.S. Census Bureau data around the actual sites were averaged to describe the six composite sites. Weather data were gathered for six actual reactors, each of which was located at one of the composite sites. The weather data from the six actual sites were used to predict the radioactive plume motion up to five hundred miles in each of the sixteen sectors around the composite reactor locations. This method eliminated consideration of actual wind direction toward actual populations. Thus, WASH-1400 arrives at a type of average prediction, rather than at a distribution of consequences over actual sites. This complicated averaging procedure was apparently used to give a net assessment without appraising the situation at specific sites. A more accurate overall risk assessment could have been made by averaging the site-by-site consequences over all the actual sites.

The meteorological portion of WASH-1400 neglects temporal variation of wind direction. While such temporal variation is extremely difficult to include in such calculations, it could be extremely important in assessing specific sites and could in turn significantly influence the averaging calculation. Another meteorological limitation is that the weather data for a particular site were collected only at the immediate site, but the weather one hundred miles downwind is generally not the same as at the reactor. The omission of correlation of wind direction with population at actual sites is a serious matter. Using the same model, it has been calculated that prompt fatalities for similar reactor accidents may vary by as much as a factor of 1,000 for different sites even if the directionality of wind patterns is ignored. Substantial differences, although presumably less, also exist for latent effects. Risk assessment of this kind, for specific sites, could be an invaluable tool for siting decisions.

Health Effects. The principal problem in the calculation of health effects from radiation exposure is the uncertainty about the quantitative effects of low levels of radiation on human cells. All human and most animal data on radiation effects are based on high-level exposures. In general, policy bodies dealing with radiation protection have adopted the conservative assumption that the risks of both cancer and genetic damage are proportional to the radiation dose, and that there is no threshold below which no damage is produced. This approach was adopted in the 1972 report of the National Academy of Sciences Committee on the Biological Effects of Ionizing Radiation (BEIR Report). The "linear hypothesis" implies that the number of latent cancers and genetic defects is the same for 1,000 people exposed to 100 rem as it is for 100,000 people exposed to 1 rem.

If one also follows the BEIR Report in assuming that no "threshold effects" exist for low radiation doses, then the question remains as to whether the same linear relationship between latent health effects and dose exists down to zero doses as has been derived from high-level exposures or whether there should be some reduction in dose-effectiveness at very low doses and low rates of dose accumulation. There are some theoretical reasons to surmise that damage is accelerated at high dose rates, at least for latent cancer. WASH-1400 therefore concluded that the BEIR Report was too conservative, and introduced for low doses and dose rates, "dose-effectiveness factors," which reduce the expected latent cancers by factors ranging from 0.2 to 1. Additionally, the BEIR Report estimates were "modified where appropriate" and used as upper-bound risk factors in the calculations. It appears that the use of these dose-effectiveness factors in the calculations reduces by about half the expected number of latent cancers that would be obtained using the original BEIR Report recommendations. Recent data, discussed in the health chapter, have cast some doubt on the justification for this reduction factor. Genetic effects in WASH-1400 are based directly on the BEIR Report recommendations. There is some evidence, discussed in the chapter on health, but not considered in WASH-1400, that genetic damage at low radiation doses may be more serious than the BEIR Report concludes.

In considering consequences, it is important to recognize that not all casualties are fatalities. Varying degrees of radiation illness occur above a prompt exposure of 50 rem. Radioactive iodine in the cloud leads to large numbers of cases of thyroid disease with many operable thyroid nodules and some cancers. Prompt ingestion of iodine tablets, however, would inhibit the uptake of radioactive iodine and substantially reduce this consequence.

Overall Consequences. WASH-1400 considers a spectrum of "release categories" ranging from small releases of radioactivity to those releasing a large fraction of the critical volatile fission product isotopes (notably cesium and iodine) contained in the reactor core. WASH-1400 does not consider accidents which, according to their analysis, have a chance of less than one of occurring in one billion years of reactor operation. Although physical mechanisms leading to even larger releases than those considered can be conceived, WASH-1400 concludes that the likelihood of even more serious events decreases very steeply with increasing extent of radioactive release. We concur with this judgment. Accordingly, we have characterized the class of accidents corresponding to most serious release categories considered by WASH-1400 as "extremely serious accidents." Without endorsing the probability estimates given in WASH-1400, we agree that accidents substantially more serious than those included in the analysis of WASH-1400 need not be included in policy considerations. Table 7-1 summarizes the consequences of the "extremely serious accident."

The most serious adverse health effect calculated in WASH-1400 would be the fatalities from latent cancer. These cancer fatalities would occur in the exposed population over a thirty-year period following the accident. It is significant to note that the detailed analysis of these potential cancers on an organ-by-organ basis in WASH-1400 indicates that 83 percent of the eventual latent cancer deaths result from exposure during the first week after the accident. This points up the need for rapid response in any evacuation plan and the inherent limitations on the value of any long-range decontamination program following an accident.

The consequences of a nuclear accident could be reduced if the population were evacuated and steps taken to decontaminate the area. This would involve very large costs and major operational problems. In serious accidents, WASH-1400 assumes that people closer than twenty-five miles from the reactor would

Table 7-1. Consequences of the Extremely Serious Accident

	Rate	*Assumed Total*
Prompt Fatalities		3,300
Early Illness		45,000
Thyroid Nodules	8,000/yr	240,000 (30 yrs)
Latent Cancer Fatalities	1,500/yr	45,000 (30 yrs)
Genetic Defects	200/yr	30,000 (150 yrs)
Economic loss due to contamination	$14 billion	
Decontamination area	3,200 sq miles	

be evacuated. The efficiency of this evacuation was derived from historical analysis of past evacuations under threats of natural disasters. In addition, models were developed for long-term evacuation and decontamination. The monetary value of the "interdicted" land and produce was used to arrive at estimates of economic damages.

The natural decontamination time for cesium-137, the principal source of ground contamination, is three to five years. It is difficult to predict how many individuals would leave their homes for extended periods to reduce their chance of eventually dying of cancer. If land contaminated in excess of current standards for permissible concentrations of cesium-137 is withdrawn from use, the economic cost is estimated in WASH-1400 at $14 billion for the accident considered. This figure depends not only on land values but on the use of contaminated land, and the effectiveness of decontamination procedures not yet developed.

A major nuclear acident with its large number of latent cancer fatalities is so different from other natural or man-made disasters that it is difficult to gain a perspective on its true impact. WASH-1400 translates this uniqueness into the surprising decision to omit latent cancer fatalities in its comparative-risk analyses, and this is in fact done in the WASH-1400 summary of consequences.[f] In contrast, some critics of nuclear power compare predicted latent fatalities over a thirty-year span with the prompt casualties of various nonnuclear events. It seems clear that these latent fatalities should be taken into account in any comparative-risk analyses, but the fatalities should not be equated directly with prompt death since they would be spread over some thirty years. The average loss of productive life due to a latent fatality is less than that of a prompt fatality.

The current rate of cancer fatalities is 1,700 per year for a population of a million. The quantitative increase in latent cancers of some 10 percent produced by the severe accident discussed above might be difficult to measure statistically although the fact of an increase would probably be apparent since the distribution of cancer types would be different from those normally occurring.

The figures in Table 7-1 describe an accident leading to extremely serious consequences. Other accidential sequences which result in core meltdown followed by breach of containment at different locations and with different weather conditions would lead to a wide range of consequences. At the lower end of the spectrum, for example, a meltdown of the core that melts through the contain-

---

[f]WASH-1400 states: "The WASH-1400 risk comparisons do not show effects such as early illness, latent illness, genetic effects and latent cancer fatalities. Such effects have been calculated for reactors and are shown in (WASH-1400) Table 7-1. Since similar data are not available for the quantification of these types of risks from other man-made activities or natural causes, no comparisons can be made between nuclear and nonnuclear risks in these areas."

ment floor might release no more than 0.1 percent as much radioactivity to the surrounding human population as the extremely serious accident disscussed above.

THE PROBABILITY OF A MAJOR LWR ACCIDENT

It has been explained that an accident releasing a substantial amount of radioactivity cannot occur unless a number of successive barriers designed to limit propagation of a malfunction are breached. This design philosophy is intended to lead to a very low probability of a really serious accident. WASH-1400 attempts to provide a thorough assessment of such probabilities for the current generation of light-water reactors. The report is frequently cited as demonstrating the extremely low risks associated with nuclear power.

The overall probability of release of radioactivity can be computed as follows: The probability of an initiating event for all possible routes to a release is multiplied by the probabilities that each safety barrier would be breached or bypassed; these product probabilities are then summed for all the routes to release. The final public health risk is then the product of (1) the probability of radioactivity release, (2) the probability that weather conditions would carry the radioactivity toward the population at risk, and (3) the probability that varying populations would be in the path of the radioactive cloud. In this manner WASH-1400 decomposed the probability for the events leading to the extremely serious accident as follows:

Probability of core meltdown leading to breach of containment	5×10^{-5}/reactor year[g]
Probability of substantial radioactivity release after containment breach	10^{-1}
Probability of unfavorable weather	10^{-1}
Probability of large exposed population	10^{-2}
Total probability of a "maximum credible accident"	5×10^{-9}/reactor year

[g]5×10^{-5}/reactor year represents a probability of 5 in 100,000 per reactor year.

The first entry is the most complex. It must take into account many initiating mechanisms and evaluate the probabilities that the various safety features will or will not operate as intended. To do this, WASH-1400 undertakes detailed analyses of "event tree" and "fault tree" sequences involving the components and systems of reactors. These sequences are grouped into classes leading to accidents of varying degrees of severity, ranging from minor mishaps to major disasters.

An "event tree" is a procedure by which the probabilities of possible outcomes induced by one initiating event are tabulated. There can be a substantial number of such outcomes, since at many stages of the accident sequence the process can proceed in more than one manner; probabilities are assigned to the various "branches" of the tree.

A "fault tree" is the reverse of an event tree. Here a final outcome is postulated and the various sequences and combinations of individual failures that can lead to that final outcome are tabulated. Again, probabilities are assigned to each branch through which the final outcome can be reached. These probabilities are derived from operational experience, from the nuclear safety research program, from judgment, or (most frequently) from nonnuclear experience with the same components, such as valves, relays, and diesel generators.

WASH-1400 used this methodology to compute the probability per reactor year of operation of various categories of radioactivity release. For instance, the probability of a core meltdown accompanied by breach of containment in the current generation of power reactors is evaluated to be 5×10^{-5} per reactor year, but only 10 percent of these meltdowns are estimated to lead to substantial radioactivity releases after containment breach.

WASH-1400 combines the probability of accidents of varying severity with assessment of a variety of meteorological conditions that would disperse the radioactivity over exposed populations. The resultant consequences fall into four main categories: prompt fatalities, latent fatalities from cancer or genetic damage, sickness, and economic damage.

This methodology faces a number of serious problems. The most fundamental is that unknown or unsuspected failure mechanisms cannot be included in the analysis. Second, the final answers are the result of the assigned probabilities at each of the branch points; while these can sometimes be based on experience, they must at times be founded on judgment. Third, the probabilities of breaching each safety barrier are not necessarily independent since "common mode" failures can increase the likelihood of failure of one barrier once another has been penetrated. The probability of such common mode failure is uncertain unless the physical mechanism coupling the supposedly independent barriers is understood. Finally, the various probabilities may be correlated in different ways for different reactors over which safety predictions are averaged.

Although WASH-1400 takes note of these limitations, we conclude that the range of uncertainty that these deficiencies imply for the final calculation of accident probability is much larger than that estimated in WASH-1400. This conclusion is based on the observations that (1) the report underestimates the basic uncertainty of common mode failure analysis in the absence of detailed understanding of the mechanism; (2) important correlation effects (e.g., the correlation of prevailing winds and population distribution at specific sites) are not considered; (3) certain other factors, identified below, are omitted. Moreover, absolute estimates of small probabilities are particularly uncertain if a single failure mechanism is a dominant factor.

Although WASH-1400 contains considerable analysis of common mode failures, in which one malfunction induces another or in which multiple failures are induced by a common cause, this has not been done either thoroughly or correctly. An example of common mode failure is the interdependence of three presumably independent control rod systems designed so that the action of any one can shut down the fission process in a section of a reactor core. WASH-1400 cites experience to assign a probability of one part in 10,000 (10^{-4}) for failure of each control rod system. If failures were independent, the overall probability would be the product of three probabilities of 10^{-4}, or 10^{-12}. A "dependent" probability can be much larger than 10^{-12} if there is some linkage between the three systems or if an external event causes all three to fail together. In the former case, the probability depends on the strength of the linkage; and, in the extreme case where the three systems were tightly linked, the combined probability of failure to shut down would be 10^{-4}, or a probability one hundred million times larger than the "independent" probability!

We gave the preceding example to illustrate the extreme range of probability assessments possible, not to give a prescription of how to handle correctly the complex control rod situation in a BWR. In dealing with this case, WASH-1400 states that 10 percent of experienced malfunctions are of the common mode variety, and that 10 percent of these result in actual failure. On face value this leads to the conclusion that the probability of common mode failure is 10^{-6}, which overrides the lesser probability of 10^{-12} of independent failure. The overall probability should be determined by adding the separate probabilities of failure. However, WASH-1400 takes the "geometric mean" (square root of the product) of the failure probabilities to obtain a value of 10^{-9} for failure to shut down. There does not appear to be any physical justification for this procedure used elsewhere in WASH-1400, where "dependent" and "interdependent" probabilities of failure, differing by three to six factors of ten, are averaged by using the "geometric mean." In the absence of an understanding of either the physical mechanism of linkage or the characteristics of the external cause controlling common mode failure, mathematical methods cannot tell us how dependent and independent probabilities are to be weighted in forming an average. However, WASH-1400 presents no such physical analysis, yet an uncertainty of a factor of only ten is assigned to the resultant probability.

The WASH-1400 methodology of calculating meteorological conditions and population at risk is also questionable. The report essentially averages conditions over sixty-eight sites, ignoring the high correlation of meteorological conditions and population densities at several sensitive sites. Therefore, the latent cancer risk to a major metropolitan population downwind from a major reactor accident is considerably larger than the risk for other sites. In fact, the assessments of prompt casualties are averages of individual situations that may differ by as much as factors of 100 or 1000. Latent casualties will also vary, but by a smaller amount. Thus, the risks to which certain metropolitan areas are exposed are considerably larger than calculated. The WASH-1400 assessment is therefore not applicable to individual reactor sites.

Another problem with WASH-1400 is that the elements in the fault tree analysis are derived from two "representative" power plants. The analysis does not consider problems associated with collocation of reactors and joint control facilities (as was the case at the Brown's Ferry reactor fire, where the control cables for two reactors went through the same room). The analysis omits the effects of component aging or retrofit. In general, WASH-1400 pays inadequate attention to uncertainty resulting from the wide dispersion in failure estimates of individual components and differences in safety-related conditions at various reactor sites.

In summary, the inherent inability to predict "unidentified" failure mechanisms suggests that WASH-1400 understates risks. Moreover, the treatment of common mode failures in WASH-1400 is inadequate and appears to underestimate the problem. Some of the factors, however, could vary in either direction. A fundamental problem with the analysis is that it consistently underestimates the uncertainty in estimates.

With this background on the probability and consequences of accidents, it is possible to analyze the following general conditions that we suggested earlier be taken into account to decide whether the risks of nuclear accidents are acceptable:

- On a predicted average rate-of-loss basis, nuclear power compares favorably with competing technologies.
- The health and property consequences of a single extremely serious accident would not be out of line with other peacetime catastrophes that our society has been able to handle.
- Despite large uncertainties, a reasonable upper limit or ceiling, that is not in itself unacceptable, can be placed on the probability of the class of the extremely serious accidents.

The average rate-of-loss predicted by the WASH-1400 analysis is extremely small, amounting to 0.02 fatalities per reactor year. The WASH-1400 extremely serious accident by itself produces a latent fatality rate of only one-hundredth this or 0.0002 fatalities per reactor year (the product of 4.5×10^4 latent cancers and 5×10^{-9}, the probability for this accident). However, to compute the average rate-of-loss, one must take into account the integrated effect of all accidents, not just the most serious one. This is calculated by multiplying the probability of a meltdown (5×10^{-5} per reactor year), by the probability of breach of containment (10^{-1}), and by the average number of cancer fatalities for all sequences of events following a meltdown and breach (4,000). By far the largest uncertainty in this calculation is the probability of 5×10^{-5} per reactor year for a core meltdown. As upper limit assumptions, one can consider the probability of a core meltdown to be 5×10^{-3} per reactor year, based on the absence of any core meltdowns[h] in the first two hundred reactor years of commercial reactor operation, and double the probability of a containment breach to give a combined probability of 10^{-3} per reactor year. While the uncertainties in siting and meteorological conditions are very large for specific sites and may have a substantial effect on the probability of the class of extremely serious accidents discussed below, a reasonable upper limit on the number of fatalities averaged over all sites and meteorological conditions probably is not more than two to three times the number estimated in WASH-1400. Consequently, an upper limit on the average rate-of-loss calculations in WASH-1400 would be roughly ten fatalities per year (based on probabilities per reactor year of 5×10^{-3} for meltdown, 2×10^{-1} for breach, and 10,000 average fatalities). It is significant that even this extremely pessimistic, upper limit calculation gives an average rate-of-loss well within the range of estimates of the average rate-of-loss associated with the use of fossil fuels.

The adverse health and property consequences of even an extremely serious accident would not be out of line with other peacetime catastrophes that our society has been able to handle without long-term impact. We have indicated that an extremely serious accident might cause the death over a few weeks of three or four thousand people and cause tens of thousands of cancer deaths over thirty years and a comparable number of major genetic defects in successive generations, as well as some $14 billion in property losses. For comparison with prompt casualties of an extremely serious accident, we note that the United States has experienced in this century many serious hurricanes, two of which have each taken over a thousand lives and others that produced physical damage in the billions of dollars. Despite the losses from these hurricanes, the United States has been able to deal with and recover from these disasters without lasting effects. In this regard it should be noted that the deaths from cancer would not be an immediate effect but would result in a 10 percent increase in deaths from cancer in the exposed population over a period of thirty years. While a major social loss, this increase in cancer deaths would not present society with any fundamentally new problems.

The concept of a reasonable upper limit or "ceiling" on the possibility of occurrence is admittedly difficult to quantify. Not only is the central value of the probability distribution very uncertain but the nature of the probability distribution itself is largely unknown. Nevertheless, precisely because there is so much uncertainty in the estimates and little expectation that fundamentally improved analyses will be available soon, it appears useful to explore in a somewhat intuitive fashion what a reasonable upper limit on the possibility of an extremely serious accident would be if all uncertainties in the WASH-1400 calculations turn out to be unfavorable to nuclear power.

[h]There have been no core meltdowns or even a loss of coolant accident (LOCA) during past commercial reactor operations. The calculation, in the spirit of an upper limit, assumes in effect that the ECCS will not function adequately to prevent meltdown.

WASH-1400 estimates that the probability of an extremely serious accident is roughly one chance in two hundred million years of reactor operation. This results in only a few thousandths of 1 percent chance that such an accident will occur in this century, assuming five thousand reactor years of operation (5×10^{-9} per reactor year times 5×10^3 years). The WASH-1400 estimate of one chance in two hundred million (5×10^{-9}) reactor years of reactor operation results from the combination of the probabilities previously discussed (5×10^{-5} for core meltdown, 10^{-1} for containment breach, 10^{-2} for high population density, and 10^{-1} for unfavorable weather). As in the above analysis of the uncertainties for average rate of risk, a reasonable upper range for the combination of core meltdown and breach of containment is around 10^{-3} per reactor year. Estimates of the upper limit of the probabilities of occurrence of unfavorable population and wind conditions, complicated by the extreme site-dependence and mutual correlation of these quantities, are difficult and uncertain as they relate to specific sites. The uncertainty in this factor was of much less significance in computing an average rate of risk over all sites and weather conditions.

For certain sites (for example, Zion and Indian Point, but not only these), there is a high correlation among site location, prevailing winds, and high population densities. Therefore, at some small fraction of the one hundred reactors in the WASH-1400 analysis, the upper limit on the probability of extremely serious consequences after release of radioactivity might be as large as 50 percent rather than 10^{-3}. At the other extreme, there are many sites where the probability of unfavorable weather impacting large populations within a few hundred miles would probably be lower than the WASH-1400 average of 10^{-3}. Nevertheless, if five out of one hundred reactors are badly sited, they might alone contribute 2.5 percent of the weather-population probability factor averaged over all sites. The additional contribution of the remaining sites would probably not more than double this factor to give an upper limit of probability of an extremely serious accident after release of some 5×10^{-2}. Accordingly, the ceiling on probability of occurrence of an extremely serious accident might be placed at 5×10^{-5} (based on probabilities of 10^{-3} for core meltdown and breach of containment, and 5×10^{-2} for unfavorable weather-population conditions). This estimate would suggest an upper limit on the possibility of occurrence of an extremely serious accident by the year 2000 of about 25 percent (5×10^{-5} per reactor year times 5×10^3 reactor years of operation).

The 25 percent upper limit figure is 10,000 times larger than the central estimate derived from WASH-1400 and represents an extremely pessimistic excursion within the range of physically permissible accident sequences. This wide gap reflects current uncertainty inherent in this estimate. We believe that the magnitude of the upper bound on the risk, while serious, is not unacceptable, considering the small chance that it correctly describes the probability of such accidents and considering the risks associated with alternative energy sources. It does, however, underscore the importance of continuing efforts to reduce the possibility and consequences of accidents by improved safety design and greater attention to siting policies.

On balance, we have concluded that the risks associated with nuclear accidents are acceptable since the predicted average rate-of-loss due to nuclear accidents compare favorably to those associated with competing fossil fuel technology; the consequences of an extremely serious accident are not out of line with other peacetime catastrophes that our society has been able to handle; an upper limit that is not in itself unacceptable can be placed on the probability of occurrence of the extremely serious nuclear accidents. However, the number of reactors is expected to grow. Even though the benefits of nuclear power will increase correspondingly, the likelihood of occurrence of an extremely serious accident should not also be allowed to increase correspondingly. Therefore, there should be a continuing effort to improve actual reactor safety. It is not enough simply to demonstrate that existing designs comply with evolving regulatory standards and to enforce strictly the standards that are now established.

SAFETY DESIGN AND RESEARCH

Safety Design Philosophy

In normal industrial practice, an engineer can compute or observe a failure point for critical components and then build a definite overall safety factor into his design. If the designer miscalculates, an accident may indeed happen, and the estimate of the failure point will have to be revised. When the consequences of an accident are not so serious as to jeopardize an entire industry, such experience together with research constitutes a learning pattern that results in continually improved safety. Generally, the safety problem is sufficiently tractable that the margin of safety and the cost of establishing it are fairly well known.

Although these methods have been applied to reactor safety design to a certain extent, the happy absence of serious accident experience with operating reactors and the complexity of reactor systems make this normal, systematic approach very difficult. Instead, safety design largely rests on a "defense-in-depth" philosophy. The following are the conceptual elements of this strategy:

- Designs that provide a succession of independent barriers to propagation of malfunctions.
- Primary engineered safety features to prevent any adverse consequences in the event of malfunction.
- Careful design and construction, involving review and licensing at many stages.
- Training and licensing of operating personnel.

- Assurance that ultimate safety does not depend on correct personnel conduct in case of accident.
- Secondary safety measures designed to mitigate the consequences of conceivable accidents.

The total effect of these measures is so complex that the designer cannot ascertain a safety factor in a quantitative way; instead, the designer is conservative in the design of individual components and replaces an overall margin of safety with a number of independent safety barriers that must be breached or bypassed before a serious accident can occur.

The combination of defense-in-depth in overall configuration and conservative design in detail has been successful. Yet despite its demonstrated value, this approach has fundamental weaknesses. The sum of many conservative design decisions may not be conservative. Even if design is conservative in each component as measured by a specific parameter, overall safety may be less, if it is controlled by another phenomenon. Penalties associated with conservative design (such as excess weight, or difficult access) can impair overall safety. Added safety features can introduce new risks of their own, as in the 1966 accident at the Detroit Fermi reactor, where a partial meltdown was caused by the breaking loose of a flow-deflecting zirconium plate that had been especially installed to reduce the likelihood of a core meltdown.

This kind of situation arises from the lack of a precise theory of operation of a reactor's primary system and safety systems. Without such a theory, the results of experiments may lose some of their usefulness, and the value of less-than-full-scale experiments is limited, since the principles of "scaling" from smaller models to the full-scale reactor are poorly understood. Moreover, the designer frequently does not know the precise cost or benefit of his conservatism, in terms of either performance or money. WASH-1400 predicts that the more severe the consequences of an accident, the less likely it is to occur. Yet in the present circumstances one cannot be certain that experience with the frequency and kinds of minor accidents will lead to reliable predictions of more serious accidents.[i]

The reliability of prediction of nuclear reactor safety performance suffers from the small scale of existing or planned experiments. To remedy this, larger and more expensive experiments would have to be carried out. These would require increased funding and considerable time for the planning and execution. However, even a full-scale reactor safety experiment would be of only limited value. While a simu-

---

[i]This limitation recalls the situation existing in the early 1960s in the U.S. *Minuteman* ICBM strategic force. At that time, the command and control system tied fifty missiles together to fire as a single unit. Such an arrangement (which no longer pertains) guaranteed that if an accident occurred it would be large. The Brown's Ferry fire also demonstrated the dangers of excessive collocation of controls.

lated loss-of-coolant accident on a full-scale reactor could be so well instrumented that much could be learned, a single destructive experiment in which the emergency core cooling system (ECCS) was asked to play a critical role could hardly provide persuasive evidence of how the ECCS would perform under a variety of circumstances.

The foregoing discussion is not intended as a criticism of current design practice but rather as an illustration that prediction of the probability of reactor accidents is an unusually difficult task. Even a more extensive research program would improve only slowly the reliability of the assessment of the safety margin in a nuclear power plant. For this reason, it must be recognized that present uncertainties—for example, the reliability of the emergency core cooling system—will persist for a long time, certainly for a decade, and that decisions will have to be made in the face of that uncertainty. Thus, proposals calling for a moratorim on reactor construction "until they have been proved safe" or "until the emergency core cooling system has been proved safe in full-scale tests" cannot be implemented within a reasonable time, if at all.

The Safety Research Program

Reactor safety research has received steadily increasing financial support since 1974. At present it is funded by government and industry at above $100 million per year. Reactor safety research serves two separate goals: to narrow the margin of ignorance in predictions of reactor safety, and to lower the risk of accident and mitigate the consequences of an accident. It is unfortunate that most research is in the former category.

The Nuclear Regulatory Commission (NRC) is currently restricted to nuclear regulatory research while the Energy Research and Development Administration (ERDA) supports safety research for new reactors and, in particular, for the LMFBR. The NRC supports independent research for its licensing responsibilities of reactors still under development. Although the Electric Power Research Institute (EPRI) has a safety research program supporting the needs of the utilities, and the vendors conduct research to improve existing reactor components, the division between NRC and ERDA creates the danger that important developments may fall between these agencies. Development of improved safety systems for existing reactors is not a clearly defined responsibility of any of the parties.

Roughly 90 percent of federally funded research is for the study of accidental sequences involving components of current design, and by far the largest share of this research is focused on the performance of the emergency core cooling system. This division of efforts reflects the conviction that current design is sound and that better calculations and measurements will bear this out. A substantial fraction of the scientific manpower analyzes the course of an accident through computation and analysis while a smaller part is in experimental work. The reason for this division is that the vendors, faced with increasing criticism and more stringent NRC regulation, as well as with the prospect of only a slowly increasing base of good safety data, are compelled to improve calculational codes so that fewer conservative assumptions must be made in the computations.

On the experimental side, the research covers a gamut from tests of separate effects of specific components to systems tests examining safety in case of failure. The largest facility existing or planned is the Loss of Fluid Test (LOFT) installation at ERDA's reactor test site in Idaho. LOFT, which is being constructed and will be operated by ERDA although the associated research is the responsibility of the NRC, will be an operating reactor with a power level of one-sixtieth that of a 1,000 MWe commercial pressurized-water reactor. This difference in scale raises the question of the transferability of the results to a full-scale commercial unit. Although an attempt has been made to match the ratio of volume to power in the essential elements of LOFT with those in a full-size commercial plant, there are essential differences in overall design. The LOFT program is twelve years old, and the reactor has yet to go critical. It is hoped that meaningful test operation can commence in 1977. LOFT will be the largest test in which actual loss of coolant accidents can be simulated in the foreseeable future, and within the limitations cited above useful information should be generated.

There are also some smaller test facilities in the program. One, called Semi-Scale, uses electric heat as a source against which emergency cooling can be tested. Another, the Power Burst Facility, is designed to test individual fuel elements in extreme conditions of stress. On an even smaller scale, there is an external separate effects program that tests and evaluates components and materials under the stress of accident.

Research dealing with actual LWR safety improvements is deficient. While studies of alternative ECCS concepts have been undertaken, no research program is now aimed at potential implementation. No substantial effort is directed at the mitigation of the consequences of a nuclear accident beyond limited improvements in the engineered safety system already incorporated inside the containment. Safety analysis of floating power plants has not reached the level of detail of WASH-1400. Little has been done to assess the value and cost of underground siting of nuclear power plants for particular locations. Finally, no intensive effort on major improvements in reactor control systems has been launched, even though such systems lag behind available technology.

We conclude that the safety research program is deficient in research leading to new and improved safety components and systems. We doubt that this deficiency can be remedied without an increase in funding, since the present regulatory requirements make it infeasible to decrease the research dedicated to evaluation of current systems. We believe that such an expanded reactor research program is fully justified. There will never be a high degree of confidence that probabilities as low as those predicted in WASH-1400 are in fact correct. One would like to be convinced that the total expectation of adverse consequences from an expanding reactor industry should steadily decrease. At a minimum, one would like to know that the safety of new reactors was improving with time. Only a more comprehensive safety research program will accomplish this. Such a program will also help to build public confidence in reactor safety.

There is no shortage of possible improvements in LWR safety. The following are some examples of improvements that might be considered for incorporation in future reactors without excessive cost:

- ECCS systems with superior reliability are clearly possible. For instance, an improved ECCS system introducing emergency cooling water directly from above is now being tested by one of the vendors, and some German reactor designs have considerably more redundancy in the ECCS systems and provide for a lesser risk of common mode failures linking the ECCS and the containment cooling spray.
- Better methods of energy absorption of fragments from a burst primary reactor vessel could be designed.
- The possibility exists of controlled venting of the containment through heat resistant filters if core meltdown occurs.
- The control systems could be extensively improved using state-of-the-art data processing techniques.
- Control rod independence could be improved.
- Underground siting could be reexplored and used where proved advantageous.

These particular examples may, on detailed examination, turn out to be less advantageous than other potential improvements. However, the decision-making process must encourage and accommodate the development of a broad range of safety improvements.

INSTITUTIONAL ASPECTS OF REACTOR SAFETY

The Safety Assurance Process and Its Conflicts

In any responsibly managed technical activity, safety considerations enter at many stages. Safety is an integral element in the research and conceptual design process. Safety is involved in the development, production, and installation of a facility. Safety has to be part of the design and execution of operational procedures, including the training of personnel.

At the same time, the safety of technological enterprises involves an inherent conflict. If the responsibility for safety is integrated with the responsibility for design, development, and installation of a new facility, there is a conflict of interest between the need to accomplish the task in the most expedient and economical manner and the need to assure safety. When responsibility for the safety and the execution of the enterprise are separated organizationally, the competence of those charged with safety tends to lag behind that of those executing the engineering. In this case, some organizational mechanism is required to reconcile conflicting positions. In recognition of this requirement, technical organizations frequently establish a safety office reporting to a higher level of management than the primary engineering organization. Such a safety office monitors the technical operation, while the regular line organization is formally responsible for safety in its own field of activity. Top management will then adjudicate any conflicts between safety and the efficient execution of the project.

The Nuclear Power Safety Institutions

The responsibility for public safety in the field of nuclear power is shared by several interdependent organizations. The power plant is generally owned and operated by a utility. The nuclear steam system is procured from one of four vendors, while the overall design of the plant is usually contracted to an architectural/engineering firm. Research and development on reactor design, including safety, is carried out by vendors, by the utility industry through EPRI, or by ERDA. Responsibility for public safety is exercised through the NRC, which also carries out confirmatory research on reactor safety.

Until 1975, responsibility for both nuclear reactor safety and reactor development was lodged within the Atomic Energy Commission (AEC). Although from the 1960s on the AEC did separate the regulatory function from those of research and development, on the detailed programmatic level the development of nuclear reactors and research on reactor safety were carried out in an integrated manner. From the outset, the AEC established the Reactor Safeguards Committee as an independent advisory body on matters of nuclear power safety.

The very structure of the AEC was something of a constitutional anomaly within the U.S. government. On the one hand, the AEC was a regulatory commission responsible for public safety in the application of nuclear energy. On the other hand, the AEC was also an executive agency responsible for promotion of the applications of nuclear energy. To counteract concern that insufficient weight was given to reactor safety, the AEC was reorganized in the early 1970s and reactor development, reactor safety research, and nuclear power regulation became three separate entities within the AEC. However, the anomaly that the AEC was both the regulator and promoter of nuclear power remained until the AEC's responsibilities were divided between ERDA and the NRC in 1975.

The NRC now has final authority in certifying the safety of domestic nuclear power. In so doing, the NRC must weigh the benefit of nuclear power against the risks. However, no guideline exists on how much weight is to be given to the risks or the benefits. Ultimately, such conflicts will have to be resolved in the political arena.

At present, no agency of the U.S. government has responsibility for assuring safety standards of exported reactors. While the NRC licenses exports, the safety of the installation and the qualifications and training of the potential operating personnel are not considered in the licensing procedures. These matters are apparently considered the sole responsibility of the importing country. There is no indication that these questions are given serious weight by the State Department or ERDA in negotiating export agreements with foreign countries. This is a serious deficiency, since, apart from the moral responsibility for the safety of exported reactors, there is the practical consideration that an accident abroad involving a U.S.-built reactor could have major repercussions on the domestic nuclear power industry.

The Regulatory Process and Technological Safety Improvement

The NRC's regulatory responsibilities for safety are organizationally separated from the reactor development responsibilities of ERDA. The NRC carries out regulatory assessment research, while ERDA is responsible for safety research as part of the development of new reactor systems. Although this arrangement preserves the NRC's independence, it results in some overlap in substance and may leave serious gaps. For instance, there is a real question where responsibility lies for developing improved reactor safety technology for LWRs of a design already licensed.

In the interest of fair competition consistent with safety, the NRC is required to establish standards against which industry can design its safety systems. As new information becomes available, the NRC can modify its standards. If such regulatory standards change too rapidly, industry complains that designers are shooting at a moving target, with increased costs and delayed schedules. On the other hand, such adjustable standards are the tools by which the NRC can react to new findings on safety and live up to its enforcement responsibilities.

The regulatory system itself can actually discourage improvements in reactor safety on the part of industry, since there is no assurance that the regulatory process will approve a new safety system without considerable delay. In addition, the introduction of a new safety system always raises the possibility that the NRC will require that it be retrofitted on existing reactors at great cost. These circumstances create a considerable incentive to utilize only designs that have passed the regulatory process. They apparently discourage, for example, the introduction of more modern data processing techniques for use in reactor control systems.

As a consequence of the regulatory process, industry, ERDA, and the NRC all tend to spend a disproportionate effort in demonstrating that existing reactor designs are as safe as regulatory requirements demand, compared to the effort toward improvements in reactor safety. It is not easy to resolve this problem, since one clearly does not want to undercut safety regulations or the independent authority of the regulatory agency. The situation might be improved by government incentives for development of new safety systems by private industry. Qualified firms, be they vendors of nuclear reactor components or systems or not, could develop superior systems without large investments of their own. There appears to be considerable opportunity in such areas as the engineering of reactor control systems that lag considerably behind other high technology areas. If firms not now active in reactor technology can be interested in this activity, it may be possible to achieve some important contributions to reactor safety engineering.

REACTOR SITING

Critics of siting regulations point to the duplication of effort and multiplicity of requirements imposed by governmental units at all levels. Only regulation of the radiological aspects of nuclear power is clearly reserved by legislation to the federal government. In other areas such as economic or environmental impact, state and federal efforts overlap. Utilities have reason to complain about this regime of multiple regulation.

In one important court case, *Northern States Power* vs. *Minnesota* (1971), the Supreme Court held that the state could not impose radioactive effluent standards more restrictive on a power reactor than those approved by the AEC. How far this doctrine of federal preemption will extend is unclear at present. Some aspects of state and local efforts to control nuclear power have been clearly discriminatory and reflect the mistaken impression that the national energy problems can be solved at the state or regional level.

Current NRC siting criteria and procedures take population proximity into account by requiring that the population density out to thirty miles not average more than 500 persons per square mile at the time of initial plant operation and not more than 1,000 persons per square mile in the plant's projected operating lifetime. Other provisions define "low population zones" which must surround reactor sites and relate these zones to local atmospheric conditions. These procedures are only loosely related to the level of radiological risk.

In principle, the type of calculations contained in WASH-1400 could provide a sounder basis for relating siting decisions to risk as discussed above. However, WASH-1400 estimates are based on calculations which aggregate and average site-specific variables such as population distribution and local meteorological characteristics. Thus the WASH-1400 calculations obscure the effects of siting on the consequences of reactor accidents and are not directly applicable to the evaluation of site-dependent risks.

In some of the potentially most worrisome cases (e.g., the New York Indian Point and the Zion, Illinois, reactors) nuclear plants are located near major population centers, and prevailing weather patterns would have a significant probability of carrying the radioactive cloud toward these centers in a serious accident. This correlation of local population distribution and meteorology increases the probability for an extremely serious accident for unfavorably sited plants. On the other hand, the same factors may decrease the probability of such an accident for favorably sited plants.

We therefore conclude that good siting decisions can decrease the overall risks associated with nuclear power. Currently such decisions involve a profusion of local, state, and federal jurisdictions. Many states play an active role in power plant siting through environmental controls, land use regulation, management of flood plain and coastal zone usages and through direct controls on power plant siting. Over twenty states have laws regulating power plant siting. Maryland's model siting law includes a state administrative agency (whose siting activities are funded by a surcharge on electricity) and a land acquisition program for suitable sites. Even local governmental units affect power plant siting through local zoning, planning, highway, health, and public safety authorities.

Notwithstanding the justified local and state interest, nuclear power poses a set of problems that must be addressed at the national and international levels. The risks of nuclear accident, environmental impacts, and the implications of nuclear theft or sabotage extend beyond state borders. Yet one can argue that states and localities should have essentially the same regulatory authority, rights, and claims in dealing with nuclear power as with fossil fuel plants. We believe that the federal regulatory process should give more critical attention to the siting problem and that considerable weight should be given to local preferences in siting decisions. However, legislation should be considered which would give the courts a basis for recognizing an ultimate federal responsibility in nuclear power plant siting decisions in case of conflicting interests.

CONCLUSIONS

Our review of the reactor safety problem has led us to the following general conclusions:

1. Neither the excellent safety record of nuclear power reactors to date nor the history of "abnormal occurrences" yields statistical data for predictions covering the rest of this century.

2. Although it is a valuable resource for study of the safety problem, WASH-1400 should not be used as a definitive guide for policy since it understates the uncertainties and has serious methodological deficiencies.

3. On an average rate-of-loss basis, nuclear power compares favorably with coal even when the possibility of accidents is included. WASH-1400 concludes that the average rate-of-loss, taking into account the full range of possible accidents, is about 0.02 fatalities per year for a 1,000 MWe nuclear power plant. This is very small compared with one fatality per year from normal nuclear operations or the two to twenty-five fatalities per year from a comparable coal

plant. Although we have not attempted to make an independent estimate of this average value, our analysis indicates that the WASH-1400 estimate could be low by a factor of as much as 500. On the other hand, it could even be on the high side as well. In the most pessimistic case, which we consider extremely unlikely, where the estimates are low by a factor of 500, the average rate-of-loss would be about ten fatalities per year for a 1,000 MWe nuclear power plant. It is significant that even in this extremely pessimistic, low probability case, the fatalities do not exceed the pessimistic end of the range of coal.

4. The adverse health and property consequences of even an extremely serious accident would not be out of line with other major peacetime catastrophes that our society has demonstrably been able to handle without major long-term impact. The most serious sequences considered in WASH-1400 might cause the death over a few weeks of three or four thousand people and cause tens of thousands of cancer deaths over thirty years and a comparable number of major genetic defects in successive generations, as well as some $14 billion in property losses. We agree that these estimates are a reasonable assessment of the probable consequences of an extremely serious nuclear accident. For comparison, the United States has experienced in this century two hurricanes that have each taken over a thousand lives and others that produced physical damage in the billions of dollars. Despite the losses from these hurricanes, the United States has been able to deal with and recover from these disasters without lasting effects.

5. The reasonable upper bound on the possibility of an extremely serious nuclear accident is not in itself unacceptable. The most serious accident considered in WASH-1400 is assigned a probability of occurrence of roughly one chance in two hundred million years of reactor operation. This implies only a few thousandths of 1 percent chance that such an accident will occur in this century, assuming 5,000 reactor years of operation. Whatever the merit of this estimate as a central value, it is extremely uncertain. Calculations based on a combination of deliberately highly unfavorable assumptions representing the upper range of the component uncertainties indicate that a single extremely serious accident could have a significant—say 25 percent—chance to occur in 5,000 reactor years of operation. This upper bound is extremely unlikely and our expectation is that the probability is likely to be *much* lower, probably at least by a factor of 10 or 100. We believe that these limits on the risk, while serious, are not unacceptable, considering the small chance that it correctly describes the probability of such accidents and the risks associated with alternative energy sources. It does, however, underscore the importance of continuing efforts to reduce the possibility and consequences of accidents by improved safety design and greater attention to siting policies.

6. On balance, we believe the risks associated with the current generation of LWRs are acceptable since, as stated in conclusions 3, 4, and 5 above, they meet the general conditions that were considered in this chapter.

7. The average risk projections are dominated by relatively few reactors located upwind from large metropolitan areas. Siting criteria specifically focused on this problem could substantially reduce the total risk to the U.S. population.

8. The LMFBR presents some major new safety problems as well as a reduction in some of the safety problems inherent in the LWR. Although a detailed safety assessment of the LMFBR is several years in the future since the design has not been completed, there do not appear to be any fundamental physical barriers to the development of a commercial LMFBR as safe as the LWR.

9. In view of the serious consequences inherent in major nuclear reactor accidents, there should be a continuing effort to improve reactor safety. To this end, greater effort should be placed on actual safety improvements, in addition to the present heavy emphasis on improving the ability to predict safety performance. Currently, the regulatory process creates disincentives to improvements in safety, and steps should be taken to change this situation.

10. The "defense-in-depth" philosophy of the U.S. nuclear power program has been the proper approach to nuclear safety, in view of the uncertain state of knowledge and the seriousness of the risks involved. In the future, however, a major effort must be made to develop better experimental and theoretical understanding of the operation of reactors under normal and abnormal conditions. This understanding will provide a basis for determining whether the totality of the conservative design features in reactor design and the defense-in-depth philosophy is indeed conservative, as well as for developing basic improvements in safety design with potential economies.

11. The present climate of doubt about reactor safety will persist for at least a decade, even if a research program to alleviate such concerns is pursued vigorously. Decisions will have to be made in the face of these doubts and uncertainties.

✳ *Chapter 8*

Radioactive Waste

Radioactive waste is the inevitable by-product of the generation of electricity by nuclear reactors. While radioactivity is encountered at most stages of the nuclear fuel cycle—in mining and milling, in fuel fabrication, in reactor operation, and in subsequent stages—the largest quantities, and those potentially of greatest concern, are removed from reactors annually in spent fuels. The radioactivity in the spent fuel results from the many different, highly radioactive fission products produced when atoms split during the fission process and form new radioactive atoms, such as plutonium, created when neutrons are absorbed by uranium or other heavy elements in the fuel.[a] The intense radioactivity of the material in the spent fuel decreases rapidly at first, being reduced by a factor of 1,000 during the first ten years and then much more slowly, reduced by a further factor of 1,000 only over the next 100,000 years. The initially high radioactivity of waste necessitates very careful handling and storage in the immediate term; the long-term nature of the waste hazard necessitates failsafe measures protecting many future generations. The interim handling and storage of radioactive material from power generation is referred to as waste

[a]Fission products are the radioactive elements formed when the nuclei of uranium or other heavy elements split after capturing a neutron. Several isotopes of each fission product element are usually present in spent fuel, each with its own radioactive decay sequence. Most of the fission product isotopes have radioactivity which is relatively short-lived— from a few minutes to a few decades; the exceptions are iodine-129 and technetium-99, which remain radioactive for many thousands of years. The heavy elements with atomic weights above actinium are known as actinides; they include thorium, protactinium, uranium, neptunium, plutonium, americium, and curium. The transuranics are the elements heavier than uranium, notably neptunium, plutonium, americium, and curium. Several isotopes of each of these are present in spent fuel. The isotopes of the transuranic elements are of particular concern because they have long half-lives and many of them decay by emitting alpha particles. The effects of this radiation are described in the chapter on health.

243

management; it is generally believed, but not universally, that this stage should have a short time horizon and that management should be followed by permanent disposal in a way which does not require human attention over the very long period of time before wastes become harmless.

While it is convenient for discussion to separate management and disposal, it should be recognized that these two problems are closely related. Since steps taken in managing wastes have an important bearing on the disposal problem, a long-range view for interim decisions is clearly indicated. Waste management and disposal involve important institutional as well as technical problems. This is in part responsible for the wide range of opinions about the difficulty of the nuclear waste problem. Critics of nuclear power point to past institutional failures to deal optimally with wastes, arguing that nuclear wastes are the basis for rejecting further expansion of nuclear power. Proponents assert that problems of waste management and disposal have solutions that would reduce risks to present and future generations to negligible levels.

Present nuclear wastes are in three general categories. Most commercially generated waste is being held as spent fuel at the reactor sites where it was produced. Some spent fuel was chemically reprocessed to recover plutonium during the period 1965–1972 at the Nuclear Fuel Services (NFS) site in western New York State. The waste remaining after these operations is still stored at the site in liquid and solid forms. Weapons wastes, resulting from reprocessing spent fuel from production reactors in order to recover plutonium, are stored in liquid and solid forms at several sites.[b] The weapons wastes and the NFS wastes presently constitute the largest management problems because of the quantities involved and, even more significantly, because of the ways in which the wastes were treated. Decisions made earlier in connection with these wastes were shortsighted and now present problems which will be expensive, and perhaps risky, to resolve.

As experience with the NFS and military waste suggests, the nature of the waste management problem depends on the choice of fuel cycle. A fuel cycle which involves reprocessing would result in the redistribution of the radioactivity from spent fuel into new physical forms. The acidic liquid waste stream coming from the reprocessing plant would contain the fission products, about 0.5 percent of the plutonium and uranium, and most of the other transuranic elements. Gaseous fission products would be released from the spent fuel in reprocessing and would either be released (for example, krypton-85 and carbon-14) or captured (iodine). Several other new categories of waste would also be formed, as discussed below. It has usually been assumed that the fuel cycle would eventually

---

[b]While the volume of military wastes is large, the concentration of radioactive material in it is small compared with that in commercial waste even if the latter is reprocessed. This has frequently led to misleading comparisons between the amounts of the military and commercial wastes. The amount of radioactivity in present commercial spent fuel waste is comparable to that in military waste and is increasing much more rapidly.

be closed by reprocessing followed by recycle of plutonium and uranium and disposal of remaining waste products. This belief has become so common that some think that failure to establish reprocessing is a barrier to management and disposal of remaining waste products. This belief has become so common that manage and dispose of the varied waste products of reprocessing operations rather than on how to deal with spent fuel.

We have concluded elsewhere in our report that spent fuel should not be reprocessed to recover plutonium for recycle in light-water reactors. If and when breeder reactors are necessary, it may be important to recover some of the plutonium in spent LWR fuel. Since the eventual need for plutonium breeders may not be known for several decades, it will be necessary to store some spent fuel retrievably for this period. The feasibility of spent fuel management and its disposal, as compared to waste problems in the reprocessing option, are thus important issues in the reprocessing and breeder decisions.

We believe that nuclear wastes can be disposed of permanently in geological formations in such a way that there is very little prospect of material escaping into the environment. Moreover, even unlikely failures of repositories in the distant future would not have large consequences to human populations. This is true independent of whether the wastes disposed of are spent fuel or the resolidified and transuranic wastes left after reprocessing and recycle. We are more concerned about the potential hazards of wastes before they are sequestered in geological formations. Experience with waste management is not encouraging. There has been a recurrent substitution of short- for long-term goals with less than adequate management practices. These include neutralization of acidic reprocessing wastes, leaks from single-walled tanks containing liquid wastes, and shallow land burial of transuranic wastes. Management of nuclear wastes as spent fuel presents fewer opportunities for failure than does a fuel cycle in which reprocessing occurs. We believe that most spent fuel should be stored retrievably in the same geological formations which would be suitable for permanent disposal, until decisions can be made on plutonium breeders.

Even measures which may appear costly, when assigned to a given amount of spent fuel, do not have much effect on the cost of power. Current estimates of the total cost of waste disposal are about $100 per kilogram of spent fuel. This is equivalent to a power cost of 0.4 mills per kilowatt hour. Even a cost of $1,000/kilogram would not make nuclear power noncompetitive. Temporary storage of spent fuel presently costs less than $5/kilogram/year, a small cost which might be reduced by large-scale facilities which would also provide greater security. Since economic constraints on waste management and disposal are not restrictive, we have focused primarily on the technical and institutional factors which bear on waste management and disposal.

In this chapter we examine the nature of nuclear waste, as it originates in the generation of nuclear power; the management problems which arise and how they depend upon fuel cycle choices such as reprocessing and recycle; possible modes of disposal and constraints on successful disposal; and institutional and programmatic approaches to waste problems here and abroad.

THE NATURE OF NUCLEAR WASTES

Nuclear waste materials are by-products of most of the stages of the nuclear fuel cycle. Mining and milling operations remove radioactive materials from the earth; the tailings remaining are a continuing source of radiation. Small quantities of radioactivity are released in normal reactor operations. By far the largest quantities of radioactivity, however, are produced by the nuclear processes which generate power in the reactor; it is removed, annually, in the form of spent fuel.

Spent Fuel

A light-water reactor discharges annually about 30 metric tons of spent fuel. Each ton contains nearly 30 kilograms of fission products, the radioactive remains of fissioned atoms, and slightly less than 10 kilograms of transuranic elements. The remainder is unburned uranium containing about 0.8 percent uranium-235. The precise composition of this material depends on reactor type and on the length of time the fuel remains in the reactor generating power; longer burnups in the reactor result in higher concentrations of fission products and transuranic elements. The intensity of the radioactivity present is very high. Immediately at reactor shutdown, a ton of spent fuel contains about 300 million curies of activity. After about ten years, this level has decayed to about 300 thousand curies.[c] Spent fuel also produces a great deal of heat: one day after reactor shutdown, 30 tons of spent fuel have a heat output of about 10,000 kilowatts; after ten years, this is reduced to about 1 kilowatt per ton.

Most of the radioactivity and heat resulting from waste during the first few hundred years after generation are due to the fission products. More than a hundred different isotopes are involved, including strontium-90, cesium-137, technetium-99, iodine-129, and krypton-85. The strontium-90 and cesium-137 have relatively short half-lives, with their activity reduced by a factor of two in thirty years. Both became familiar to the public during the era of atmospheric nuclear weapons tests, with strontium-90, which behaves biologically much like calcium, appearing in milk supplies in some areas. Iodine-129 and technetium-99 have much longer half-lives and thus have lower levels of radioactivity spread

[c]Some sense of these magnitudes is given by considering that a person living 100 meters away from material with 1 curie of radioactivity for a year will receive a radiation dose of about 1 rem, or about five to ten times the dose he will receive from natural background radiation during the year. Whole-body exposures higher than 500 rem have a high likelihood of causing death within weeks; lower doses increase the long-term risk of cancer and genetic disease. As discussed below and in Chapter 5 (Health Effects), however, the potential for harm is different if radioactive material enters the body by ingestion, inhalation, or other routes.

out over longer periods. Both are biologically active, with iodine concentrating in the thyroid and technetium in the gastrointestinal tract. Krypton-85 is an inert gas which is released to the atmosphere during reactor operation. The total activity of the fission products is reduced by a factor of about ten million 700 years after generation. By this time, the dominant source of radioactivity in the waste comes from the transuranic elements.

The transuranic elements in waste include neptunium, plutonium, americium, and curium. The long half-lives of iostopes of these elements, and their health risks, make the long-term disposal problem difficult. Plutonium-239, for example, has a half-life of more than 24,000 years. The amount of plutonium-239 in reactor waste is reduced by radioactive decay only by a factor of twenty over a period of one hundred thousand years. The toxicity of plutonium-239 is known to be very great; that of the other actinide elements is not yet as well known. The principal health dangers arise from the fact that these elements emit alpha particles. These heavy particles cannot penetrate skin. However, alpha-emitters present a cancer risk even in quantities as small as ten-millionths of a gram if inhaled so that they come in contact with lung cells or other living tissue by transport in the blood stream. The ingestion hazard is considerably lower since the gastrointestinal tract is protected by a mucus layer. Several isotopes of plutonium are present in commercial reactor wastes. For example, plutonium-238, which accounts for about 2.5 percent of the plutonium in spent uranium fuel, has a half-life of only eighty-six years and is several hundred times as radioactive as plutonium-239. Most research on the biological effects of plutonium has been done with plutonium-239 and most human experience, in weapons work, is with this isotope.

Commercial waste is presently being stored as spent fuel assemblies, most of it in water-cooled facilities at the reactor sites where it was generated. Subsequent steps taken to deal with this waste will have a bearing on the nature of future waste management and disposal problems. A decision to reprocess the spent fuel and to recycle uranium and plutonium in light-water reactors would result in changes in the radioisotope composition of waste from reactors and in the creation of different physical categories of waste, each of which requires separate treatment. A decision to defer reprocessing, as recommended in this report, would require that consideration be given to methods to store spent fuel securely but retrievably until it is clear whether the plutonium contained will be necessary for breeder reactors and to permanent direct disposal of spent fuel if it is not. These alternatives are discussed below.

Reprocessing and Recycle

It has generally been assumed in the past that spent fuel would be reprocessed to recover and reuse uranium and plutonium and, for this reason, it has been allowed to accumulate at reactor sites. Reprocessing and fabrication operations put the radioactivity in spent fuel into new physical forms; recycle of plutonium in reactors as a fuel supplement to uranium-235 changes the quantities of various radioactive species in waste. Reprocessing results in an acidic liquid stream bearing the fission products and perhaps one-half percent of the plutonium and uranium present in the original spent fuel. This liquid high-level waste also contains nearly all of the other actinide elements, notably americium and curium. Reprocessing also liberates radioactive gases and volatile radionuclides from the original waste material. Provision for recapturing most of these, including isotopes of iodine, would be included in reprocessing plants, such as that at Barnwell, South Carolina. The exception appears to be krypton-85, which would be released in relatively large amounts from reprocessing plants presently contemplated but which could be captured using available technology, at some cost. The effects of the release of krypton-85 are described in Chapters 5 and 6. Reprocessing also results in the creation of new categories of waste contaminated with the radioactivity in the original spent fuel: cladding and fuel assembly materials; solid waste consisting of process materials, filters, containers, tools, rags, and so forth; and a large volume of liquid waste containing only very small concentrations of fission products. Some of the cladding and solid waste would be contaminated with transuranics and other long-lived radioisotopes. Fabrication of fuels bearing plutonium also results in the creation of solid process trash contaminated with plutonium.

Reprocessing and recycle of plutonium provide a way to reduce long-term risks by reducing the amounts of transuranic elements in wastes. However, the magnitude of this effect is not large. Recycling plutonium in reactors results in higher concentrations of americium and curium in spent recycle fuel than in ordinary spent uranium fuel. These elements present risks similar to those of plutonium and would go out in the waste stream; some of their isotopes eventually decay to plutonium. Spent fuel also has higher plutonium concentrations; larger quantities of plutonium therefore go out in the reprocessing waste stream since the percentage recovery remains about the same. Thus, while reprocessing might result in removal of all but 0.5 percent of the plutonium in a given load of spent fuel, the amount of plutonium in the high-level waste from a reactor system utilizing the recycle option will be equal, after about thirty years, to about 5 percent of that contained in ordinary spent uranium fuel which has simply been disposed of in unreprocessed form.[d] Moreover, waste from multiple recycled fuel will have higher concentrations of other long-lived transuranics than does spent uranium fuel. The net reduction in transuranic content of waste is less than a factor of 10 rather than by the factor of 200 which might appear to be implied by the 99.5 percent recovery rate in reprocessing. The consequent reduction in risk after permanent disposal achieved by reprocessing and recycle

---

[d]Quantities in this section are based on the *Final Generic Environmental Statement on the Use of Recycle Plutonium in Mixed Oxide Fuel in Light Water Cooled Reactors*, (GESMO), NUREG-002, August 1976.

must be balanced against the new risks entailed in using plutonium in the active fuel cycle and in processing large quantities of radioactive materials.

The high-level acidic liquid waste from reprocessing must be resolidified in a form safe for disposal. A calcine process for acidic waste is preferable in that it is possible to add material and make a borosilicate glass containing the wastes by melting. The resolidified high-level waste has a volume about eight times smaller than the original spent fuel. Unfortunately, some acidic high-level waste has been neutralized and some of it converted to a salt cake. Neutralized waste is difficult to treat successfully for long-term disposal.

The volume of solid waste (cladding hulls, process trash, and so forth) resulting directly from reprocessing and fabrication of mixed oxide fuel (reactor fuel containing uranium and plutonium oxides) would be rather large and would contain quantities of plutonium and other transuranics comparable to those in the much smaller volume of resolidified high-level waste. The total volume of transuranics-contaminated solid waste (TRU waste) would be comparable to the volume of the original spent fuel. In the past, TRU waste has been buried in commercial sites on state or federal land. New regulations should require much more careful disposition.

Liquid waste is predominantly water from processing operations with small concentrations of short-lived fission products. This waste would be discharged into holding ponds and perhaps subsequently diluted into large volumes of natural water bodies where the resulting concentrations would be much below harmful levels. Low-level solid waste, containing low concentrations of radioactivity decaying to innocuous levels in a few hundred years, would be buried at commercial facilities on state or federal land as at present.

Spent Fuel as Waste

The alternative to early reprocessing is either to store spent fuel until it is evident whether the plutonium contained in it will be needed for breeder reactors, or to dispose of it permanently. Light-water reactor fuel consists of sintered uranium dioxide fuel pellets assembled as fuel elements clad in sealed zircaloy tubes. Roughly 0.1 percent of the fuel elements develop small cladding leaks during their life in the reactor. Despite the high operating temperature and aqueous environment of the reactor, only small quantities of gaseous and volatile fission products migrate out of fuel elements. Routine holdup reduces emissions from the site to low levels (the exceptions are the noble gases xenon and krypton, of which 5,000–10,000 curies are released per gigawatt-year). After removal from the reactor, spent fuel assemblies are stored in water-cooled stainless-steel-lined pools. Zirconium-clad spent fuel has been stored successfully in this way for more than ten years at Savannah River.

Since reprocessing has always been assumed to be the next step in the fuel cycle, virtually no technical consideration has been given to longer-term secure storage of spent fuel or to the possibility of permanent disposal of spent fuel.

The issues which must be considered include the feasibility of interim storage for periods up to several decades, the feasibility of disposal in geological formations as compared to resolidified high-level waste, and the long-term leachability of spent fuel compared with treated high-level waste if the multiple barriers which isolate the waste are breached.

WASTE MANAGEMENT

Between its generation and permanent disposal, nuclear waste must be managed carefully to prevent accidental releases. Waste management problems are both technical and institutional, and they depend on other decisions taken regarding nuclear power. Waste management practices also have a bearing on the chances for successful long-term disposal. Past experience with nuclear waste from the military program and from an unsuccessful commercial venture has not been encouraging. Because of the importance of the cautions suggested by this experience and the illustration of the technical problems, a brief review of this history is useful.

The Present Waste Problem and Its Management

High-Level Waste at Federal Installations.   As a result of the nuclear weapons program beginning in World War II, the Atomic Energy Commission (AEC) generated some 205 million gallons of high-level radioactive liquid waste (as of June 30, 1974), primarily as a by-product of the production of plutonium in special reactors. Weapons work currently generates waste at the rate of about 7.5 million gallons annually. The total amount of waste has been reduced in volume by partial solidification to about 81 million gallons total in liquid and solidified form. It is stored at three sites: at Idaho Falls, Idaho (3 percent); at the Savannah River plant near Aiken, South Carolina (25 percent); and at the Hanford Reservation near Richland, Washington (72 percent). Solidification converts neutralized high-level liquid waste to salt cake at Richland and Savannah River and acidic waste to calcine (which is a dry granular material) at Idaho Falls. Salt cake is a neutralized high-level waste with most of the water removed. The process used to produce calcine consists of spraying acidic liquid waste into a bed of heated fluidized particles. The calcine form is preferable to salt cake because it allows for greater volume reduction, relative ease of handling, and because technology exists for further immobilization in glass or ceramic form. At present, neutralized waste, which is alkaline and contains considerable amounts of water, cannot be calcined or put into glass form except on a laboratory scale.

Experience with the storage of high-level liquid waste has not been encouraging. From 1958 to 1974, eighteen leaks, totaling 429,400 gallons, were detected at Richland. In 1973, a leak involving the loss of 115,000 gallons went forty-eight days before being noticed. Fortunately, the nature of the soil around the storage tanks has prevented the material from migrating significantly. Rainfall rates in the area are presently so low that the ground water penetrates only a few tens of feet before drying rapidly. The water table is far below, and there has been no contamination of wells (which are monitored) to date. The tanks at Richland are of the single-walled carbon steel construction which were considered acceptable if acidic wastes were neutralized. They are not all fitted with automatic leak detection systems.

There has been some degree of complacency at Richland because of the arid land and the deep water table. In the period 1956–58, 31 million gallons of radioactive waste (with removal of strontium-90 and cesium-137 but not plutonium-239) were intentionally dumped. In addition, liquid wastes from the Plutonium Finishing Plant have been released, until recently, directly to subsurface trenches not isolated from the surrounding dirt. One of these became of concern when it was discovered to contain as much as 100 kg of plutonium. Monitoring has shown no significant migration (no more than a few feet); however, on the time scale on which plutonium remains a hazard, the experience to date can give little indication of future risks.

Experience with liquid waste storage at Idaho Falls and at Savannah River has been much better. At Savannah River the tanks are also carbon steel but most have double containment and all have automatic leak detectors. The Idaho Falls tanks, which contain acidic waste, are stainless steel with double containment and leak detectors. There has been one very small (100 gallon) leak at Savannah River and none at Idaho Falls.

At Richland most of the waste should be solidified in the form of salt cake by about 1982. Not much more waste will be generated there because only one reactor is in operation. Two approaches to immobilizing the salt cake are being studied. The first calls for removal of the cake from the tanks for disposal or storage elsewhere, and the second would treat the salt cake and leave it in the tanks for the time being. Removal of the salt cake is costly and risky from a contamination standpoint. Hydraulic removal would dissolve self-sealed leaks and cause waste leakage. Mechanical mining could release material into the atmosphere unless some containment method were devised. Having removed the cake, one would still have to decide what to do with it. The second alternative would be to add material to the tanks and cover them with concrete, say, to prevent animal or human intrusion which could cause dispersal. Considering that the Richland site contains waste tanks, reactors, burial grounds, and other facilities spread over more than five square miles so badly contaminated that the land may never be cleaned up, immobilization of the waste in place might be the most practical alternative; it is also the least expensive. Whether these wastes really can be sealed off for hundreds of thousands of years and the land be removed from man's use for this period is far from clear. A study of the Richland problem is being undertaken by a National Academy of Sciences committee.

At Savannah River the alkaline waste evaporation to salt cake is continuing.

Conversion to an acid waste process for new waste seems economically un-desirable according to a duPont study. Neutralization of waste may be short-sighted, however, since it may make long-term disposal much more difficult: conversion to forms suitable for disposal is presently not possible on the scale necessary.

Commercially Generated High-Level Waste

The waste stored at the Nuclear Fuel Services (NFS) commercial reprocess-ing plant at West Valley, New York, presents problems similar to those at Rich-land and Savannah River. Closed since late 1971 because of increasing difficulty in meeting more stringent standards for releases of radioactivity, the facility has been abandoned because expansion was not economically justifiable. Re-sponsibility for the waste presently rests with the State of New York, though requests have been made to transfer this responsibility to the federal govern-ment. There are 600,000 gallons of neutralized waste (of which 30,000 gallons are sludge) stored in a single 750,000 gallon carbon steel tank and 12,000 gal-lons of acid thorium waste stored in a 15,000 gallon stainless steel tank. It is now recognized that the waste cannot remain in these tanks indefinitely. Be-cause of obstructions and limited tank access, it is unclear how much of the neutralized waste sludge can be removed by a combination of pumping, hy-draulic jetting, and chemical flushing. Pumping of the acid thorium waste poses no special problems.

The easiest thing to do after removal of the neutralized waste from the tank would be to convert it to salt cake. It is believed that a suitable facility could be built in six years to do the job. Another possible method is already used for permanent disposal of intermediate-level waste at Oak Ridge: a grout made up of waste, cement, and other additives is pumped down a well and injected into an underground shale formation that has been first hydrofractured. As the crack propagates, it is filled with grout which sets after a few hours and fixes the waste permanently in the shale. Provided the West Valley shale is suitable, such a process is thought to be possible in five to six years. However, serious questions about the long-term geological suitability of this process at the NFS site must be answered before it is attempted. Other suggestions are said to require at least ten to fifteen years to be carried out and involve undeveloped technolo-gies and research specific to the NFS problem. In a recent study for the Nuclear Regulatory Commission (NRC), the Battelle Pacific Northwest Laboratories estimated costs ranging up to several hundred million dollars for removing and disposing of the neutralized waste.

Solid Waste Management

Solid waste, including cladding hulls and other material contaminated with plutonium, has been traditionally disposed of by shallow land burial. The ade-quacy of this practice has recently been addressed by a committee of the Na-tional Academy of Sciences (June 1976) and in a report from the Government

Accounting Office (January 1976). We shall simply summarize some of the findings and recommendations of these studies.

There are six licensed commercial and five federal burial sites. Although some of these facilities have been in use for over thirty years, it is not yet known what hydrogeological characteristics and engineering features give the greatest assurance against migration of radioactivity. Since additional land burial sites will be required, particularly if reprocessing should occur, it is important to establish sound site-selection criteria. The Environmental Protection Agency (EPA) estimates that with reprocessing all six sites would be full by 1998 and two would reach capacity by 1985.

There is evidence that some of the sites are releasing radioactivity into the environment. In December 1974 a report issued by the Kentucky Department for Human Resources concluded that the disposal site at Moorhead, Kentucky, known as Maxey Flats, which was opened in 1963, was "leaking," but at levels that did not pose a health hazard. The difficulty is apparently associated with water in the trenches (this is a heavy rainfall area), and corrective measures are said to have been taken. In January 1976 a report from EPA's Office of Radiation Programs reported the migration of plutonium through some hundreds of meters in less than ten years. Analysis of this occurrence is as yet incomplete. An important but unresolved issue is whether the plutonium moved through the ground (which it is not supposed to do according to ion holdup theories) or was rather bodily transported by excess water. Some radioactivity has been detected in streams at the West Valley, New York, disposal site and a plan is being developed by the state and the licensee to stop water infiltration into some of the trenches. Levels have not reached those considered hazardous to health. There have also been problems at the Oak Ridge facility in Tennessee where the trenches have intercepted the water table and material leaches into a creek feeding into the Clinch River. Although the activity exceeded maximum permissible concentrations in the creek, it was down to 1 percent of these limits at the river. Remedial action has been undertaken. Though none of these examples has yet been proven to have appreciable health or environmental impacts, it is clear that past solid waste management practices have been less than adequate and that existing facilities must be carefully monitored.

Since 1974, solid wastes containing transuranic elements (TRU wastes) may no longer be sent to disposal sites, pending a ruling on whether such waste must be segregated and turned over to ERDA for ultimate disposal. ERDA is also planning to exhume and rebury solid TRU waste in existing sites where possible. This effort is not without hazard to employees and perhaps the public, since corroded containers may rupture and uncontained material may be scattered.

Management of Future Waste

At the end of 1975, roughly 1,200 tons of spent fuel were held at reactor sites and in the storage pools at reprocessing facilities. While this storage technology is well proven, relatively inexpensive, and easily expandable, it is at best a temporary solution. In the past, it has been assumed that spent fuel would move directly from these facilities to reprocessing plants as soon as there was enough accumulated to justify operating a large reprocessing plant. Indeed, this belief has become so widespread that many regard the failure of reprocessing plants to operate as the principal barrier to successful management of nuclear wastes and some have even suggested that reactors may have to shut down if reprocessing plants do not operate soon. While there is a need for better planning and more flexible storage arrangements for spent fuel, the choice of fuel cycle option does not itself impose essential constraints on waste management or disposal or on use of nuclear power.

The alternative decisions on reprocessing lead to two different waste management scenarios. If reprocessing occurs, the radioactive elements in spent fuel would be separated into new physical categories (high-level, TRU or transuranic, low-level, gaseous effluents, and so forth). Research programs have focused almost exclusively on the management of these new categories of waste. The timetable for waste management, if this is done, is constrained by regulation to a decade. Wastes, in their various forms, must be dealt with or transferred to ERDA for disposal within ten years of generation or five years after reprocessing. This timetable puts a tight deadline on the resolution of disposal problems.

The other alternative, which is recommended in this study, is that spent fuel be stored retrievably until it is decided whether or not the plutonium contained in it will be used. If this no-reprocessing option is chosen, spent fuel must be stored securely for two or three decades. While efforts to provide for this are only now getting underway, there appear to be no insurmountable technical barriers to such storage. Some spent fuel has been kept for more than ten years without appreciable degradation, and present fuels are of high integrity in aqueous environments. An important step is to provide facilities which offer additional protective barriers against accidents and sabotage. Spent fuel could be stored retrievably in the geological repositories presently planned. It should be possible to do this in such a way as to constitute permanent disposal should the plutonium component not be needed.

A delay in reprocessing offers an advantage in keeping the management problem simple and allowing more time to develop and demonstrate technologies for permanent disposal of reprocessing and spent fuel wastes. It will be some years before ERDA disposal plans can be implemented and even longer before assessments of the efficacy of particular methodologies can be made with confidence.

PERMANENT DISPOSAL OF WASTE

Public attention has focused most critically on the risks of permanent disposal of nuclear wastes. If there is even a small risk that large quantities of radiotoxic waste might some day find their way into the human environment with consequences which threaten the future of civilization, most would question the right of contemporary society to create this hazard. Assessments of the risks entailed in permanent disposal of nuclear wastes are thus important to policy choices. This section reviews suggested disposal mechanisms, discusses the features crucial to successful disposal, and considers the risks involved.

Modes of Permanent Disposal

Many methods have been suggested for the permanent disposal of nuclear waste. These proposals generally fall into the following categories:

- Ocean and Sea Bed Disposal
  - Deep ocean dumping
  - Sea bed disposal in stable ocean floor
  - Ocean trench disposal
- Disposal in Geological Formations on Land
  - Liquid injection into hydrofractured rock
  - Disposal in very deep holes
  - Burial in deep geological formations
- Ice Sheet Disposal
- Extraterrestrial Disposal

Dumping waste in the deep ocean would be feasible only if the integrity of the container could be guaranteed. Given the corrosiveness of sea water, this is not possible. Since the residence time of the deep water is only of the order of 100-1,000 years, long-lived wastes from broken containers would mix with the upper, biologically active, layers and thus enter the food chain. Emplacement of canisters in the thick clay materials of the deep ocean is subject to many uncertainties about possible thermal currents and sediment behavior as well as risks associated with extended sea transport and emplacement in water of 5 kilometers depth. The recent British Royal Commission report recommends a major study of disposal of vitrified waste in 1,000 meter deep holes drilled into stable areas of the bed beneath the deep ocean (4,000 meters). The advantages of the scheme include security from inadvertent retrieval and a very stable geological environment. Containers placed in an ocean trench might actually be drawn down into the earth by the descending plate. However, the rate of this process is very slow, about 2 to 4 centimeters per year. Until a great deal more is known about the phenomena involved and the associated risks, this method cannot be exploited.

Disposal into Antarctic ice, even if it were not forbidden by the Antarctic Treaty of 1959, is not attractive for a number of reasons. The ice sheets are not very old, most of the ice being deposited less than 100,000 years ago. The sheets are also known to undergo surges on the order of once every 10,000 years during which periods there is relatively rapid movement. The heat given off by waste containers might even trigger such surges. The possibility of emplacement near ridges in the underlying bedrock where the ice is much older has been mentioned, but far too little is known about the presence of water or other factors. To this must be added the hazards of emplacement in an environment in which all human activity is difficult.

Extraterrestrial disposal of solidified high-level waste or spent fuel would be very expensive, something in excess of $2,000/kg of payload. Such expense would be possible only for the smaller volume of long-lived elements, such as iodine-129 and the actinides, assuming that they could be separated from other waste at low enough cost and risk. An additional and important source of risk is that of malfunction of the space vehicles carrying waste, with re-entry of waste packages into the atmosphere. Considering the large numbers of launches required each year, the chance of failure and its consequences are too great to make this an attractive approach at this time.

If retrievability is unimportant, disposal in deep holes is a promising alternative. Holes might be drilled to depths of 10–20 kilometers and wastes placed at the bottom. The hole would then be filled. Eventually, the waste would melt and become part of the rock structure deep in the earth's mantle. The drawback to this method is the expense of drilling to such depths holes of large enough diameter; the technology necessary may be beyond present capabilities. Since it would be good to have a backup to present plans for repositories in accessible geological formations, the deep hole concept is worthy of further study.

There are a number of accessible deep geological formations which could be used as repositories. These include bedded salt, granite deposits, some shales, carbonates, and metamorphic rock such as that in the Precambrian shield. Among these, bedded salt has long been considered ideal; however, other environments may meet the criteria for repositories equally well or perhaps better. These criteria include geological integrity and stability and a confirmed absence of appreciable contact with ground water in adjoining strata. In what follows we will discuss waste disposal in the context of salt, with the understanding that most of our conclusions would apply to other geological environments satisfying the above criteria.

Disposal in Bedded Salt

The fact that salt formations have survived is ample testimony to their relative lack of contact with ground water. The technology of salt mining is highly developed, and salt beds have several favorable characteristics. First, salt beds are widespread in the United States, underlying more than 100,000 square miles. Second, salt has a high thermal conductivity which facilitates the transport of heat away from the waste canisters. Third, it has high structural strength and in experiments withstood both heat and radiation effects well. Fourth, salt flows plastically under pressure, and any cracks that form have a tendency to heal. A deficiency of salt is that it has local inclusions of brine, about 0.5 percent by weight, which would tend to migrate toward hot waste canisters and would eventually corrode them. Brine, however, does not provide a mechanism for transport of radioactivity away from the site. Realistically, such transport could only occur through ground water intrusion; intrusion could only occur through massive faulting or through other diversion of aquifers in surrounding formations. The possibility of this clearly depends on the site chosen, but there is evidence that sites can be found which have negligible risk of intrusion. Oak Ridge scientists, for example, studying a prospective site for weapons waste near Carlsbad, New Mexico, have concluded that if all the water flowing in aquifers above the salt beds there were diverted into them, 40,000 years would be required to dissolve the salt surrounding the waste and expose it. Diversion of these aquifers would require a 350 meter vertical fault; the probability for this to occur is judged, on the basis of geological evidence, to be so small as to be negligible even on geological time scales. It is clearly important that sites be studied carefully to be certain that human activities have not and will not compromise the integrity of the repository. A primary threat comes from past exploratory drilling for oil. Frequently, no record was made of the location of these exploratory holes.

Past efforts to select repositories have not been successful. After a considerable amount of research beginning in 1963 on the properties of salt mines, the AEC in 1970 announced plans to construct a full-scale pilot disposal plant near Lyons, Kansas. State geologists and citizens' groups argued that there were serious problems with the site. Concerns centered on bore holes into the mine, and the fact that 1,800 feet away another salt deposit was going to be solution-mined with a danger of breakthrough. This episode contributed to the general lack of confidence in how the waste disposal program was being managed by the government.

Disposal in Geological Repositories

Disposal will probably ultimately involve sequestering wastes deep in geological formations chosen for their stability and isolation from ground water. The soundness of this choice will be the ultimate insurance against release. Thus detailed analysis of prospective areas for peculiarities of geology, lithology, and hydrology will be vital before final choices are made. Because of the long times involved and the inevitable degradation of engineered facilities, long-term disposal schemes should be planned to involve very little dependence on nongeological factors.

As noted earlier, the waste problem has two characteristic time constants. For perhaps 700 years the radioactivity present is very high and is dominated by short-lived isotopes. Thereafter, the longer-lived isotopes, including the actinide elements, provide most of the radioactivity with little change in activity for tens of thousands of years. Because of the large quantities of radioactivity present during the first few hundred years, the potential consequences of repository failure will be higher than when radioactive decay has appreciably reduced the quantities present. During the early years after disposal, however, engineered barriers in waste repositories, such as stainless steel canisters, will provide additional barriers to leakage. Beyond this time, after man-made barriers have been degraded, risks depend upon natural physical factors. These include the presence or absence of water, the leachability of the waste, the permeability of the environment, and ion exchange delay.

While it is possible to avoid formations which have appreciable contact with ground water, even the presence of such water does not imply an efficient mechanism for transport. Estimates have been made of the time required for a repository to leak; water moves through aquifers slowly, with water at the 500 meters depth of a repository requiring 100 to 1,000 years to reach the surface. However, the rate at which waste ions in the water would move is much less, for most of the radioactive isotopes involved, because of ion exchange. Waste ions would exchange with ions in rock or other material and are thus held up until the reverse exchange occurs. This phenomenon cuts the transport rate by a factor of about one hundred for strontium-90, by one thousand for cesium-137, ten thousand for americium and plutonium, and one hundred thousand for radium. Thus, even if water were present, it might take well over ten thousand years for waste to reach the surface. This time delay could probably be increased considerably by optimal siting. These figures are derived from laboratory experiments and are consistent with theoretical estimates. These estimates could be improved by using more sophisticated models, and modeling could also be used as a way to evaluate and select potential repositories. Measurements should be made in aquifers in the neighborhood of any proposed disposal sites to establish the actual water flow rates as well as actual ion exchange holdup factors.

There are two important waste products which apparently do not participate in ion exchange and are thus not held up relative to ground water movement. These are the long-lived and biologically active isotopes technetium-99 and iodine-129. These elements are a relatively insignificant fraction of initial fission product activity; however, they dominate the long-term risk from fission products and their level of radioactivity approaches that of the transuranic elements over many thousands of years. Ion holdup therefore cannot be relied upon to provide protection from important sources of risk after the first thousand or so years.

The results discussed here, it should be noted, all require contact with ground water. Such contact has not occurred in salt formations, for example, for millions of years. One would thus ordinarily expect that, even if wastes were exposed to water already in the repository, it would remain immobilized in the formation. Some local dissolution and migration is possible, particularly in the case of salt deposits which have some brine included in the salt itself, but migration is very unlikely to extend beyond the repository.

The primary barriers against release of radioactivity are thus those due to the geological character of the repository itself and to the lack of transport mechanisms away from the site. Given proper site selection, the risk of release appears to have limited dependence on the precise character of waste material emplaced after the first few hundred years. This observation is central to the debate over the value of reprocessing and partitioning of plutonium or other elements. Proponents of reprocessing have contended that the resolidification of high-level waste as a glass after reprocessing is necessary for permanent disposal.

Under proposed plans, glass cylinders of high-level waste from reprocessing would be placed in stainless steel canisters for disposal in a geological repository. Transuranic waste would be accorded similar disposal. While the volume occupied by fission products and actinides in the high-level waste would be about a factor of eight less than in spent fuel, the heat generated by the smaller volume would be about the same. Moreover, the volume of TRU waste would be comparable to that of spent fuel. Thus the volume and heat load requirements on repositories for reprocessing and spent fuel disposal options would be comparable. In salt deposits, at least, heat load appears to govern the amount of waste that can be placed in a given area.

While it may be possible to immobilize high-level reprocessing waste in a glass with initial low leachability, it is not known how this material ages. Some believe that it crumbles in as little as one hundred years under radiation and heat. Since stainless steel canisters are expected to provide some protection initially, it is not the initial leachability which is important but that characterizing the contained material if and when the canister leaks. It is not known how long it takes for canisters to deteriorate—estimates range from tens to hundreds of years for disposal in salt. The time to failure could be extended by additional measures. Moreover, leachability is relevant only if the repository fails and ground water is encountered.

While little is known about the long-term leachability of spent fuel, experience with its service in reactors and storage in water suggests that it initially has high integrity. Disposal techniques for spent fuel have been little studied. The GESMO comparison of disposal options assumes that spent fuel assemblies would be placed in canisters similar to those anticipated for high-level reprocessing wastes. The fuel element cladding would thus provide an additional barrier to release, compared with high-level waste from reprocessing.

Partitioning of long-lived isotopes has been much discussed. These elements include technetium-99, iodine-129, as well as the actinide elements plutonium, americium, and curium. Presently contemplated reprocessing plants would separate only plutonium. Many additional separative stages involving major capital investments and technical developments would be required to remove all of the long-lived elements. Even if this were accomplished, the question remains of what to do with the isotopes removed. Partitioning does not automatically reduce long-term hazards, and the operations involved increase intermediate-term hazards by potentially exposing current generations.

It has been suggested that after partitioning, reactors or special facilities could be used to transmute the dangerous long-lived elements into more tractable forms. While this is an appealing technical possibility, it may prove difficult, expensive, and involve larger risks than geological disposal. The rate at which the isotopes involved may be burned up in present reactor types is very low. Some of the actinides might require tens to hundreds of years of continuous treatment to be burned up completely. Dedicated facilities might reduce this time, but only at considerable expense and some risk to safety in operating systems processing large quantities of these elements. Thus, while partitioning is much discussed, there is little realistic prospect that its use would simplify the waste problem.

Risk Assessment

Given the present status of waste disposal technology, only theoretical calculations can be done to evaluate the risk of failures of geological repositories. Both natural phenomena and human intrusion or disruption must be considered. To the water transport mechanisms, discussed earlier, must be added the possibilities of earthquakes, meteorite impact, and explosions of gases trapped or generated in the repository. All of these are very unlikely. For example, the probability of a meteorite large enough to disrupt a repository 500 meters underground has been estimated on the basis of past experience as being of the order of 10^{-14} per square kilometer per year. The earthquake hazard can be reduced to negligible levels by judicious choice of sites. Repositories could be situated in material, such as salt beds or the Precambrian shield, which would suffer little damage even if an earthquake did occur.

The need for isolating waste for such long periods has often caused much concern in the public mind about accidental or even purposeful intrusion of disposal sites by humans. Following a period of a disruption of society and a disappearance of all markings, one might imagine inadvertent drilling into the burial ground by a future civilization looking for minerals. However, it is hard to believe that a society technically capable of drilling to such depths (even if they were unlucky enough to hit one of the rather widely dispersed canisters) would at the same time not be able to recognize a waste depository and take appropriate action. Even if the danger were not recognized, the consequences of inadvertent encounter with aged waste could hardly have catastrophic consequences. There appears to be essentially no chance that saboteurs or terrorists could penetrate a sealed salt bed 500 meters below ground; if they ever did, it is hard to imagine what they would do there. It is our conclusion, therefore, that human intrusion is not a problem.

Eliminating the highly improbable and inconsequential risks, one is still left with some uncertainty regarding the long-term hazards. It is difficult to predict in advance all the mechanisms by which wastes might leak from repositories thousands of years in the future. Moreover, the assessment of the risk involved depends not only on the probability that some release might occur but also upon the prospective consequences of such a release. However, only if the potential consequences of releases could be truly catastrophic to future generations or civilizations should the waste problem be singled out as having unique significance in making decisions concerning nuclear power.

A number of hypothetical situations have been considered by those concerned by this question. Most calculations show that the results of gradual leaks from repositories would be an increase in background radiation, locally, of less than a factor of two.[e] Others, notably Cohen,[f] have computed the deaths which might result under various conditions of release. Many of these calculations are unrealistic, assuming, for example, that all the waste is mixed uniformly in all the rivers in the United States. While these calculations might appear to give the worst consequences of a waste accident, there are numerous assumptions required about biological effects of material ingested or inhaled, food chain concentration factors, and long-term effects of waste material in the environment. Many of these effects are neglected in available studies or are highly uncertain. Despite these deficiencies, such calculations do suggest strongly that even large releases would not threaten future civilizations with large numbers of deaths.

A sense of how large a failure is required to have noticeable consequences is provided by a simple but somewhat more realistic calculation. Suppose that a river runs near a waste repository and that waste is somehow transported by ground water from a repository to the river after a thousand or more years in the repository. If the river has a flow of 10^9 cubic meters per year (about that of the Hudson River) and dilution is uniform, maximum permissible levels (for ingestion) in the river will be exceeded (using current standards) if more than 1 percent of the waste in a repository (containing the waste from 1,000 reactor-years of power generation) reaches the river each year. Even drinking this water would result in only a small health risk to individuals. During the first one hundred years, the repository must be much more secure with less than one part in ten million reaching a river of this size, if the river water is to be used for drinking without special treatment. The slow rate of transport by ground water alone would appear to guarantee against this possibility even without ion holdup attenuation or engineered barriers within the repository.

Models leading to larger effects can, of course, be constructed. One might imagine, for example, that river water would be used to water crops without anyone noticing that it was radioactive, thus initiating food chain concentration of certain isotopes and perhaps increasing the risk of inhalation by farmers. Higher concentrations would also occur if waste seeped into a lake, though

[e]References to these calculations may be found in "Environmental Survey of the Reprocessing and Waste Management Portions of the LWR Fuel Cycle," NUREG-0116, U.S. NRC, October 1976.
[f]B. Cohen, *Reviews of Modern Physics,* to be published.

monitoring and cessation of water use could reduce the consequences. Since water eventually finds its way to oceans, the long-term risk of repository failure could involve a general increase in background radiation levels, and in the presence of actinides, iodine-129, and technetium-99, in the environment or in particular food products. None of these effects, however, can be regarded as catastrophic. Moreover, the chance that the rate of release could reach the levels above is very small. Virtually all of the waste would have to be dissolved and swept out of the repository over a short period of time and transported very efficiently into the watershed of a river in order to achieve the concentrations discussed above. Anticipating such possibilities in the choice of a repository is the best way to avoid them.

While calculations of this type obviously cannot anticipate all possibilities, it appears to be very difficult to postulate realistic failure mechanisms which would imply consequences of catastrophic, or even serious, proportions to future generations. Additional studies of such failure mechanisms are not only essential to testing this conclusion but would also assist in selecting repositories and insuring against failure. We believe that the probabilities and consequences of failures of waste repositories are very small for properly chosen repositories and that this will be substantiated by the detailed site-specific studies necessary to make such choices. Since the present rate of generation of waste will no more than double the magnitude of our current waste problem over the next five to ten years, there is time to conduct such studies without increasing the ultimate risk appreciably over what it already might be. However, our interest in the waste problem extends beyond optimism about technical feasibility. There are important questions about the abilities of institutions and individuals to recognize and fulfill responsibilities for implementing successful management and disposal practices.

INSTITUTIONAL FACTORS AND WASTE PROGRAMS

It is clear from the history of waste management, and from the rigorous demands placed on future waste management and disposal efforts, that institutional factors will play an essential role. At present, responsibilities for waste are split several ways. Up to the point of permanent disposal, most wastes are managed by the commercial entities which generated them (utilities) or last processed them (reprocessing plants), under regulation by the Nuclear Regulatory Commission. For solid wastes, which in the past included transuranics, disposal is made in burial sites on state or federal land by commercial firms. For high-level waste (or the equivalent, spent fuel), responsibility for disposal has long been seen as belonging to the federal government. Transuranic waste now appears destined to become, and should become, a federal responsibility; repository plans now include provision for disposal of this material. Mill tailings are also under commercial control and appear to have been virtually unregulated in the past. Notice was given in 1976 that mining and milling operations will be the subject of a forthcoming NRC generic environmental statement.

There has been little incentive, on the part of the government or commercial interests, to deal with waste on more than a short-term basis. In the military program, short-term goals have all too often displaced long-term federal responsibilities, with the results described in the preceding sections. In commercial power, the natural tendency of short-term interests to dominate has been aggravated by a bifurcation of responsibility between private concerns and the government and by slow regulatory development. While successful waste management is strongly dependent upon an integrated series of actions, the present arrangement, in which early management steps are a commercial responsibility and disposal a federal one, creates incentives for each sector to try to transfer to the other responsibility for the intermediate steps. It is essential to establish the necessary integration through a comprehensive regulatory framework.

The uncertainty over when, and now whether, the fuel cycle would be extended to include reprocessing and recycle has not helped this situation. Government and industry have assumed that these stages would be added when economic factors permitted. Thus utilities have been content to store spent fuel at reactor sites, and virtually all government research and development has centered on the recycle option. In part because of the delay in recycle, internal pressure to develop disposal technology has been low. It was assumed that reprocessing and recycle fuel companies, as new commercial entities, would have an economic incentive (under regulations) to prepare waste for disposal by the government.

The Nuclear Fuel Services experience reveals the difficulty of instituting a comprehensive waste management policy by regulatory action alone. Pressure to develop a commercial reprocessing industry in the mid-1960s led to adoption of the shortsighted practices used for some weapons wastes, including burial of transuranics and neutralization of liquid high-level waste, despite the fact that waste management practices at NFS came under AEC regulatory control. Recognition that interim management decisions must be keyed to disposal requirements has been slow to develop.

The NRC has recently begun a program to develop a comprehensive regulatory approach to waste management and disposal. This effort includes definition of waste goals and performance criteria—technical, economic, social, and environmental standards against which programs and strategies can be measured. Generic criteria are particularly useful in those areas in which technology is not well developed and where it must be developed by ERDA. An example is that of selection and acceptance criteria for geological repository sites. The NRC intends that specific regulations will be made as part of a larger framework which specifically confronts tradeoffs between short- and long-term goals. The NRC will also define criteria for acceptable risk for proposed waste management, handling, transportation, storage, and disposal methods and will develop an independent capability for performing risk assessments for various waste systems, including geological disposal. Most of these efforts are scheduled for completion in the next two years. Provision of the technical and economic resources necessary to carry out this ambitious program is essential.

Several specific problems will receive early NRC attention. These include assessments of present land burial practices, a generic environmental statement on management of uranium mill tailings, analysis of requirements for decontamination and decommissioning of nuclear facilities, and regulations governing the disposition of the 600,000 gallons of high-level waste at the Nuclear Fuel Services site. The NRC may also rule on methods for handling waste contaminated by transuranics; however, there is considerable industry opposition to the level of contamination (10 nanocuries of transuranics per gram of material) proposed as the basis for categorizing this waste. Finally, there is uncertainty about the extent to which NRC's responsibility extends to ERDA's efforts to deal with weapons wastes. These wastes involve many of the same public issues which arise in connection with commercial wastes, and efforts to deal with them may be prototypic of techniques applicable to commercial wastes. For these reasons and since public concern about commercial wastes may be traced, in part, to experiences with weapons wastes, NRC involvement in the weapons waste problem could contribute to improved public confidence in the resolution of both waste problems.

The ERDA nuclear waste program has recently been unified by integration of previously separate efforts dealing with military and commercial wastes. Budget authorizations for waste management and disposal activities have been increased by a factor of nearly ten in the past two years. Plans to develop a salt disposal site for weapons waste near Carlsbad, New Mexico, are well advanced, with site evaluation having begun in 1972. The focus on geological repositories for commercial waste was made central to the ERDA program in 1975. (The earlier program, involving the Lyons, Kansas, facility had been abandoned in favor of the retrievable surface storage facility concept.) Present plans call for up to six geological repositories with the first two in salt deposits. The earliest wastes could be deposited in 1985. Provision will apparently be made for disposal of spent fuel as well as solidified waste (high-level and TRU) from reprocessing; moreover, repositories must be licensed by the NRC under the regulations it is now developing. This has removed an ambiguity as to NRC's role in federal waste demonstration projects. A generic environmental impact statement on the waste management program will be issued in 1977, to be followed by specific impact statements for particular repository sites. ERDA now plans to demonstrate all components of waste management technology by 1978. Geological studies will be conducted in thirty-six states in 1977 to determine potential sites.

The timetable and philosophy of the new government effort are oriented toward early demonstration. However, the concept of meaningful demonstrations of disposal technology is elusive since the disposal problem has a characteristic time of tens of thousands of years. This implies that confidence in repositories will rest on conservative measures, geological investigations, and analytical projections. Public confidence will be enhanced by efforts to subject this analysis to the scrutiny of the wider technical community, perhaps through commissioned reviews by independent groups.

Waste Problems Abroad

All countries using nuclear power must deal with waste management and disposal problems. There are differences in attitudes toward these problems, however. England, for example, has distrusted permanent geological disposal and preferred management in engineered storage facilities, believing that provision for the next few hundred years is an adequate undertaking for the present. This attitude, which has been criticized by the recent British Royal Commission report (September 1976), is in contrast to that in West Germany, where disposal in salt deposits is a well-advanced technology. Italian scientists are considering disposal in thick clay beds; Belgium contemplates mixing the liquid wastes from its past reprocessing operations with bitumen and putting them into engineered storage; Japan is storing spent fuel and, if reprocessing proceeds, will store liquid wastes in tanks; France stores reprocessing wastes in tanks and has an experimental solidification program but faces a decision on whether to proceed with permanent disposal or engineered storage.

A reasonable conclusion from such a review is that other countries, with the possible exception of West Germany, have not really settled on long-term solutions to waste problems despite research programs and plans. One difference between present U.S. attitudes toward wastes and those abroad concerns the role reprocessing plays. Most foreign efforts still regard reprocessing as essential either for future breeder programs or as part of the waste management process. Germany and Japan, for example, have legal requirements for reprocessing as part of the treatment of wastes. Thus, research efforts focus exclusively on the handling and disposal of waste products after reprocessing and do not deal with spent fuel as a waste. This is in contrast to the recent shift in U.S. emphasis and to the conclusions of our analysis, both of which suggest that early reprocessing is not essential and may in fact complicate waste management more than it helps in reducing permanent disposal risks.

Waste management and disposal will present particularly difficult problems for countries with small nuclear deployments or limited technical capabilities. Comprehensive waste management and disposal facilities, many of which are characterized by large economies-of-scale, would require investments not immediately justified by small nuclear deployments, without major economic penalty. Moreover, many countries will lack geological formations appropriate to permanent disposal. Since the long-term hazards of nuclear wastes easily extend beyond borders, the success of waste management and disposal practices in one country is a matter of concern to its neighbors.

Nations with major waste programs and suitable geological storage or disposal sites can assist these countries by handling their nuclear waste for them. The cost of this assistance would be small and would not significantly change the magnitude of waste problems already faced by countries with large nuclear programs. If the United States, for example, accepted all of the rest of the world's commercial nuclear wastes, it would only increase its waste problem by a factor of two. The wastes from less developed countries with their small nuclear programs would add only 10 or 20 percent to U.S. wastes.

The coupling of reprocessing and waste disposal also complicates the nuclear proliferation problem. Nations that develop independent facilities for this purpose will have the capability to produce plutonium. Even if fully safeguarded, these facilities would provide all but the final stage in a weapons program. Management of the wastes as spent fuel, for either permanent or retrievable storage in supplier nations with suitable geological formations, would eliminate this proliferation problem and at the same time reduce the hazards and costs associated with the management of separated wastes. Our confidence in the ability to handle nuclear wastes in the long term is sufficiently high that we believe the United States should be willing to take back spent nuclear fuel for permanent or retrievable storage if this will help control proliferation.

SUMMARY AND CONCLUSIONS

Disposal of waste in stable geological formations which are isolated from appreciable contact with ground water appears to provide adequate assurance against the escape of consequential amounts of radioactivity even over long periods of time. While it is evidently impossible to demonstrate the security of such repositories on the relevant time scales experimentally, present geological evidence and analytical techniques applied to site selection and evaluation can provide the necessary confidence in the system. Reasonable scenarios of repository failure lead to the conclusion that they would not lead to catastrophic or even substantial consequences. Since the security of permanent disposal would depend primarily on repository integrity, there is little difference in the risk in storing spent fuel as compared with waste from reprocessing which has had plutonium partially removed. Even if partitioning of other actinides and of long-lived radionuclides could be accomplished economically, this would substitute intermediate-term management problems, with potentially high risks, for a relatively insignificant change in the long-term disposal risk.

Problems in the management of wastes are potentially of more serious concern than the purely technical aspects of disposal. The past history of management practices and the problems left by them—nonretrievable wastes, leaky tanks, buried solid wastes containing transuranics—suggests the need for greater care in future decisions. Some problems, notably emissions from tailings, releases of krypton-85 from reactors, and improper treatment of material contaminated with transuranics, can and should be resolved by appropriate regulatory decisions.

The magnitude and complexity of future management problems depend in part on fuel cycle decisions. Reprocessing would complicate waste management by broadening the spectrum and potential difficulty of problems. A decision to defer reprocessing relieves pressure on the development of more complex waste regulations and practices and relieves some of the urgency of seeking an immediate solution of the permanent disposal problem. Without reprocessing and recycle, there would be time for a more orderly and assuredly more error-free process in both the management and disposal aspects of waste.

If plutonium is to be held in spent fuel for several decades as part of the insurance against very high energy costs provided by breeder reactors, storage facilities must be provided for some spent fuel. Spent fuel could be stored retrievably in the same geological formations that would eventually be used for permanent storage. In any case, flexible temporary storage facilities should be provided until more secure ones are available.

The institutional basis for managing and disposing of wastes is important. Past experience has revealed the defects of split or misplaced responsibilities, of improper tradeoffs between short- and long-range goals, and particularly of the failure to demonstrate the adequacy of proposed solutions to the wider technical community and to the public. We are encouraged that new NRC and ERDA programs are moving toward a more unified treatment of waste management and disposal problems.

In addition to the measures already proposed, we would urge that plans for the disposal of military waste be integrated, insofar as possible, with research and development and regulatory programs dealing with commercial waste problems. If it should ultimately prove necessary to reprocess and recycle plutonium for breeder reactors, the experience gained with reprocessed military waste would be useful in developing technical capabilities in the handling of the waste problems associated with this more complex fuel cycle.

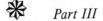 *Part III*

Nuclear Proliferation and Terrorism

Chapter 9

Nuclear Power and Proliferation of Nuclear Weapons

The consequence of nuclear power that dominates all others is the attendant increase in the number of countries that will have access to the materials and technology for nuclear weapons. At the beginning of 1976, fifteen non-nuclear weapon countries had operational power reactors, each generating as a by-product enough plutonium for a few to a score or more bombs annually. When reactors now on order or under construction are completed in the mid-1980s, the number of countries will double and amounts of plutonium will increase rapidly. In addition to the danger of peacetime "proliferation" of announced and demonstrated weapon capabilities, there will be a spread of potential capabilities that could be activated in a crisis. Having a potential weapon capability is, however, not the same thing as "proliferating"—having actual weapons—as has been demonstrated by Japan, West Germany, Sweden, Canada, East Germany, Spain, Switzerland, and others.

The link between nuclear power and weapon capability has been recognized since the end of World War II, as have the potential for nuclear proliferation and its attendant political and security risks. A network of agreements and safeguards has accompanied the international development of nuclear power. These arrangements, the principal ones being the Treaty on the Non-Proliferation of Nuclear Weapons (NPT) and the International Atomic Energy Agency (IAEA), provide incentives and reassurance to states willing to forego nuclear weapons status without thereby foregoing the benefits of nuclear power. In this sense, nonproliferation arrangements constitute a "bargain" between weapon states and nonweapon states, the latter renouncing nuclear explosives and accepting safeguards in exchange for assurances of access to nuclear materials and technology, and as to pursuit of nuclear disarmament.

A worldwide nonproliferation regime prevails as a result. It is imperfect, but it is a uniquely constructive element in the current international system and a principal asset in efforts to make manageable the difficult task of living with nuclear power.

The contrast with the conventional arms trade is instructive. It has repeatedly been suggested that a suppliers' club could put a brake on conventional arms sales. However, the fact is that the United States and the Soviet Union use arms sales for political influence. Other suppliers have little more common ground. The developing countries have made it clear that they value access to diverse arms suppliers and view controls as contrary to their security. Supplier co-ordination faces resistance from both suppliers and purchasers.

For nuclear weapons, the difference is striking. There is widespread, though not universal, consensus that the spread of nuclear weapons is dangerous. As early as 1963, the Limited Test Ban, with broad acceptance, put partial limits on nuclear weapons testing. The Soviet Union and its allies cooperate in the IAEA and the London Suppliers Group (an informal group of nuclear exporting states which meet regularly to coordinate their export and safeguards policies in support of NPT objectives and provisions). The NPT provides a consumer-endorsed basis for this supplier coordination; it obligates each party to the NPT to supply nuclear materials or equipment only under safeguards. IAEA safe-guards apply to almost all nuclear power activities of nonNPT parties as well as to those of NPT parties.

Uncertainties, strains, and limits exist for the NPT and other measures to contain proliferation. Yet the common understanding and accord are remark-able, an invaluable base on which to build in seeking to deal with nuclear pro-liferation, which, in the last analysis, is a political problem.

At the present moment, the nonproliferation venture is under test. The am-biguous Indian "peaceful" explosion of 1974 was a reminder of the nuclear weapons potential of a growing number of states. At international tension points, rival insecure or ambitious states confront each other. If one or more states overtly "go nuclear" in the near future, there is risk of a chain reaction of nuclear commitments by imitators or rivals. At this crucial time, nuclear power is spreading and bringing with it growing basic nuclear capabilities. In the United States and in other countries with nuclear power programs, tech-nical decisions are impending whether to proceed with plutonium reprocess-ing and recycle and with breeder development, which could constitute a water-shed for proliferation.

NPT nonweapon parties currently undertake not to make or acquire nuclear explosives; they do *not* undertake not to acquire the prerequisites for doing so—materials and basic technology. The current active spread of nuclear power in-dustries thus could be attended by widespread availability of separated pluto-nium or highly enriched uranium usable in weapons and the fuel cycle facilities for producing, separating, and handling them. Many nations would be only a step or two from a nuclear weapons competence if they decided to go that route. The nonproliferation regime which currently prevails, and indeed international stability, could be subjected to new and perhaps intolerable strains.

The challenge now is to create arrangements controlling the sensitive nuclear fuel cycle. This is a difficult but by no means impossible venture. The dangers attendant on power systems fueled with and generating increasing quantities of a bomb material, plutonium, can imperil smaller and less powerful states no less than they do the United States and the Soviet Union. Perceptions of the risks have led to the acceptance of the NPT by most developing and industrial states—as in their interest, not as an imposition by the nuclear weapon states. A diplomatic effort embracing weapon and nonweapon states, nuclear suppliers and consumers, advanced and developing states thus has precedent for controlling the fuel cycle internationally. The original "bargain" would be extended to safe and reliable supply of nuclear fuel and related services. Supplier coordination would become easier to work out among suppliers, and easier to accept for consumers.

Interest in national facilities for separating plutonium has been sustained by expectations about the value of plutonium as a fuel. There are increasing doubts, discussed elsewhere in this study, about the soundness of these expectations. If advanced countries recycle plutonium in LWRs, treat breeders as inevitable and urgent, and develop complete national fuel cycles, forebearance by emerging industrial nations cannot be expected. Our study (see Chapters 11 and 12) has persuaded us that plutonium recycle in LWRs has little economic value to large nuclear programs and even less to smaller programs and that the breeder is not as urgent as official plans now project. A realistic appraisal of economic merit, safety, and environmental and proliferation risks would afford time for working out international arrangements governing the fuel cycle.

For nuclear fuel cycle arrangements, as earlier for NPT adherence, the most basic responsibilities of nations are at issue—their security and economic well-being. The importance of coupling assurance of fuel supply with control of the fuel cycle has already been identified. No less crucial will be the security environment in which nations make their nuclear decisions. South Korea's deferral of a reprocessing plant reflects South Korean dependence on U.S. commitments, troop deployments, and arms. The absence of concern over reprocessing or enrichment capabilities in West Germany and Japan is a consequence of their internal stability and even more their security relationship with the United States. Avoidance of hostilities in the Middle East or South Asia will be a prime determinant of nuclear weapon policies there.

Containing nuclear proliferation and controlling the fuel cycle involve a range of political and technical complexities. We shall examine:

1. The prospects for nuclear proliferation; the pressures for and against it.

2. What can be done to reduce its likelihood.

3. The bearing of U.S. nuclear power policies, domestic and international, on the control of proliferation.

NUCLEAR NONPROLIFERATION: ORIGINS AND STATUS

"Nuclear proliferation" can be defined with reference to the Treaty on the Non-Proliferation of Nuclear Weapons (NPT). Nonnuclear weapon states parties to the treaty are "not to manufacture or otherwise acquire nuclear weapons or other nuclear explosive devices" (Article II). Proliferation in this context is not a matter of degree, but of whether a state has or has not built or otherwise obtained at least one nuclear explosive device (whether or not tested).

Five states (United States, Soviet Union, United Kingdom, France, and China) had demonstrated their nuclear weapon status by 1964. The NPT was negotiated in the mid-1960s; was opened for signature July 1, 1968; and entered into force March 5, 1970. India, not a party to the NPT, detonated a nuclear device on May 18, 1974; though asserted to be for peaceful purposes, the act constitutes proliferation under the NPT, which applies to all nuclear explosive devices whether for military or "peaceful" purposes. Manufacture or acquisition of a device without testing could be a less conspicuous or even clandestine form of proliferation. There have been allegations, for example, that Israel, not a party to the NPT, may have some assembled or unassembled nuclear weapons that have never been tested.

The spirit of the NPT is endorsed by the majority of nations. There are ninety-eight parties (including major nonweapon states such as West Germany, Italy, Japan, and the industrial states of Eastern Europe) and thirteen additional signatories that have not yet ratified (including Egypt, Indonesia, Switzerland, and Turkey). France has not signed, but has declared it will act as though it were a party and has done so in the IAEA, in the London Suppliers Group, and in its bilateral activities. China has not signed and has questioned the objectives of the NPT but does not in practice encourage or aid proliferation. Thirty-nine states are not signatories; noteworthy are Argentina and Brazil, India and Pakistan, Israel, South Africa, Spain, and North Korea. Isolation, insecurity, local rivalry, ambition, and regional or global status pretensions: one or more of such motivations is evident in each of these cases.

Virtually all the nuclear facilities in the territory of nonNPT parties are subject to nonweapon pledges and safeguards as a condition for obtaining fuel or equipment from suppliers. Exceptions are the small Indian, Argentine, and Spanish reprocessing facilities; the large Indian and Israeli research reactors; the fifth and sixth Indian power reactors under construction; and the South African uranium enrichment plant under development. All of these are in countries where other materials or facilities are under IAEA or bilateral safeguards. De facto nonproliferation and safeguards thus supplement formal NPT adherence.

Support of nonproliferation nevertheless remains incomplete and conditional. States facing potential regional conflicts will look at rival defense and nuclear programs for signs that nuclear renunciation may no longer be tolerable. Insecure states will ask themselves whether they can continue to rely on their conventional military strength, or on particular great power protectors, or on the good offices of the international community. States in relatively secure situations may assess the status they might gain from weapons, particularly if nuclear weapons retain their roles as symbols of world power and prestige. Desperation or ambition may move national leaders to threaten or exercise the nuclear option. Some states that fit these categories are not parties to the NPT; for parties, the NPT withdrawal clause is available if events jeopardize "supreme national interest."

For the United States, the root of concern over proliferation does not differ basically from that of other countries—fear of enhanced international instability and of the risk of nuclear war. Indeed, the risks may be greater for other countries, since U.S. nuclear and other military strength make it relatively secure and it is not exposed to adjacent rival nuclear powers. In unstable countries, there is the added threat of seizure and possible use of nuclear weapons or materials by dissident or rival groups, in time of military revolt or civil war.

A special U.S. concern is escalation in the event of a nuclear clash between regional states to which the United States and the Soviets may have commitments. A broadening nuclear war, drawing the United States and the Soviet Union into confrontation, is one conceivable way in which deterrence could fail and a U.S.-Soviet nuclear exchange could be triggered.

The opposition of the United States to nuclear proliferation goes back to the period just after World War II, when the Baruch plan called for internationalization of nuclear energy and nuclear weapons. The proposal was not adopted, and the United States attempted to maintain its monopoly by strict protection of its "secret" of the atomic bomb; by termination of nuclear cooperation, peaceful as well as military, even with close allies; and by a governmental monopoly of all domestic atomic energy activities.

The sharp change from monopoly and secrecy that was ushered in by the Eisenhower "Atoms for Peace" proposal of December 1953, and the revised Atomic Energy Act of 1954, was a recognition of changed circumstances, not an abandonment of concern with proliferation.

Secrecy and autarky could no longer preserve monopoly. The Soviet Union and United Kingdom had tested weapons. Civilian programs were launched in other countries. If the United States wished to shape the direction in which such programs moved, it was concluded that cooperation rather than isolation should be the choice.

The Atomic Energy Act of 1954 and later revisions authorized international cooperation, bilaterally and through the new IAEA. But cooperation was constrained; fuels and equipment could be exported only under a formal agreement submitted in advance to the Congress. The recipient would have to pledge not to use the assistance for weapons or any other military purpose, and to accept inspectors from the United States, the IAEA, or (later in the case of members of the European Community) Euratom.

While the abandonment of autarky and secrecy was timely, the motivations of Atoms for Peace were complex and led to excesses which accentuated proliferation dilemmas. Unrealistically optimistic projections of the development schedule and economic benefits fed vigorous promotion of nuclear power in the United States and abroad, including a number of developing countries. This promotional spirit was officially encouraged by the Eisenhower Administration for foreign policy reasons. U.S. nuclear weapon programs, particularly nuclear testing, and overseas weapon deployments were encountering widespread opposition. Fears of nuclear bombs were intensified by nuclear fallout controversies and by movements to "ban the bomb." Atoms for Peace was promoted to emphasize the beneficent side of the atom and to give other countries a stake in nuclear energy. Nuclear science and technology, with the exception of bomb design and uranium enrichment, were declassified and disseminated. Training of foreign scientists and engineers was fostered in the United States. Thus, while nonmilitary commitments and safeguards were firmly promoted, basic, and at times advanced, nuclear capabilities increased rapidly as a result of Atoms for Peace. The dual approach persisted through the 1960s and early 1970s. On the one hand, the Limited Test Ban and then the NPT were negotiated, while at the same time nuclear exports and nuclear activities abroad were actively encouraged. Only in the past two years have the contradictions in nuclear policy and the need to resolve them been officially recognized.

Nations are not the only conceivable agents of proliferation. Terrorists, criminal groups, and revolutionary or military groups seeking power may see nuclear weapons as instruments. Currently, nuclear weapon stockpiles are in the possession of relatively stable and well-organized governments. Seizure by terrorists or dissidents should be avoidable provided alertness and efficiency of custodial forces are maintained. Weapons deployed on foreign territory in U.S. custody should similarly be defensible against nongovernmental groups. If national proliferation occurs, however, weapons might be less well protected against terrorists, and in case of internal conflict or attempted coup might be seized by rival groups, brandished as instruments of political pressure, even used. A different route would be seizure of plutonium or highly enriched uranium, actual (or alleged) fabrication into crude nuclear explosive devices, and threatened use for terrorist or other purposes. Such "subnational proliferation," and measures to deal with it, are examined in Chapter 10.

REQUIREMENTS FOR NUCLEAR
WEAPONS MANUFACTURE

The requirements for making nuclear weapons can be reduced to three:

1. Uranium.
2. Trained personnel and information to build or operate facilities, and design and fabricate weapons.
3. Facilities to produce highly enriched uranium or plutonium from natural uranium.

Uranium

Although uranium is widely dispersed in the earth's crust, major deposits have so far been found in only a few countries. Accelerated exploration may turn up extensive deposits in additional countries, but at present the principal sources (outside the Communist states) are the United States, Canada, Australia, South Africa, and France, with its associated states of Gabon and Niger and the Central African Republic. Smaller amounts or very low-grade deposits are more widespread. Chapter 2 discusses current and prospective supplies in detail.

Uranium has rarely been available commercially on an unsafeguarded basis. Among nonNPT parties, Argentina, India, and Spain have exploited small deposits but have obtained most of their reactor fuel as safeguarded imports. South Africa is the only major supplier not an NPT party. But, while uranium is unevenly distributed and most countries are dependent on imports, relatively little is needed for a small weapons program.

South Africa, which is not an NPT party, occupies a special place, in view of its uranium resources and its potential role as supplier of enriched uranium fuel. At present South Africa requires safeguards on exports. Continuation of this policy is important for nonproliferation.

Technological Base

If uranium is available, trained personnel are the key requirement for weapons. The requisite people are at the disposal of an increasing number of countries outside the advanced industrial states; India, Pakistan, Taiwan, Israel, South Africa, Spain, Brazil, and Argentina are examples. Research and power programs provide opportunity for training and experience.

Most of the technology necessary for a weapon program is readily available to scientists and engineers. Although detailed information on isotope separation and bomb design are closely held, the general concepts are well known. Successful application in a weapon program is a major undertaking, requiring qualified personnel, substantial funds, and time. Such a program nevertheless is certainly within the capabilities of developed countries, and of many developing countries, if they put in sufficient resources.

Reactor and chemical separation technology has been widely diffused since 1954. Enrichment technology has in contrast been tightly held. Seven countries (United States, Soviet Union, United Kingdom, France, China, and West Germany, and The Netherlands in cooperation with the United Kingdom) have working enrichment facilities. South Africa has a plant under construction, and several countries (Japan, Australia, and Italy among others) have done substantial studies and preparatory work. Hitherto the large size and high capital cost of gaseous diffusion plants, coupled with the ready availability at attractive prices of U.S. slightly enriched uranium fuel, have been important disincentives. The UK-FRG-Netherlands centrifuge project has restrictions on technology diffusion, in coordination with the United States. If laser isotope separation proves successful, it could be highly sensitive for proliferation since this method requires smaller scale and investment, is easily adaptable to high enrichment, and appears likely to be very difficult to safeguard in operation. Control of information can delay dissemination for a time.

Another kind of technology transfer is "know-how." It is one thing to obtain scientific principles or plant design; it is another thing to make them work. Training military or civilian defense personnel in nuclear science and engineering, or giving scientists and engineers of another country the opportunity to participate in construction and operation of a multinational reprocessing or enrichment facility, can add markedly to the quality and confidence of an indigenous national capability.

Facilities

Weapons can be made from either highly enriched uranium or plutonium. A very large investment has so far been required for uranium enrichment. Current enrichment facilities, located in the five nuclear weapon states, are gaseous diffusion plants, which inherently require large capital investments and consume large amounts of electricity.

New methods on a much smaller scale are in various stages of development. At a cost of tens of millions of dollars, any of a dozen advanced nonnuclear states could build a centrifuge plant adequate for a small weapons program; other countries would need to import the main equipment. Laser separation, which may prove feasible in the 1980s, would in principle require less investment and electric power than a centrifuge plant. Most countries would have to import the critical equipment for the foreseeable future. The aerodynamic nozzle process, which is being developed by West Germany and by South Africa, and which is a component of the West German-Brazilian agreement, is more analogous to gaseous diffusion in its requirements for investment and electric power.

Plutonium is generated in a nuclear reactor and separated from discharged fuel elements in a chemical reprocessing plant. Various reactors are possible:

1. Special plutonium production reactors fueled with natural uranium, maximized for output of weapons-grade plutonium.
2. Large natural uranium research reactors yielding enough plutonium annually for one or a few weapons (the Israeli Dimona reactor and the Indian Trombay research reactor are examples).
3. The light-water reactor (LWR), which is fueled with slightly enriched uranium and designed for power production. These reactors, which comprise the vast majority of the power reactors now being operated or constructed in the world, in sizes of 1,000 MWe produce some 200–300 kgs of plutonium over a year's operation. When operated efficiently for power purposes, these reactors produce plutonium with a high concentration of plutonium-240 which, while less desirable than low concentration plutonium-240, is usable for fabrication of weapons.
4. The Canadian CANDU power reactor, which burns natural uranium and is moderated with heavy water, is continuously fueled. It can be operated without undue economic penalty at low fuel irradiation to produce plutonium with a low concentration of plutonium-240 which is more suitable for reliable weapons. It is in operation or under construction in Argentina and India as well as Canada.

The process to separate plutonium from spent fuel elements is much simpler than uranium enrichment and is well known. Although a plant to handle large amounts of heavily irradiated material from a power program, subject to stringent environmental protection standards and quality control, would cost hundreds of millions of dollars, a small plant for weapons could be built at a cost of tens of millions of dollars. Preventing such facilities is an urgent task.

ALTERNATIVE ROUTES TO NUCLEAR WEAPONS MANUFACTURE

Manufacture to nuclear weapons could be based on several alternative approaches:[a]

1. Unsafeguarded facilities, constructed specifically for production of enriched uranium or plutonium for weapons, or for research.
2. Clandestine diversion of weapons material from nuclear fuel cycle facilities ostensibly devoted to peaceful purposes and under safeguards.
3. Overt use of weapon materials produced in nuclear fuel cycle facilities in

[a]Acquisition of nuclear explosive devices by any other means (e.g., seizure or purchase or gift) is forbidden by the NPT. These alternatives raise different problems and are not treated here, though seizure is treated briefly in the chapter on terrorism.

the absence of, after withdrawal from, or in disregard of international agreements to the contrary.

4. Fabrication of nuclear weapons labeled "peaceful nuclear explosive devices" (PNEs), using materials from the previous categories.

Special Facilities

Special facilities, including plutonium production reactors, have been the route of choice by the five nuclear weapon states. India (and Israel, if it has weapons) have not used power reactors as a source of plutonium, but have produced it as a by-product of large natural uranium research reactors.

States determined to acquire weapons openly are likely to find that construction of specialized facilities for weapon materials will be cheaper and quicker than relying on power reactors. To produce plutonium for weapons, there is little if any extra cost, and possibly an advantage, in spending $50-$100 million on a small plutonium production reactor rather than interfering with a power reactor that costs between $500-$1,000 million and is used as a base-load generating plant. Early removal of fuel from LWRs for the sake of a low concentration of plutonium-240 would be uneconomical for power generation, since less efficient use would be made of the fuel. Moreover, the LWR has to be shut down for fuel removal. The Canadian CANDU reactor, however, can produce weapons-preferable plutonium with far less cost of interference with operations, since spent fuel (which is natural uranium) is discharged without shutdown.

A plutonium separation facility serving a nuclear power industry could be used for separating plutonium for weapons without comparable operating penalty. If such a facility were not available, or its use precluded by agreement and safeguards, a small special-purpose facility would not be expensive to construct.

Hitherto the plutonium route to weapons has appeared more plausible than the uranium enrichment route because of the high investment and power requirements for gaseous diffusion plants. Centrifuge and laser separation techniques could change this if they prove to be of manageable expense and difficulty for advanced countries. Enrichment to around 90 percent would be necessary, as contrasted with about 3 percent uranium-235 enrichment for LWR power reactor fuel. Small dedicated plants for uranium enrichment for weapons could be preferred to diversion from safeguarded nuclear fuel facilities.

Clandestine Diversion of Fissionable Material

Clandestine diversion of fissionable material is not a very plausible path to weapons for a state. Slightly enriched uranium, the fuel of LWRs, is not usable for weapons. Until plutonium separation facilities are available, plutonium in discharged fuel is not accessible. Even with separation capabilities, the plutonium from LWRs, though usable for weapon purposes, would be inferior unless its production was allowed to interfere with normal power operations.

IAEA safeguards are by no means certain to detect diversion, but they pose a substantial risk of detection, over time if not immediately. A weapons program based on safeguarded facilities would, when discovered, precipitate a major international incident and lead to suspension of fuel deliveries or other retaliation by suppliers.

Abrogation of Commitments or Safeguards Agreements

NPT or bilateral agreements could be abrogated. This course, though followed by suspension of fuel supply or political countermoves, nevertheless will be an option for countries where one or more large reactors have generated a substantial amount of plutonium. In adverse circumstances, such as initiation of a weapons program by a hostile neighbor or heightened international isolation, a nation would have the stark option of abrogation of existing restraints and utilization of the plutonium for weapons. If uranium fuel had been stockpiled, indigenous sources of uranium and enrichment facilities developed, or plutonium recycle perfected, these measures would reduce the impact on nuclear power generation from sanctions which might follow.

Peaceful Nuclear Explosives

Promotion of the alleged economic benefits of so-called peaceful nuclear explosives (PNEs), initially led by the now disillusioned United States, and more recently by the Soviet Union, has complicated nonproliferation efforts. PNEs do not differ from nuclear weapons, save in packaging and function. They can therefore serve as a convenient cover for a weapon program, and provide a state with a de facto weapon capability. They are banned by the NPT equally with nuclear weapons (and subject to the same yield limitations under the recently negotiated U.S.-USSR Threshold Test Ban Treaty). Some states have expressed concern that they not be excluded from the presumed benefits, if not by their own capabilities then at least through PNE services the nuclear powers might perform on contract. A few (India and Brazil) argue that PNEs should not be treated as weapons, and that the ban on them is unwarranted and discriminatory and a reason for withholding accession to the NPT. It would help if the Soviet Union would abandon its interest in PNEs and join the United States in renouncing them. The discriminatory element in asking nonweapon states to forego them would then be ended, and arguments against them on safety and economic grounds would be reinforced. Until such total renunciation is possible, a fifteen-year moratorium on PNEs, with reassessment after that period, might be an acceptable course for all.

CURRENT NUCLEAR PROGRAMS AND
PROLIFERATION CAPABILITIES

There are currently no explicit aspirants for nuclear weapon status. Major non-NPT parties like India and Brazil stress the peaceful character of their nuclear programs and explain nonadherence to the treaty in terms of its banning of peaceful nuclear explosives for nonnuclear weapon parties and other asserted discriminatory aspects of the NPT. Israel takes the official position, when asked about indications that it may have obtained nuclear weapons, that it will not be the first to introduce nuclear weapons into the Middle East. Official circles in Turkey, Pakistan, Taiwan, and South Korea have talked of circumstances in which recourse to nuclear weapons would become imperative—e.g., weakening of U.S. security commitments, refusal of the United States to supply conventional arms. For the immediate future, these hints tell more about interest in U.S. arms and security backing than about firm military plans. Individual national leaders have on occasion spoken grandly of the eventual acquisition of nuclear weapons as their growing nation takes its rightful place on the international scene. All of these clues tell something about future risks of proliferation, but there is no professed weapons program outside the five nuclear powers.

Capabilities are more important in the long run than present intentions. Table 9-1 exhibits for the relevant countries the nuclear programs that seem fairly firmly established. The categories within which states are grouped are arbitrary, but highlight some political and economic factors which are important for nuclear power spread, or for ultimate national decisions on nuclear weapon programs or on adherence to the NPT. Key states for nonproliferation strategy are those in the categories reflecting isolation or insecurity or ambition, and their rivals. If proliferation gathered momentum, domestic and international pressure on major nonnuclear states such as Japan and West Germany to reconsider their nonproliferation commitments would build up.

The tabulation sets forth nuclear power plants operating, under construction, or on order. Even these data have substantial uncertainty: there may be slippage in completion, some planned plants may be deferred or canceled, etc. These data are better for present purposes, however, than more general forecasts of energy demand and nuclear power capacity. Even though large increases in total nuclear power capacity may be delayed, the plants in Table 9-1 appear likely to be built and operated.

From a proliferation perspective, they establish a nuclear base of experience and competence that is relevant to a nuclear weapons potential. Table 9-1 thus provides a rough forecast of the number of countries with an early option. Other developing countries are unlikely to undertake nuclear power programs in the next decade or so because of the size and capital cost of nuclear reactors, the lack of an integrated grid to exploit such large generating units, and uncertainties as to reliability and capacity utilization factors. Some may buy a single plant, however, for experience and a general nuclear base. Some of the more advanced and affluent developing countries (Brazil, Iran, Spain) are considering, and very well may build, quite extensive nuclear power capacity and related fuel cycle facilities. While this possible future expansion does not show up in Table 9-1, the proliferation situation would not be markedly changed.

These power reactors are not fueled with, nor do they produce, material directly usable for bomb fabrication. Small-scale chemical reprocessing facilities to separate out plutonium have been built locally in a number of countries. Those in Spain, Argentina, India, and presumably Israel are not subject to safeguards except as imported fuel elements may come to be authorized for local reprocessing. The South African enrichment plant would process local uranium and would only be subject to IAEA or other safeguards if South Africa voluntarily offered to accept them since South Africa is not an NPT party or a member of the London Suppliers Group. South Africa was, however, a founding member of the IAEA and has acted responsibly as a uranium supplier.[b] Brazil has contracted with West Germany for uranium enrichment and plutonium separation facilities; the plants and their output would be safeguarded under the covering West German-Brazilian agreement and associated Brazilian-West German-IAEA safeguards agreement, which contain exceptionally stringent provisions.

The presence of plutonium separation or uranium enrichment facilities will shorten the lead time between the decision to acquire weapons and the achievement by perhaps five to ten years in some of the less developed countries, but probably not more than one to three years in countries with substantial nuclear power programs. A country with several thousand megawatts capacity in 1985, such as Taiwan, might produce a nuclear device within a year of a decision to do so—if at that time it had its own enrichment or reprocessing capacity. If on the other hand such "sensitive" facilities were not available at the outset, the lead time would lengthen by no more than a couple of years, especially if local personnel had prior experience. Such experience could come from involvement in the design, construction, and operation of a multinational reprocessing facility. If the country chose to test an early device for political or confidence purposes, the detonation might be the first public notice of a program, since most or all of the preparatory activities could be carried out clandestinely.

Subnational proliferation presumes the existence of separated plutonium or

[b]"Sales of uranium by South African producers to nonnuclear weapon states are made subject to the safeguard provisions of the International Atomic Energy Agency or the equivalent safeguards arrangements . . . in order to ensure that the nuclear material will be used for peaceful purposes only." Statement by representative of South African Atomic Energy Board, February 11, 1976, p. 127, *Nuclear Fuels Policy*, Report of the Atlantic Council's Nuclear Fuels Policy Working Group.

Table 9-1. Nuclear Power and Nuclear Proliferation Capabilities

(1) Country	(2) NPT Status	(3) Nuclear Power Reactors (MWe) Operational	(4) Under Construction or on Order	(5) Forecast Capacity Mid-1980s	(6) Annual Bomb Equivalent
Nuclear Weapons States					
United States	Party	37,600	170,800	208,400	4,168
USSR	Party	4,600	9,800	14,400	288
United Kingdom	Party	5,300	6,500	11,800	236
China	Nonparty	?	?	?	?
France	Nonparty	2,800	18,500	21,300	426
Insecure States					
Israel	Nonparty	0	(a)	?	?
South Africa	Nonparty	0	(a)	?	?
South Korea	Party	0	1,800	1,800	36
Taiwan	Party	0	4,900	4,900	98
Yugoslavia	Party	0	600	600	12
Status-Seeking States					
Brazil	Nonparty	0	3,200	3,200	64
India	Nonparty	600	1,100	1,700	34
Iran	Party	0	4,200	4,200	84
Spain	Nonparty	1,100	7,200	8,300	166
Rivals to States in Preceding Categories					
Argentina	Nonparty	300	600	900	18
Egypt	Signatory	0	(a)	?	?
North Korea	Nonparty	0	?	?	?
Pakistan	Nonparty	100	?	100	2
Politically Constrained Major States					
Czechoslovakia	Party	100	1,800	1,900	38
East Germany	Party	900	1,800	2,700	54

Italy	Party	500	4,700	5,200	104
Japan	Party	5,100	10,400	15,500	310
Poland	Party	0	400	400	8
West Germany	Party	3,300	20,000	23,300	466
Other Developed Countries					
Australia	Party	0	?	?	?
Austria	Party	0	700	700	14
Belgium	Party	1,600	3,800	5,400	108
Bulgaria	Party	900	900	1,800	36
Canada	Party	2,500	9,300	11,800	236
Finland	Party	0	2,200	2,200	44
Hungary	Party	0	1,800	1,800	36
Luxembourg	Party	0	1,300	1,300	26
Netherlands	Party	500	0	500	10
Romania	Party	0	400	400	8
Sweden	Party	3,200	5,200	8,400	168
Switzerland	Signatory	1,000	4,800	5,800	116
Other Developing Countries					
Chile	Nonparty	0	0	0	0
Greece	Party	0	0	0	0
Indonesia	Signatory	0	0	0	0
Mexico	Party	0	1,300	1,300	26
Philippines	Party	0	1,400	1,400	28
Thailand	Party	0	?	?	?
Turkey	Signatory	0	0	0	0

Notes:

Columns (3), (4), (5) derived from "World List of Nuclear Power Plants, December 31, 1975," *Nuclear News*, February 1976. Many countries' programs have undergone change in the past year, but the overall picture is reflected here.

Column (6), "Annual Bomb Equivalent," is a rough approximation which assumes each 1,000 MWe in Column (5) is operated in such a way as to produce 200 kg/plutonium annually as a by-product, and 10 kg/plutonium are required for one bomb.

(a) indicates that a country has formally announced that it has one or more nuclear plants on order or under construction since the data for this table were compiled.

highly enriched uranium as objects of seizure. As separation facilities and enrichment plants come into operation, stringent security will be required at the facilities, during transport, and during storage.

ENCOURAGING DECISIONS TO FOREGO WEAPONS

Fabricating nuclear explosives is within the capacity of an increasing number of countries. Nevertheless, proliferation is not inevitable because it is possible. It is remarkable that, since the Chinese detonation in 1964, no other nation has avowedly launched a nuclear weapons program. The widespread consensus for nonproliferation is a fact to which Indian professions of peaceful objectives for their first nuclear explosion are an ironic tribute. The political considerations behind this consensus, the strains to which it is subject, and ways in which national nonproliferation decisions can be encouraged will now be examined. These national decisions are critical, whether in the form of adherence to the NPT, or that of acceptance of ad hoc commitments and safeguards against use for explosives as a condition of supply of imported nuclear materials and facilities. To deprecate these commitments because they can be repudiated is to devalue the most basic proliferation asset.

What stands between a nonweapon state and a nuclear weapon program may be a lack of full capability, in addition to or instead of a political decision to forego weapons. The focus in the NPT is on political commitments; the treaty does not limit capabilities short of direct weapons and PNE activities. Even these latter limitations are linked to exchange of political commitments, in a consensual document, as part of a bargain among nuclear "haves" and "have nots."

It is now timely, in addition to efforts to broaden acceptance of nonproliferation, to consider extending the NPT "bargain" to enhance stabilizing limitations on national nuclear capabilities. Such limitations in nonweapon states, if freely accepted and if not impairing efficient and reliable use of nuclear power, can strengthen the nonproliferation regime and its political base. The absence of key fuel cycle facilities under national control, for example, can lengthen the lead time between a decision to acquire weapons and the acquisition. The absence of such threshold facilities in the territories of neighbors or rivals may reassure uneasy nations and help avoid reciprocal weapons efforts. This lead time can be a deterrent to launching a program which may be conspicuous and be accompanied at its outset by tensions and perhaps risks and vulnerabilities. A national decision to launch a weapon program could be sufficiently close that possession or nonpossession of a uranium enrichment or plutonium separation facility would tip the balance. Finally, the greater the lead time, the greater the likelihood that a weapon decision would be aborted in response to internal or international changes or pressures. Thus measures to limit certain capabilities can reinforce nonproliferation; such possibilities will be examined after review of political approaches.

To see what might encourage nations to renounce weapons, and to buttress those which have already done so, it is useful to look both at why countries might turn to nuclear weapons and why they might not.

The principal barrier to further proliferation is probably the perception by an increasing number of states that nuclear weapons will not advance their security and that a regime in which they and their neighbors refrain from weapons can make them more secure. The principal thrust of nonproliferation policy in that case would be to do the things which will make national decisions to forego weapons demonstrably in the interest of the security and welfare of the countries which so decide.

In a turbulent world in which nuclear weapons have a symbolism not only of danger and catastrophe but also of international power and status, countries approach so fundamental a decision from a variety of perspectives. For many, any significant nuclear program is infeasible. For others, the burden of a military nuclear program would be heavy and the opportunity to forego it, by mutual renunciation and mutual openness, a welcome escape. Other states are exposed and isolated (South Korea, Taiwan, Israel) or located in areas of confrontation or conflict (Japan, India, Pakistan, Iran, Central Europe). For some of them, security assurances from a major ally may be a continuing prerequisite to foregoing nuclear weapons.

Foregoing weapons is not irrevocable. Indeed, NPT signature or deferring purchase of a plutonium separation plant can be moves to buy time while technical capabilities and physical facilities enabling quick weapons development are built up. But if the duality of peaceful programs has this inherent risk, it also strengthens NPT commitments: the capabilities built up by peaceful programs are a "hedge," an assurance that countries which try the nonproliferation option will not be permanently disadvantaged if it fails. Time will be bought for broadening adherence to the NPT and easing tensions and settling disputes, so that the weapons option will not have to be taken.

There will be states for whom the discrimination of the NPT will be difficult to accept indefinitely. The trends in the developing international system, and the recognition accorded their economic and political weight, will affect their continuing ability to accept a nonnuclear international role.

The prospects for nonproliferation will be imperiled if the thirty-year taboo is broken and nuclear weapons are used. Not only would a firebreak against escalation of violence, maintained since 1945, be breached, but also, by making recourse to nuclear weapons in regional situations less unlikely, indefinite renunciation of weapons by nations resolved to assure their own defense and to play a major regional role will become more difficult.

Promotion of doctrines that emphasize nuclear weapons for limited objectives, including political leverage, stimulates the belief that such weapons are acceptable and necessary tools of modern warfare and diplomacy.

For nations which out of a sense of peril have spoken of or hinted at recourse to nuclear weapons—South Korea, Pakistan, Taiwan, Israel—the option is a matter of desperation, not exuberance or optimism. Nuclear weapons bring dubious benefits. For key states, the decision is a security one, with political overtones. Yet against a nonnuclear rival, nuclear weapons may offer no advantage if a state is militarily superior. India might be wise to consider whether a weapons capability could fail to be followed by a Pakistani capability. Pakistan, now badly outmatched, would then be able, even if at great cost, to hurt India in a way it now cannot. For a state militarily inferior or faced with erosion of its superiority, nuclear weapons may simply trigger its rivals' acquisition and lead to no net advantage or even further deterioration. Israel must weigh these considerations in its changing balance with the Arab world.

Against a major nuclear power, being nonnuclear may offer better protection than a nuclear arsenal, which risks being an invitation to preemptive attack. Association with a nuclear protector and ally has been a surer way of neutralizing any nuclear threat.

For precariously situated rulers of instable states, weapons pose a different kind of risk. In the event of a military revolt or civil war, nuclear weapons or fissionable materials may be tempting targets for seizure and use by dissident or rival groups.

One weighty cost of nuclear weapons may be alienation of a crucial ally and guarantor. South Korea, Israel, Taiwan, even Japan and West Germany, must weigh this risk to their relations with the United States. Presumably this was central in South Korea's decision in 1976 to defer acquisition of a plutonium separation plant in the face of U.S. opposition.

There are also more general considerations that weigh against nuclear weapons. There is widespread sentiment against nuclear weapons. U.S.-Soviet agreement and often concerted action on nonproliferation, while at times resented as superpower dictation, has undercut ideological rationales for going nuclear on anti-American or anti-Communist grounds. While North-South antagonisms may weaken the force of this unusual U.S.-Soviet convergence, Third World consensus has been more anti-nuclear weapons than anti-superpowers. In this atmosphere the prestige or status gains of nuclear weapon gestures have been ambiguous, as in the case of the Indian explosion. It is easier to keep open the option, or maintain a figleaf such as the PNE cover, than to launch an overt weapon program.

Not only have there been no new acknowledged nuclear weapon states since 1964, but the growth of an international political acceptance of nonproliferation goals has been striking. In contrast with the bitterness of much of the NPT negotiations, the five-year NPT Review Conference in 1973, while occasionally contentious on narrow issues, did not shake or even question the foundations of the NPT. Criticisms were not of the NPT but of failure to implement or strengthen it, with the nuclear powers asking for more effective safeguards and security measures and the nonnuclears seeking more aid and disarmament. Major holdouts like Brazil and India feel compelled to profess lack of weapons interest and to excuse themselves for nonsignature on ground of political or PNE discrimination. Other holdouts have special regional problems (Argentina and Pakistan vis-à-vis Brazil and India; Israel and Egypt) or political problems (South Africa, Spain).

The United States and the Soviet Union have kept in step, in the negotiation of the NPT, at the IAEA, and in the London Suppliers Group. France, while not an NPT signatory, has acted "as if a signatory." It has been cooperative both in the IAEA and in the suppliers' group. China does not assist or encourage nuclear weapons acquisition by other countries. It has not changed its stated opposition to the NPT as a fraud, designed to maintain the superpowers' monopoly, but it no longer maintains, as in the early 1960s, that proliferation is not only inevitable but even desirable on the part of Socialist or developing nations.

Most basic of all, the isolated and insecure states (Israel, South Korea, Taiwan) with growing nuclear potential, as well as the advanced industrial states with demonstrated potential (West Germany, Italy, Japan, some of the East European states) have had the nuclear umbrella and security guarantee of an ally or protector as a more powerful and less risky defense than a nascent nuclear force.

There are a number of situations in which weapons may appear tempting. There are corresponding security actions which can facilitate retention of nonnuclear status. The most important are external to the NPT system. They relate to the security and prestige of key nations (whether parties or nonparties to the NPT). Less decisive but also important will be how the NPT works, and whether the "bargain" appears indeed to give the nonnuclear weapon states what is promised, not only in security from nuclear war, but in arms control and in access to the benefits of nuclear power.

The United States has acquired a central place in the security calculations of many countries. As a result, the United States has influence on proliferation postures. Future U.S. policies will affect the immediacy of interest of other countries in the nuclear option.

For a number of NPT parties, the U.S. security guarantee is essential to nonnuclear commitments. This is true of South Korea and Taiwan; it is true also of NATO, Japan, and Australia. While U.S. responsibility is burdensome, the U.S. stake in forestalling proliferation is great. The United States accrues leverage on allied decisions, as in the recent South Korean deferral of purchase of a reprocessing facility.

U.S. ties are also critical for isolated or threatened states which are not NPT parties (Spain, Israel), enabling them to remain nonweapon states with the option of future NPT accession. The Spanish case, in which political isolation and the desire for a prominent international role are more controlling than security, is an illustration of the role of political action in nonproliferation strategy. Spain, which has a major nuclear power program, has built its own small reprocessing facility, so that separating plutonium is within its capability.

It is not an NPT party, though most of its facilities are under IAEA safeguards as a condition of import of fuel and equipment. Yet, the United States has not withheld nuclear cooperation and exports. In June 1976, a large Senate majority approved a U.S.-Spanish Treaty of Cooperation and Friendship under which the United States would obtain certain limited base rights, and in return would among other things provide $450 million in Export-Import Bank credits mainly for U.S. nuclear power exports to Spain. The effect was calculated to minimize chances of a Spanish weapon decision. The purpose of the treaty goes beyond U.S. bases; the objective is to help Spain move toward a modern democratic state away from a military regime and into Atlantic institutions. The treaty thus counters political isolation, which is the decisive force for a Spanish weapon option, rather than the capability for reprocessing. The alternative strategy, use of leverage to impose constraints on Spanish nuclear power programs, could gain some time, but at the cost of heightened isolation.

In other cases (Pakistan, South Africa, Egypt, Yugoslavia), U.S. influence is less direct and unique. U.S. action must be concerted with that of other countries, in regional settlements, easing of tensions, and resolution of disputes in the Middle East, Southern Africa, and South Asia.

Nuclear-free zones are a method to achieve regional mutuality of weapon renunciation. They can add elements not in the NPT, like absence of nuclear deployment by outside powers, and provide a framework for regional cooperation in nuclear facilities. But a background of regional cooperation is indispensable to a nuclear-free zone, and one absent in areas of tension.

Such approaches are not the key to decisions by nations such as India or Brazil, where status rather than security is the crux. Drawing them and others such as Iran and Spain into a responsible international role not only makes good sense but provides status. Trying to force, rather than persuade, such countries to renounce a full fuel cycle is likely to be futile.

A dilemma arises out of India's "peaceful" detonation and the indications that Israel may have bombs. Treating them as nuclear weapon states would be unwise. The way back to renunciation of weapons ambitions by these states should not be blocked.

ARMS CONTROL

Article VI of the NPT commits the parties to pursue negotiations toward cessation of the nuclear arms race and nuclear disarmament. Article VI is part of the NPT bargain; it is one of the promises of the major nuclear weapon states in return for the commitments of the nonnuclear states. It is directed at the discriminatory aspect of the NPT structure—the division of the world into nuclear weapon "haves" and "have nots"—and holds out hope for long-term equalization in a peaceful international order moving toward deep disarmament. Critics and hold-outs of the NPT have pointed to the limited and halting progress of the nuclear powers on SALT, on a comprehensive test ban, and on other nuclear measures as evidence that the treaty has not worked fairly. Uncontrolled continuation of U.S. and Soviet nuclear weapon programs, a check in SALT negotiations, would underscore the discriminatory aspect of the NPT. Minimal measures such as the Threshold Test Ban may be judged as cynical and diversionary. Thus vigorous and productive arms control negotiations, on SALT and other nuclear measures, are important for nonproliferation success.

SALT and other nuclear arms control negotiations should be pursued on their merits, for U.S. security and a more stable world. A stable NPT is unlikely over the long term if progress in such efforts is not substantial. But in the short term, so long as good faith efforts are pursued and the major powers show responsibility in their management of their nuclear arsenals, the fate of nonproliferation will be decided on other grounds. No nuclear arms control agreements foreseeable in the near future are likely to be decisive in moving significant hold-out states to NPT adherence. Major SALT agreements, a comprehensive test ban, might however make it more difficult for a hold-out actually to start nuclear testing.

Another measure contributing to ease any sense of NPT discrimination is the longstanding U.S. proposal for a cutoff of further production of fissionable materials for weapons. This proposal, still on the books but muted recently, would not turn the clock back on fissionable material already committed to nuclear weapon stockpiles. However, by freezing the quantities of fissionable materials allocated to weapons, it would establish a base from which the nuclear weapon states could subsequently proceed to negotiate reductions of nuclear weapon stockpiles. Immediately, it would counter discrimination arguments. A cutoff would put all countries on equal footing as far as future production of fissionable materials for weapons is concerned. IAEA verification (adequate for U.S. security, in view of the huge existing stockpiles of the nuclear superpowers) within the territory of nuclear weapon states would remove the discrimination presently attributable to application of NPT safeguards only to nonnuclear weapon states. Soviet resistance is predictable, but no reason for the United States to cease pressing a useful proposal which existing and prospective SALT agreements should make easier to accept.

Attention to reducing the political and security symbolism of nuclear weapons as marks of power and status is also important. Stress on the nuclear arms race between the United States and the Soviet Union, or on the vital need not to appear "inferior," is not helpful. Nuclear war-fighting strategies, whether strategic counterforce or limited tactical nuclear war doctrine, and weapons developments or deployments related to them, imply that nuclear war may be a deliberate and rational national policy, and thus something any state with major power aspiration may need to prepare for.

Formulas guaranteeing nonnuclear weapon states against use of nuclear weapons against them can also be evisaged and might increase the acceptability of nonnuclear weapon status.

SAFEGUARDS

Safeguards over the use of indigenous or imported nuclear material or facilities play an important but limited role in preventing proliferation. They do not in themselves limit the capability to make nuclear weapons; they do not prevent misuse of nuclear capabilities or prevent proliferation. Nevertheless, they do help deter proliferation and create an international atmosphere in which a non-proliferation regime is possible.

Safeguards presuppose the operation of a system of national accounts and reports for nuclear materials. IAEA inspectors have access for verification to accounts, materials, and facilities. Safeguards are thus an alarm and a reassurance. As an alarm, they deter by the threat of timely discovery. And they offer the opportunity to demonstrate observance of commitments, and thereby to reinforce arrangements and build mutual confidence.

Safeguards are not foolproof. There are technical problems in maintaining and verifying accounts even within a national program. IAEA safeguards do not protect facilities or material from attack or seizure. International safeguards operate within painfully negotiated and rigidly defined procedures, constrained by sensitivities of national sovereignty. Efforts to install better technical and procedural measures and improve the skill and access and number of inspectors are important, are worth additional funding, and should yield results. Opening of extensive U.S. and United Kingdom nuclear power facilities to IAEA inspectors, now being worked out with the IAEA, will undercut resistance to inspection and safeguards justified on grounds of discrimination; similar action by the Soviet Union should continue to be sought.

Safeguards over sensitive fuel cycle facilities such as reprocessing plants, enrichment plants, and storage of separated plutonium (including that in mixed oxides) or highly enriched uranium pose particularly difficult problems. Even with a staff of full-time inspectors, there could be little confidence that diversion of enough plutonium for a single explosive could be detected, although material for a significant number of weapons would be more detectable. However, full-time inspectors could probably assure that a low-enrichment facility was not being reconfigured or operated to produce highly enriched uranium. International operation of such facilities would be more effective in preventing diversion at the facility itself.

Under the NPT, safeguards are to be applied to *all* peaceful nuclear activities in the territories of nonnuclear weapon NPT parties. In the case of nonparties, safeguards are applied only to materials or technology imported from NPT parties or others requiring safeguards. This element of discrimination between parties and nonparties has been troublesome to some states. The United States and other NPT parties should seek to have all states become parties to the NPT, and in the interim to have all activities subject to safeguards.

Should exports of materials or technology for peaceful programs be withheld unless one or the other of these conditions is accepted? There may be cases where the wiser course is to work with nonNPT states, to insist on IAEA safeguards on all supplies of materials and technology and their by-products or derivative applications, and to draw these states and their programs as extensively as possible into safeguards and commitments against explosives use. A de facto nonproliferatoin regime would then apply. Violation or abrogation of arrangements would then be subject to penalties. A pattern of peaceful cooperation and of interdependent safeguarded activities might come to prevail throughout their atomic energy programs.

IAEA sanctions are limited. In case of violation, the agency can (1) notify nuclear countries and the UN Security Council; (2) suspend the state in question from IAEA membership; and (3) demand the return of agency-supplied materials. Members are reluctant, however, to invoke even disclosure and investigation because of the political opposition which might ensue. For the latter deficiency, provisions for resolution of unclear cases—perhaps through use of a nonpunitive investigative and conciliation committee of the IAEA Board—would reinforce confidence in the efficiency of the safeguards system. For actual violations, action by major members (who will be major suppliers and thus have substantial leverage) will be imperative.

ATTITUDES AND EXPECTATIONS

In the developing countries, the complexities of nuclear power may not be fully appreciated. Active promotion of nuclear power, first under the U.S. Atoms for Peace program and also by the IAEA, has led to unrealistic expectations of benefits. Nuclear power plants appear increasingly to be competitive only in large units (1,000 MWe) operated at high capacity. Market estimates based on smaller units (300-500 MWe) and on previous optimistic capital cost projections have overstated both the number of countries and the aggregate nuclear generating capacity that might be installed in the developing world. Supplier countries and the IAEA can perform a valuable (if at times unwelcome) service by conveying caution to potential clients. Expectations as to the merits and imminence of plutonium recycle, the breeder, and PNEs also need a stiff dose of realism.

There is currently a "mystique" of nuclear power which is not warranted on objective grounds. It appears, in addition to ways already mentioned, in such gestures as the offer of reactors to Israel and Egypt by President Nixon, and in the heavy involvement of the U.S. Export-Import Bank in financing of exports of nuclear facilities as contrasted with conventional power facilities.

Gestures of this kind make it awkward for the United States to remonstrate should other suppliers use nuclear assistance to establish special or political relations.

Extending International Control of the Nuclear Fuel Cycle

If plutonium and highly enriched uranium become increasingly available, the route to national or subnational explosives capabilities will become easier and quicker, and safeguards will lose much of their ability to provide reassurance. In current reactors, fresh fuel is not usable in weapons without further enrichment, and the discharged fuel contains plutonium which still must be separated from large quantities of highly radioactive material. If countries lack the facilities to perform these functions and do not have stocks of plutonium or mixed oxide fuel returned from out-of-country reprocessing, there is a gap still to be closed before reaching a weapon capability. Moreover, the possibility that weapon-grade materials can be seized by terrorists will be avoided.

The United States exports neither of these facilities. Agreement that separation and enrichment facilities should only be exported and operated under safeguarded commitments against use for nuclear explosives is unanimous among NPT parties, nuclear and nonnuclear, and is not contested by nonNPT importers. However, an embargo on further national facilities beyond those of present supplier states, even under such commitments and safeguards, is urged by some. The suppliers have discussed, but have been slow to agree on, a policy of refraining from sales of such facilities. West Germany has agreed to help Brazil build a reprocessing facility and an enrichment plant, and France has contracted to sell a reprocessing plant to Pakistan; it is not clear how firm plans for actual import and construction now are. At the end of 1976, France and West Germany announced a policy of refraining from further sales. The IAEA is studying the possibility of regional fuel cycle centers—which might be multinational, or national facilities serving a regional market under international supervision.

The case for avoiding additional national facilities is that only a brief interval would be required between a decision to make weapons and their fabrication, once these facilities were in operation. It is plausible to suspect that the real motive for seeking such facilities is weapons. There can, however, be other motives for such interest. Developing nations may be genuinely interested in nuclear power as an energy source. It has been widely believed that recycle of plutonium and enriched uranium from discharged fuel elements is necessary to achieve the full economic potential of nuclear power. If these countries have uranium, they can hope that fuel facilities will give them self-sufficiency and improve their balance of payments. Countries without uranium may seek diversification of suppliers to reduce dependence on uncertain energy sources or supplier countries.

This scenario depends on the feasibility and economics of recycle and the breeder. While the economics of both technologies are questionable, as discussed in Chapters 11 and 12, expectations are high in many places. These expectations have been stimulated by the public statements and heavy national commitments of funds and efforts to the plutonium cycle in the United States and other supplier nations. Countries projecting nuclear power programs totaling several thousand MWe in the 1980s can readily rationalize enrichment and reprocessing technology. Familiarization with novel and complex technologies, experience to make choices among alternative technologies, expanding indigenous energy options—all have appeal, and the additional investment may seem small.

Like reactors, fuel cycle facilities are high technology symbols of industrial and technical status. The appeal may be enhanced by the awareness that the rich countries want to withhold these facilities from developing nations. Irrational as this proposition may appear, it is hard to discount completely when sensitive facilities are being developed and built in advanced countries and when development of plutonium recycle and breeder reactor systems there are supported by industry and government.

A policy limited to embargo, or penalty against countries importing or exporting sensitive facilities, can subject the nonproliferation regime to severe strain. Countries accepting nonnuclear explosive commitments and safeguards may consider such additional inhibitions to be outside the NPT provisions and implicit bargain, and view them as arbitrary. New discrimination will appear between members of the suppliers group "permitted" to have sensitive facilities and those outside it. Tensions between advanced and developing states may occur in the IAEA where they have hitherto been muted. NonNPT parties that are developing their own facilities (Spain, India, Israel) rather than importing them will not be affected by the restrictions. The charge of discrimination can be undercut if suppliers are not themselves reprocessing for their own nuclear power programs.

A common supplier policy of embargoing exports of reprocessing technology and mixed oxide fuels has encountered obstacles. Soviet attitudes on atomic energy activities are not always the same as those of the United States—on PNEs, or safety practices, for example. West European initial reserve involves more than reluctance to lose exports or the desire to gain a competitive edge over the United States. An embargo not founded on the NPT or other international understandings is difficult to justify to clients with whom there are good political and commercial relations and who will accept safeguards over sensitive facilities.

Nevertheless, other suppliers have begun to show readiness to act to limit the diffusion of fuel cycle technology. The political problems hampering supplier coordination can be finessed if supplier cooperation is embraced in a broader approach, in which consumers as well as suppliers join in an effort to strengthen the nonproliferation regime by international fuel cycle arrangements which both reduce proliferation risks and provide assurance of availability of fuel supplies and related fuel cycle services when needed.

If control of the nuclear fuel cycle is vigorously pursued as an extension of the nonproliferation regime, based on consent and common interest and equitable access to fuel and services, broader cooperative measures along with supplier coordination become feasible. These can take various forms—either including suppliers, who might offer incentives to developing states to forego sensitive fuel cycle facilities, or among recipient states, who might participate in and use multinational rather than national facilities.

Reprocessing and enrichment plants are technically complex and costly, with economic uncertainties and risks. As a practical matter, they are a burden rather than an advantage to an embryonic nuclear power program. Instead of arbitrarily deciding that developing countries must forego them, implicitly or explicitly because they cannot be trusted to resist weapon ambitions, the United States and other supplier countries might concentrate on making alternatives attractive and national facilities a wasteful course difficult to justify. Nuclear fuel and services, on this approach, would have to be competitively priced, reliable, and efficient, with some diversification in source.

Advanced countries should be able to produce slightly enriched uranium more cheaply and reliably than less advanced states in small plants. Long-term supply contracts, opportunity for "toll enrichment" of natural uranium, sales for stockpiling, investment participation if desired, could reduce incentives to construct new plants. Since a number of enrichment facilities are in operation or in prospect in supplier countries, customers wary of dependence on a single source should be able to diversify their sources; and supplier cooperation should not be so tightly structured that purchasers do not feel they have the benefits of choice. Encouragement of this kind of "buyers' market" would not respond to all the motivations for indigenous plant construction, but it would satisfy some countries' needs and probably complicate and delay decision-making in those where political motives are stronger.

Offering reprocessing services is more debatable. If the result were to be return of plutonium to the customer, it would increase rather than reduce diversion risks. Return in a form still requiring chemical separation (e.g., mixed oxide fuel elements) would leave only a small barrier in place, since chemical separation is far simpler than in the case of spent fuel. Contract separation and return of plutonium should in any case not precede the basic decision on whether there is an economic need for recycle.

Disposition of plutonium is, in any event, a major problem for the plutonium cycle. Safeguarding of national stockpiles on current surveillance patterns would have limited value. International custody of plutonium (whether in spent fuel elements or after reprocessing) may be the solution. Arrangements for international custody might apply to nuclear power programs of supplier as well as recipient states.

Better than reprocessing would be retention of plutonium in spent fuel. Storage (interim or permanent) could be safeguarded or managed by the IAEA.

Return of spent fuel to the supplier, with credit for unused uranium or contained plutonium, redeemable if and when economically warranted, might be an interim solution.

Multinational or other regional fuel cycle facilities have been proposed. They might offer economies and facilitate safeguarding. But organizational, siting, and transport problems are evident, as well as selection of host countries. The crucial drawback is the diffusion of technology. Modern fuel cycle facilities under multinational operation might serve as training schools for national cadres. In the case of plutonium separation facilities, return of plutonium to participating countries (presumably the reason for their participation) would present the basic problem of susceptibility to diversion. In the case of plutonium reprocessing, then, any steps toward a multinational or regional facility should not precede the decision on plutonium recycle. If reprocessing comes into substantial operation, technical fixes (spiking, denaturing, supplying plutonium fuel in mixed form) would provide some safeguard against diversion, though more against theft than against subsequent national diversion.

Plutonium storage and surveillance deserves urgent attention quite aside from their bearing on multinational separation facilities. Internationally operated or supervised storage might be practical, and the statute of the IAEA authorizes such a function.

Waste disposal is another area which may be not only susceptible to, but in need of, international cooperation. Feasibility and acceptability of suitable sites will be a controlling consideration.

Low-enrichment facilities should be more adaptable to the multinational or regional approach than plutonium reprocessing, and might offer a means of diversification of supply. Surveillance by resident inspectors could assure that a plant was not configured or operated to produce weapons-grade uranium. The product, so long as limited to slight enrichment, would be usable only as reactor fuel and not as bomb material.

SUMMARY AND CONCLUSIONS

A strategy to constrain proliferation must be complex and comprehensive. U.S. nuclear power policies and programs can be shaped to support such a strategy, but they can be only partially effective unless they are meshed with political actions and with broader international arrangements.

Some of the fundamental elements of a nonproliferation strategy are broader than nuclear power. They include:

• A foreign policy in support of international security, peace, and stability; patient efforts to contain and resolve conflicts which might lead to recourse to nuclear weapons.

- Relations with the Soviet Union that will foster continuing cooperation on nonproliferation.
- Aherence to security commitments serving the interests of the United States and allies and certain threatened states, so that isolation will not make nuclear weapons appear necessary.
- Where necessary, use of U.S. influence to discourage states from preparatory moves toward a nuclear capability.
- Vigorous efforts toward nuclear arms limitations which would implement Article VI of the NPT and build additional barriers (e.g., a comprehensive test ban) against proliferation.
- An active program of international cooperation in energy development and exploitation, so that the full range of energy resources will be available in international commerce and national planning, without need for excessive reliance on nuclear power.
- Deemphasis on nuclear weapons in military policy, particularly doctrines that present nuclear weapons as acceptable necessary armaments for a limited application or political pressure.

Other international measures are directly related to nuclear power.

- Support of the NPT and encouragement of present nonparties to adhere.
- Support of the IAEA, in particular, generous financial and technical help for IAEA safeguards, standards, and practices relating to safety and physical security.
- Leadership in supplementing the NPT with arrangements controlling the spread and utilization of sensitive nuclear fuel cycle facilities.
- Coordination with other suppliers to handle nuclear exports and associated safeguards uniformly and in support of both current NPT requirements and new fuel cycle arrangements. Classification of the technology of laser isotope separation would be one useful area of coordination.
- Interchange of objective data and analysis on recycle and the breeder, in the interests of a realistic appreciation of costs, benefits, risks, and time scale.
- Support of coordinated actions by suppliers of low-enriched uranium to assure supply of reactor fuel, with special attention to needs of countries cooperating with international arrangements for control of the nuclear fuel cycle and related facilities.
- Leadership in exploring arrangements for storage of spent fuel and for disposal of nuclear waste.
- A clearer stand on the doubtful practical value and ominous proliferation dangers of "peaceful nuclear explosives," and greater effort to persuade the Soviet Union to discontinue its PNE activities.

Actions and policies regarding the U.S. domestic nuclear power program, discussed in other sections of this study, would affect proliferation in important ways. The following measures would have major nonproliferation significance:

- A clear decision to defer plutonium reprocessing and recycle.
- Deemphasis of the breeder program with deferral of the early date for commercialization.
- Reduced priority for nuclear power in energy research and planning, in a framework giving equal weight to coal in the short term and alternative replenishable energy sources over the longer term.
- Avoidance of promotion of nuclear power both at home and abroad.
- Orderly expansion of enrichment facilities to correspond to realistic projections of future demand at home and abroad.
- Continued refusal to export plutonium separation and enrichment technology, coupled with efforts to achieve similar action by other suppliers.
- Approval of nuclear exports only where consistent with U.S. security interests and obligations and nonproliferation policy.

✳ *Chapter 10*

Nuclear Terrorism

The past few years have seen an upsurge in the size, sophistication, and capabilities of terrorist groups around the world. Terrorist activities offer strongly motivated political or dissident groups a way to dramatize their causes and influence their adversaries. Modern communications enable terrorists to project their messages widely. Weapons of increasing sophistication are available to terrorists, including automatic weapons, modern high explosives, and even rockets and antiaircraft missiles.

For some years the possibility has existed that terrorists might attempt to steal nuclear weapons. The worldwide development of civilian nuclear power provides additional opportunities for terrorists—whether revolutionaries, nationlists, dissidents, or criminally motivated—to employ nuclear energy as a weapon. One is to acquire, from somewhere in the nuclear fuel cycle, the material with which to fashion a simple nuclear explosive. The second is to use a reactor itself as a potential radiation weapon, by threatening to take actions leading to core meltdown and breach of containment.

Terrorists might choose the nuclear industry as a target to exploit the mystique that surrounds nuclear energy and nuclear weapons. The threat of nuclear terrorism may be used to extort money, secure the release of prisoners, or publicize a particular cause. Whether terrorists are actually in a position to carry out their threats is probably less important than their plausibility. A group may be able to achieve its objectives with a hoax such as the one perpetrated by an Orlando, Florida, teenager whose claim to have a bomb was taken seriously for a time by local officials.

Some incidents have already occurred. In 1973, the Argentine Atucha reactor was temporarily occupied during its construction by a Trotskyist urban guerilla group. In separate 1975 incidents, bombs were detonated at French nuclear plants at Fessenheim and Mt. D'Arree by as yet unidentified groups. In 1974, a meteorological tower at a proposed nuclear reactor site in Massachusetts was toppled by a saboteur protesting the proposed plant. On the other hand, no terrorists have yet credibly claimed to possess stolen nuclear materials, and no actual weapons have been stolen during a military coup, although U.S. nuclear weapons were located in Greece at the time of two successive military revolts.

The implications of terrorism are international in scope. Nuclear materials might be stolen in one country, fabricated into a weapon in a second, and used to threaten a third, in different parts of the world. Precipitation of a nuclear incident anywhere in the world could affect public acceptance of nuclear power in many countries. Radioactive clouds would not halt at national borders. The security of nuclear facilities everywhere is therefore of interest to the United States.

The possibility of nuclear terrorism poses difficult questions. What can the United States do, and what is it worth to keep nongovernmental groups from possession of nuclear weapons or materials? How should the United States respond to the threat or occurrence of terrorism? What legal and constitutional issues are thereby raised?

Decisions about civilian nuclear policy have an important bearing on the terrorist problem. If plutonium is not recycled, the opportunities for plutonium theft in civilian industry are essentially eliminated. Similarly, if the high temperature gas-cooled reactor is not commercialized, the amount of highly enriched uranium in commerce will be very small. Technologies which introduce weapons-grade materials into commerce are clearly undesirable from a security viewpoint.

As presently operated, the security system even in the American nuclear industry does not appear adequate to meet the potential threat of terrorism. Measures to strengthen security, at the local, national, and international levels, need to be carefully developed and vigorously carried out.

This chapter addresses the security of civilian nuclear facilities and those nuclear materials (i.e., plutonium, highly enriched uranium-235 and uranium-233) that might be used by terrorists to make weapons. It addresses theft of materials, the vulnerability of reactors to terrorist attacks, technical and institutional problems in prevention of terrorism, and the constitutional implications of governmental action in response to the terrorist threat. This chapter is closely linked to the chapter on reactor safety, since terrorists might attempt to cause a serious reactor incident, and to the chapters on nuclear proliferation, plutonium reprocessing, and the breeder, since the possibility of terrorists diverting material to fabricate weapons depends on the availability of plutonium or enriched uranium in the commercial fuel cycle.

THEFT OF NUCLEAR MATERIALS

Criminal or terrorist groups might seek to obtain nuclear materials to construct a crude nuclear explosive or conceivably to use the materials as radiological poisons.

The potential sources of such materials in the United States are under strict government control. These include government facilities for the production of highly enriched uranium and plutonium, and a handful of licensed commercial facilities involved in the transportation, storage, and use of materials in research, and in the nuclear weapons and nuclear submarine programs. Nuclear materials suitable for weapons are not now available in the commercial market and do not flow in commercial channels. Decisions on plutonium recycle, the future of the breeder, and construction of other reactors of all types have important implications for increasing or decreasing the opportunities available to unauthorized groups of individuals to gain access to dangerous nuclear materials.

Adequate security at U.S. sites alone will not eliminate the threat. If nuclear materials are in commerce in any country, their security will be of concern to the United States. It may prove simpler for terrorists to steal materials in one country and use them to threaten another.

Availability of Fissionable Material

The vulnerability of nuclear installations to the theft or capture of materials by terrorist groups depends on the reactor types and the fuel cycles in use. The low-enriched uranium used in most commercial power reactors cannot be made to explode or be fashioned into a radiological weapon. Theft of spent fuel is also unlikely since it is extremely radioactive and can be handled only with special shielding and equipment. The heavy casks (30–100 tons) in which it is shipped further complicate theft. Thus, with the nuclear power technology presently in use and under construction in the United States, the danger of terrorist bombs from stolen American materials is minimal.

Plutonium reprocessing and recycle would increase the opportunities for theft. In a reprocessing or mixed oxide fuel fabrication plant, large quantities of plutonium will be on hand. Because of time lags and uncertainties in the inventory measurement system, employees might be able to steal small quantities of materials over an extended period without arousing suspicion. (To get the materials out of the plant, the thieves would also have to evade existing security systems including sensors capable of detecting small quantities of nuclear materials.) Another possibility is forcible theft by outside groups. Under current NRC security procedures and industrial practice, one cannot have confidence in the ability of plant security personnel to defend against a well-armed group until help arrives.

The LMFBR would pose problems like those of plutonium recycle for LWRs, but with much greater quantities and higher concentrations of plutonium. Less than 100 kilograms of fresh LMFBR fuel would yield enough plutonium oxide for an explosive.

An HTGR fuel plant would use large quantities of highly enriched uranium. But once fabricated into the HTGR fuel stacks, the uranium is highly dilute (about 0.4 percent by weight) and difficult to separate for use.

With plutonium fuels, or high enriched uranium, the weakest link in security would be transportation. A dispersed national and international nuclear industry will have large flows of nuclear materials within nations and across national borders. In addition to flows in military and experimental programs, terrorist targets could include shipments of highly enriched uranium from enrichment plants to fuel fabrication plants, shipments of highly enriched uranium or mixed oxide fuel from fuel fabrication plants to reactor sites, and shipments from reprocessing plants to storage sites and fuel fabrication plants.

In 1969–72, several incidents occurred in the United States where shipments of nuclear materials were misrouted in transit, lost for periods of time, or delivered to the wrong facilities. As a result, the AEC tightened regulations governing their transport (Title 10, Code of Federal Regulations, Section 73.30). These regulations apply to shipments of enriched uranium-233 and plutonium larger than two kilograms and highly enriched uranium-235 in quantities of five kilograms or more. There are no security procedures for shipment of smaller quantities of nuclear materials.

The 1973 revisions to transport regulations reduced the likelihood of, for example, a lone truck driver diverting a shipment of nuclear materials. The use of two-man driver teams, escorts, regular check-ins en route, etc., have reduced considerably the probability of employee theft in transportation. The regulatory changes did not eliminate the possibility that a small armed group could successfully hold up a truck and hijack its load of nuclear materials.

Neither the NRC nor any other federal agency directly regulates the carriers and transporters of commercial nuclear materials. NRC requirements are levied upon licensees who, in turn, have responsibility for assuring that common carriers and contract shippers meet the NRC standards. Transport procedures are not vastly different from those employed by armored car companies and other carriers that ship high-value cargo. Some criminal and terrorist groups have already demonstrated capabilities to defeat such precautions.[a]

---

[a]Nuclear weapons and ERDA-owned materials are transported under more strict security procedures in specially designed tractor trailers. The ERDA trucks have special armor plating, bulletproof windows and have HF and VHF communications and a radio telephone. The trucks are also fitted with special immobilization features. The trucks are escorted en route by a separate vehicle, which has redundant communications. Typically, there are four armed driver guards (two in each vehicle), who are ERDA security personnel.

Clandestine Fabrication of Weapons

To fabricate nuclear weapons a terrorist group would have to obtain either plutonium or highly enriched uranium. The plutonium would have to be obtained in the form of metal or plutonium oxide since separation of plutonium from irradiated fuel would almost certainly be beyond the capabilities of a small group. Separate plutonium will only become accessible in commercial channels if plutonium is reprocessed and recycled.

The difficulty of designing, planning, and constructing a crude weapon from reactor-grade plutonium should not be underestimated. The final device would most likely consist of one or more subcritical masses of plutonium metal or oxide surrounded by other machined metal parts and a ton or more of high explosive. Mechanisms would be necessary to assure simultaneous detonation of the high explosive at various points so that the plutonium could be symmetrically imploded to form a supercritical mass. This requires a good design, very precise work, and microsecond-scale synchronization of detonation of the high-explosive charges. The process of fabricating the weapon under less than ideal conditions would present serious risks from accidental criticality leading to low-level nuclear explosions in addition to the general health hazard of working with plutonium without special equipment. These problems are not insoluble, but they do require substantial knowledge, planning, and extraordinary care in execution. A small group of even highly intelligent people is unlikely to have all the skills needed to carry out such a program successfully.

The yield of a simple device using reactor-grade plutonium would be substantially impaired by the presence of the isotope plutonium-240. Plutonium-240 builds up in reactor fuel used for extended periods as is customary in commercial reactors. Thus, a device which otherwise would have a yield equivalent to ten thousand tons of TNT would probably yield a few hundred tons.[b] This reduced yield is a consequence of the fact that plutonium-240 undergoes spontaneous fission to produce neutrons. The presence of these neutrons would initiate the chain reaction in the device as soon as it reaches criticality and cause it to disassemble before it became sufficiently super critical to produce a full yield. Although sophisticated designs might produce full yield despite the presence of plutonium-240, the complexity of such a device puts it beyond the capability of a small group.

The simple bomb that a group of terrorists might be able to construct with reactor-grade plutonium might still have a yield equivalent to a few hundred tons

[b]The availability of weapons-grade plutonium would not necessarily improve the terrorist's chances of achieving a high yield. Unless the designer were able to construct a mechanism to introduce neutrons into the plutonium core at the exact moment of maximum criticality, the device would have a substantially reduced yield, or no yield at all. Development of such a mechanism would additionally burden the capabilities of any small group. A continuous source of neutrons would result in a low yield for the same reason as would a high concentration of plutonium-240.

of TNT, unless some element of the system actually malfunctioned. This would be a factor of at least ten less than a comparable military weapon but would still be able to do terrible damage in a densely populated area. One has only to recall that the largest conventional bombs of World War II contained about ten tons of high explosives.

A weapon employing highly enriched uranium would present fewer fabrication problems than a plutonium weapon since it could make use of the so-called gun design in which two subcritical masses of uranium are brought together rapidly by gun powder within a container similar to a gun barrel. Such a weapon might have a yield equivalent of around ten kilotons (ten thousand tons of TNT). As with the implosion weapon, to obtain this yield it would be necessary to develop a complex neutron-injecting mechanism, a task of very different nature from the rest of the weapons problem. Simpler approaches to design might result in zero yield or an unpredictable yield anywhere between zero and ten kilotons. At present, there is little highly enriched uranium suitable for weapons in commercial channels, and there are no plans to deploy a reactor, such as the HTGR, that would utilize highly enriched uranium. However, there is considerable interest in the HTGR abroad, and this situation could change in the future.

The success of such an attempt to construct a nuclear weapon clandestinely would be critically dependent on the technical skills of the terrorist group. The chances of success would be increased if individuals with backgrounds in nuclear materials or weapons design, construction or handling, or individuals with experience with high explosives could be recruited. After extensive planning, many months of intense work would be involved to produce a weapon. Even in a well-planned effort, there is a good chance that the weapon would fail to detonate or that the group would suffer fatal accidents during its construction. Prospects for success would be somewhat enhanced if terrorists could operate freely enough within a society to test the high explosive parts of the weapon.

ATTACKS ON NUCLEAR REACTORS

Attacks on nuclear power plants pose a spectrum of potential consequences. At the low end are symbolic acts, perhaps even by disaffected employees. Such actions cause property damage but may not hazard lives, threaten radioactive release, or pose risks to the public. At the other end, a group that wanted to cause or threaten widespread damage could seize a reactor and damage it to precipitate a core meltdown and radioactive release. While other dangers are conceivable, such as an attempt to disperse spent fuel at storage sites, the consequences would be considerably less than a serious reactor accident.

The NRC has recently reviewed how terrorists might set in motion the sequence of events leading to a major reactor accident. The NRC concluded that the safety characteristics of reactors (massive containment structures, redundancy of safety systems) make it difficult to induce a radioactive release and that actions which might endanger the public could be carried out only by knowledgeable people with technical competence.

While it is true that safety features reduce the likelihood of a major incident, they cannot reduce it to an inconsequential level. In contrast with an accident where "defense-in-depth" deals with chance coincidence of malfunction, probabilities here must take into account *deliberate* simultaneous sabotage of reinforcing safety measures.

It is also true that it would require techanically sophisticated and knowledgeable commandos to have a high probability of causing a large radioactive release. However, this does not pose an insuperable barrier to a group with time, resources, and determination. The flow of personnel through military nuclear programs and the growing international civilian nuclear industry provide a large pool of experienced manpower from which a group could seek assistance. Reactor personnel held as hostages might be forced to assist their captors under duress. The technical problems in blowing up a reactor would be easier than those in designing and constructing a nuclear explosive. Explosives could be carried by a few people into a reactor or other facility and could cause major damage. Shaped charges could severely damage main inlet pipes for cooling water. Automatic control and safety equipment could be destroyed. Even primary containment could be ruptured with conventional explosives.

One serious deterrent to nuclear sabotage is the likelihood that the terrorists would be killed or captured during the operation. The terrorists would have to operate openly since there is little chance they could clandestinely emplace explosives at the critical points necessary to disable a reactor. Having initiated a meltdown, they would either be the first victims of prompt radiation or would face likely capture if they tried to escape.

The NRC does not impose rigorous security requirements on reactor sites, fuel fabrication plants, or other facilities that do not possess strategic quantities of special nuclear materials. Industry is not required to protect such facilities against threats of the type that could be posed by "foreign enemies of the United States" i.e., international terrorists with paramilitary capabilities or by extension against similar threats from domestic groups.[c]

Reactors are usually protected by armed contract security guards who may be little more than watchmen and could not be counted on to hold off a well-armed team of attackers until assistance arrived. If local police do not arrive until the attackers have secured themselves in the reactor building (and perhaps

[c]Licensees are not expected to defend against foreign enemies of the United States. The Atomic Safety and Licensing Board in a 1974 decision on the Indian Point reactor (Docket number 50-247) extended this exemption to include well-armed domestic threats. The Board concluded that licensees have no realistic means to discriminate between domestic terrorists and foreign enemies, and thus are not responsible for defending against either.

taken hostages as well), the terrorists have time to set explosives or perform other preplanned acts. Local police units are rarely equipped or trained to deal with such contingencies.

The probability of such events can be reduced through increased security at reactor sites and more effective coordination of response at the local, state, and national level. Explicit measures should be prescribed for actions that threaten widespread damage. While this is by no means a likely threat, the potential damage is such that a sizable investment is warranted.

If terrorists succeeded in defeating the "defense-in-depth" safety technique and initiated a meltdown, the consequences would be like those estimated for accidents, discussed in Chapter seven (Reactor Safety) and Chapter five (Health Effects), although more severe consequences are possible for sabotage. In safety calculations, fission products are assumed to decay for several hours between core melt and containment breach. This essentially eliminates many short-lived fission products. Thus, if containment were breached at the outset as part of the sabotage operation, more radioactivity could be released, with possible increases in the number of prompt casualties.

RESPONDING TO NUCLEAR TERRORISM

Many methods for upgrading security of nuclear facilities have been suggested. Some are technical improvements—inventory of materials, or defense of stages in the fuel cycle that might become targets. Others are improved security procedures and institutional arrangements. Although none of these proposals will solve the problem completely, some of them will make nuclear terrorism much more difficult to accomplish and thus hopefully deter it. Some proposals may create more problems than they solve.

Technical Measures

The security system can be alerted to potential thefts by improvements in the inventory and accounting systems to keep better track of nuclear materials flowing through facilities and reactors. If spent fuel is reprocessed, the currently used techniques to measure the plutonium and uranium content have a margin of error of about 1 percent. This discrepancy is termed "material unaccounted for" or MUF. MUF does not mean that material has been stolen or lost, but simply indicates the imprecision inherent in the measure of the inventory.

Other sources of MUF include scraps when plutonium or enriched uranium is compacted and machined into pellets for insertion in fuel rods. A small fraction of this scrap, about 0.5 percent, is unavoidably lost during processing. When nuclear materials are processed in liquid form, small amounts may remain in vessels and process lines. The NRC requires that these losses and discrepancies be within the NRC's estimated "limits of error" of material accounting techniques (LEMUF).

Current regulations require that a facility's inventory books be balanced to LEMUF standards every six months for reprocessing plants and every two months for all other facilities holding strategic quantities of nuclear materials. If small quantities of plutonium or highly enriched uranium were diverted over the course of several weeks, discrepancies might not be noted until months later. There are many difficulties with such scenarios. Thieves would still have to remove the material from the facility through various technical detection barriers that can be designed to be sensitive to extremely small quantities of plutonium. Such a barrier to physical removal would be equally sensitive, regardless of the amount of plutonium in the facility or the size of the MUF.

To combat this problem, the timeliness of material inventories and the accuracy of measurements can be improved. Real-time or near real-time computer-based systems might allow material balances at the end of each shift before workers had left the plant. Improved measurement could reduce the margin of error reflected in MUF. Improved timeliness in accounting and inventory could also supplement international safeguards for nations utilizing plutonium or highly enriched uranium in their fuel cycles.

A second countermeasure is to attach radioactive sources, emitting hard-to-shield radiation, to material shipments or certain areas in plants to discourage attempts at theft. This process, known as "spiking," could be utilized for detection, location, or denial of access.

Spiking could be applied to facilitate the detection of materials in a plant; to enhance their detectability if stolen; and intense sources could be attached to shipping canisters to produce extremely high-exposure doses that would incapacitate unprotected persons within a short time.

In a plutonium plant, spiking for detection would be feasible at little increased cost or hazard to workers. Spiking for location is also feasible, though this involves more substantial cost and some increased hazard to workers in an accident. Spiking shipments to deny access is feasible, but poses a dilemma. To pose a threat to potential thieves, the source would be so intense as to endanger the public in event of accident. If the current one hundred ton shipping cask was employed for shipping plutonium-bearing fresh fuel, the radioactive inventory (1–2 megacuries) of the spiked cask would be similar to a load of irradiated reactor fuel. One assessment of an accident involving spent fuel assumed that about 1 percent of the fuel would be scattered in a nonurban area (population density one hundred per square mile), causing 0.6 early deaths and three hundred latent cancer fatalities. If a spiked fresh fuel cask was opened and some of the cobalt sources were somehow dispersed, it could result in many times as much damage in a comparably populated area. Because of the attendant risks of accident or deliberate dispersal, spiking should probably not be the preferred method for the protection of nuclear materials.

There are many other improvements such as advanced security systems for identification and access control, and systems for continuous communication.

At present, transporters of nuclear materials may be out of radio contact for periods of up to several hours during shipments. Continuous communications could be obtained by add-on capabilities to planned earth-orbiting satellites.

In general, all of these improvements are inexpensive. The total investment for the list of security options identified in the draft GESMO is a few hundred million dollars for the entire industry. Annual operating costs for the industry might be in the tens of millions of dollars, adding 1 or 2 percent to the cost of nuclear-generated electricity.

Institutional Considerations

The institutional response to terrorism encompasses preventive measures such as improved guard capabilities at reactor sites, better alerting and intelligence, and activities associated with response to a crisis once it is underway. Recognition of the need for upgraded security, and action to achieve it, has been slow. For industry, security and safeguards are additional burdens and expenses and there are few incentives to accept these costs unless required. The NRC has not moved rapidly on these issues and has been reluctant to specify in detail what the utilities should do about security. The NRC usually operates through the lengthy and sometimes indirect process of proposing rather general standards for public comment, modifying them, promulgating them, and finally inspecting and evaluating their implementation. The NRC does not have direct authority over critical elements in the security system, such as the contract guards employed by most utilities or the transport services used to ship nuclear materials.

A similar problem arises among federal, state, and local law enforcement authorities. The NRC has not expected industry to defend against armed threats, but government responsibility to do so is not clearly assigned. The NRC requires utilities to establish liaison with local law enforcement authorities, but these forces are often unresponsive or inadequately equipped. The NRC has little authority for dealing with other federal agencies. It is trying to alert local FBI agents to the potential hazards of nuclear terrorism, but such a campaign hardly assures an adequate federal response.

Federal authority to deal with nuclear terrorism is diffuse. Theft of nuclear materials is a federal crime; but an attack on a reactor might not automatically involve federal authorities, since it is not a crime under the Atomic Energy Act. Federal authorities might become involved if national defense statutes or the interstate commerce clause of the Constitution could be invoked.

Security could be improved by federal training, equipping, and regulation of privately employed guards. Legislative and administrative action is required to fix institutional responsibility, set adequate standards, and see that they are met. At minimum, the NRC or the Justice Department should have regulatory authority over the operation and training of guard forces at nuclear facilities. Local and state law enforcement cooperation with utilities and federal officials could be encouraged through mechanisms such as Law Enforcement Assistance

Administration (LEAA) planning grants. In any event, an attack on a nuclear facility should be clearly designated a federal crime.

A more comprehensive measure would be the creation of a federal Nuclear Protective Force. The advantages of a federal force include uniform training, access to and authority to use a wide range of weapons, and a clear conception of mission and responsibility. The disadvantages include infringement on police powers of the state, layering of security bureaucracies, and the possible expansion of federal police powers at the expense of civil liberties. If a federal force was created, its mission should be limited to local, on-site security against criminal or terrorist action. The force should not have investigative or intelligence functions. A model might be the Executive Protection Service (EPS) which protects the White House grounds and foreign embassies in Washington, D.C., and has circumscribed powers.

Statistics on most terrorist activity (bombing, kidnappings, etc.) show a distinct increase in recent years. In airplane hijacking, however, the trend is sharply downward. International diplomatic efforts, coupled with security improvements at airports, appear to have reduced significantly the incidence of air piracy. Modest increases in security have bought a lot, even in the highly accessible airline industry. Though not all of the early hijackers were international terrorists, improved security has raised the threshold from something that a lone gunman could accomplish to a small-scale paramilitary operation. Moreover, while many hijackings had political overtones, nearly all nations found a common interest in countering a threat to international commerce, whatever the affiliation of particular hijackers. By the same token, modest improvements in security at nuclear facilities may substantially raise the threshold for theft, and there is a similar basis of mutual dependence and common interest among nations in preventing nuclear incidents.

The federal government appears to be poorly prepared to respond to an act or threat against a reactor or the theft of nuclear materials. The responsibilities of the NRC, the FBI, ERDA, Civil Defense, and other federal agencies are not clearly defined, nor are the liaison procedures with state and local authorities in the event of an emergency. It is not even clear where the focus for information and decision in Washington would be.

A modest effort has been started under the auspices of the Cabinet Committee on International Terrorism chaired by the Secretary of State. However, nuclear terrorism, whether perpetrated by domestic or foreign terrorists, is different from other terrorist activities and merits separate attention. It involves a different set of agencies and should have a separate organizational focus.

Resolving this set of problems is independent of improvements in security for nuclear facilities and materials. Whether site security is provided by federal, local, or private guards, it is essential that the federal government be able to respond if local defense fails. There must be a recognition of the federal interest and role in a nuclear crisis and an effective chain of command. There must also be clarification of responsibilities and authority of many federal, state, and local agencies in a nuclear crisis. Finally, there should be contingency planning, including compilation of information on nuclear sites and transportation routes, inventories of federal, state, and local equipment and forces in different areas, what terrorists or saboteurs could and could not do, and the associated consequences.

U.S. actions are only part of the story. If plutonium or highly enriched uranium become items of international commerce, terrorists or criminals will go where they can get them. Materials or bombs will not necessarily be used against the country from which they are stolen. The United States should therefore use available bilateral and multilateral channels such as the IAEA to encourage adequate security standards and practices for nuclear activities and facilities abroad. The IAEA has published advisory guidelines on security at nuclear facilities but has not made acceptance of the guidelines a condition for IAEA assistance. The United States should urge a reconsideration of this policy by IAEA and propose security standards as an integral part of all future international agreements for peaceful future cooperation in nuclear energy.

SECURITY AND CIVIL LIBERTIES

Preventive or responsive actions may impinge on civil rights and liberties of those employed in the nuclear industry, those living or working near nuclear facilities, and the general public. Civil liberties may be affected by activities designed to prevent terrorist incidents or by activities undertaken in response. Preventive measures could affect civil liberties by invading the privacy of employees as a result of personnel security check, and physical search, and by surveillance and domestic intelligence dealing with or anticipating antinuclear threats.

Restrictions on employees of the nuclear industry are part of any effort to increase security against terrorism. Such measures must be subject to constitutional guarantees. For example, physical search is subject to Fourth Amendment limitations. Use of devices such as magnetometers, which do not require intrusive physical search, has been upheld by courts in searches for airline and courtroom security. But frisk-type, "hands on" searches (absent probable cause) are of doubtful constitutionality.

Security clearances and personnel screening are subject to restrictions, e.g., membership in particular organizations may not be sufficient grounds to deny employment. The NRC may soon require security clearance for 30,000 employees in the nuclear industry. This should be compared to five million persons already subject to the government's security clearance program. As long as the procedures and standards for nuclear industry employees are carefully drawn, there need be no new cause for concern about civil liberties. Substantial protections for employee rights exist.

Problems could be posed by domestic surveillance to identify potential terrorists. Likely targets would include criminals, terrorists, and possibly domestic dissidents. Surveillance of foreign nationals can be conducted under the national security authority of the President. Where U.S. citizens are involved, techniques such as electronic eavesdropping are subject to the Fourth Amendment requirement for a warrant based on probable cause. Wiretapping in domestic security cases is also regulated by the Omnibus Crime and Safe Streets Act of 1968, which spells out surveillance procedures and enumerates specific crimes in which surveillance is permitted. Wiretapping is permitted with a warrant in crimes relating to the Restricted Data provisions of the Atomic Energy Act and for crimes such as extortion and sabotage. However, theft and unauthorized possession of special nuclear materials, while a federal crime, are not crimes for which a warrant can be issued under the Act.

The use of informants or undercover agents is less well regulated in U.S. law. Use of informants per se has never been held to threaten constitutional rights. However, informants or undercover agents cannot contravene a citizen's constitutional protection in the course of their activities. Though in some cases recognizing a potential "chilling effect" on free speech, the courts have been reluctant to interfere in this area.

If a crisis is threatened or underway, the urgency may subject civil liberties to different pressures. The character of the risk can vary enormously. At the low end of the scale would be a company's response if a quantity of plutonium appeared to be missing at the end of a shift. Employees might be detained, searched without probable cause, or otherwise subjected to abuse of their liberties. At the other extreme, if terrorists had stolen nuclear materials, there might be calls to subject hundreds or thousands of citizens to blanket search, warrantless surveillance, forced evacuations, and detention and interrogation without counsel or probable cause.

Once the crisis is past, there is the risk that some tactics employed in the crisis might be carried over into routine operations or extended to other law enforcement problems. This tendency would be exacerbated if there was strong public sentiment that such a crisis should never be allowed to happen again, and widespread fear of society's vulnerability to nuclear terrorism. It is essential to protect against such spillover effects on civil liberties. There are tradeoffs between physical security measures for the protection of facilities and materials and activities that might affect civil liberties. Where this is the case, additional investments in site protection and local security forces are preferable to expanded surveillance.

Well thought out and well understood guidelines and contingency plans for federal, state, and local law enforcement officials would minimize the confusion and panic in which ill-advised actions infringing on civil liberties might be taken. Uniform response procedures should be developed and subjected to realistic testing by utilities in conjunction with appropriate authorities at all levels. Only through careful planning, testing, and evaluation in advance of a crisis can law enforcement officials assure that response procedures will not affect civil liberties.

CONCLUSIONS

The likelihood of nuclear terrorism is impossible to quantify, but the possibility must be taken seriously. If terrorists were to obtain reactor-grade plutonium, a small group of technically trained people might be able to build a bomb that might have a few hundred tons of explosive yield. If highly enriched uranium were available, the design and fabrication would be somewhat less complicated than for a plutonium bomb but the resulting yield would be unpredictable, and might well be zero. In any case, even in a well-planned effort, there is a good chance that the weapon would fail to detonate or that the group would suffer fatal accidents during its construction.

If plutonium is not reprocessed for recycle in LWRs or breeders, plutonium would not be available in the fuel cycle in a form suitable for use by a small terrorist group since reprocessing fuel elements would be beyond their capabilities. Highly enriched uranium is not now generally available in commerce for nuclear power reactors although this could change if new types of reactors are introduced. At present, both plutonium and highly enriched uranium would have to be diverted from national nuclear military programs.

An armed group could seize a reactor, damage it in such a way as to overcome safety measures, and precipitate a meltdown and radiation release. While this would not be a simple operation, it would not be beyond the capabilities of a knowledgeable, well-trained group willing to risk perishing. The consequences of such an act would be similar to those of a nuclear reactor accident.

Current physical security arrangements for both nuclear facilities and materials require strengthening. At a minimum, federal regulation and training of private security forces are required. Ultimately, the federal government may need to assume responsibility for physical security of materials and facilities against theft or sabotage.

Public authorities are poorly organized to respond to a crisis involving terrorist acts involving nuclear facilities or materials. The federal government should provide leadership and fix responsibility for dealing with terrorist acts. This should include coordination of action and information and assurance that local security forces (public and private) are properly trained and understand their roles. New legislation may be desirable to establish reactor attacks as a federal crime and provide authority for dealing with it.

Improved security measures can be introduced without endangering civil liberties. A potential problem does exist, however, and in establishing new procedures, care should be taken to avoid infringing on civil liberties. An actual crisis could create precedent-setting problems. This possibility puts an additional premium on improved physical security over key nuclear facilities and transportation activities to forestall such incidents.

Nuclear terrorism is a problem international in scope. The security of materials and facilities elsewhere has direct importance to the security of the United States. The United States should press for effective international standards and measures in nuclear export policies, discussions with other supplier countries, and in bilateral and multilateral arrangements with other countries.

✳ *Part IV*

Issues for Decision

✳ *Chapter 11*

Plutonium Reprocessing and Recycle

The spent fuel removed annually from a commercial light-water reactor contains about thirty tons of uranium and 250 kilograms of plutonium. Since the beginning of nuclear power development, it has been assumed that these materials would be recovered and used to make fuel for LWRs or for breeder reactors. If used in the growing U.S. LWR power system, recovered uranium and plutonium could reduce uranium needs by 22 percent and enrichment requirements by 14 percent. The spent fuel from about twenty LWRs operating for a year would produce the inventory of plutonium necessary to start one breeder reactor.

To recover uranium and plutonium, spent fuel must be reprocessed. Spent fuel would be dissolved and the uranium and plutonium separated by solvent extraction. The plutonium would be converted to a solid and sent to a plant making mixed oxide fuel (consisting of a blend of oxides of uranium and plutonium) for LWRs or for breeders; the uranium, which has about the same fuel value as natural uranium, could be sent to an enrichment plant.[a]

The reprocessing and fuel fabrication operations change the nature of the nuclear waste problem: most of the radioactivity in the spent fuel is redistributed into several new material forms—high-level liquid waste, contaminated process materials, and so forth—each of which would have to be treated for disposal.

Until recently in the United States, expectations were that reprocessing would begin as soon as there were enough LWRs to justify the large-scale facilities needed for economical operation, and that plutonium would be recycled in LWRs until breeders were introduced. The belief that reprocessing was imminent led to

---

[a]The steps which reprocessing and recycle would add to the present uranium fuel cycle are shown in the Appendix in Figure A-7 and discussed in Chapter 8.

provision for storing only limited amounts of spent fuel at reactor sites, rather than for more secure interim storage. Consequently, reactor storage pools are filling up, leading some to conclude, incorrectly, that failure to reprocess is a barrier to waste management and disposal and a threat to continued operation of LWRs.

As reprocessing and recycle have moved closer to reality, cost estimates have escalated until there are doubts about its economic merits. Simultaneously, concern has intensified about stimulus to nuclear weapons proliferation due to separation of plutonium and the possibility that terrorists might steal plutonium for weapons. The potential health hazards of plutonium and the impact of reprocessing on risks from nuclear wastes have also become important issues. This process has stimulated assessment of the benefits and costs of reprocessing and recycle and has reversed earlier optimism about the value of plutonium in the fuel cycle.

Abroad, the case for reprocessing rests on the beliefs that it reduces long-term risks from nuclear wastes and that plutonium stockpiles will be needed at an early date to achieve energy independence through use of breeder reactors. Reprocessing has not, however, gone ahead on a large scale, although plants in France and the United Kingdom originally built for weapons work are being converted and expanded to handle LWR spent fuel and a few small plants are planned or under construction in Europe and Japan. A number of countries with very small nuclear reactor deployments, and hence little need for commercial-scale facilities, are building or attempting to purchase pilot-scale reprocessing plants. The threats of weapons proliferation and plutonium theft are raised by these efforts and by the possibility that supplier countries will perform toll reprocessing and thus introduce plutonium into world commerce. In some countries, the economic and social benefits and costs of reprocessing may appear to be of less importance than the possibility of establishing a measure of autarky in energy matters or a technical base which might underlie a future energy technology. A few countries appear to regard competence in plutonium technology as a hedge against future security threats or as a mark of national prestige.

In this chapter, the issues involved in the U.S. decision on reprocessing and recycle are analyzed in the light of conclusions of other parts of this report. The conclusion is reached that there is little or no economic incentive to use plutonium in the LWR fuel cycle and that wide use of breeders is so far in the future and uncertain that reprocessing for this purpose is unnecessary for many years. The value, if any, of reprocessing and recycle in reducing the long-term hazards of nuclear wastes is small compared to the complexity and contemporary risks which it would introduce in the fuel cycle, including increased opportunities for failure in the interim management of wastes.

These conclusions apply in other advanced countries as long as the world market in low-enriched uranium develops as expected. They apply even more strongly in countries with small nuclear deployments: reprocessing and recycle introduce a complexity into the fuel cycle that cannot be justified on economic grounds and that makes waste management problems particularly difficult in those countries. Even if the complicated logistical problems of recovering and using plutonium are solved, plutonium recycle would have only a small impact on energy needs. Small countries will also be dependent on supplier countries for breeder fuel cycle services for many years after breeder introduction, and introduction on a competitive basis, even in advanced countries, is still far in the future.

Our net conclusion is that reprocessing and recycle are not essential to nuclear power, at least during the remainder of this century. In addition, there are potentially large social costs, including proliferation and theft risks in proceeding. A U.S. decision to proceed despite disincentives would induce other countries to follow suit and undermine efforts to restrain proliferation. We believe that the reprocessing of spent fuel, even on a demonstration basis, should be deferred as a matter of national policy, until it is clearly necessary on a national scale.

Historical Experience

Plutonium recycle in light-water reactors is not being practiced in the United States, and the only reprocessing being done is for weapons. This has not always been the case. In the early 1960s, the belief that plutonium recycle would soon be profitable led to an early effort to claim the fissile values left in spent fuel. In 1966, Nuclear Fuel Services (NFS) obtained a provisional operating license from the Atomic Energy Commission (AEC) to operate the world's first private reprocessing plant. This plant was built in West Valley, New York, on land leased from the state. It cost over $30 million and had a design capacity of 300 metric tons of spent LWR fuel per year. The plant operated until 1972 but reprocessed only 630 metric tons of spent fuel. About 60 percent of this was from U.S. government sources. Since the reprocessing price to commercial customers was only about $30/kg, the plant did not even recapture its capital costs. Moreover, the NFS facility was coming close to exceeding limits on radioactive contamination in effluents and in exposure to personnel when the plant was shut down in 1972 for renovation and enlargement. Operations over six years resulted in 600,000 gallons of neutralized high-level waste, now stored in a large steel tank on the site. The neutralization of the waste yielded a sludge that is not amenable to further treatment for terminal disposal. The State of New York presently has responsibility for the waste though efforts are being made to transfer this responsibility to the federal government. New health and safety requirements, particularly regarding seismic criteria, have added to these economic problems and have resulted in commercial failure of the NFS venture.

The subsequent effort by General Electric to build a regional scale plant at Morris, Illinois, fared even worse. General Electric attempted to use a modification of the conventional Purex process which would lead to more economical operation in a smaller-scale plant, a plant that would have advantages in regional markets. The plant failed to operate as anticipated and work was suspended in 1974 after investment of $64 million.

The Allied General Nuclear Services plant at Barnwell, South Carolina, was to have begun reprocessing at a rate of 1,500 tons/year in 1975; however, it is not yet operational. Its problems have to do with development of ancillary waste technology, changes and uncertainties in regulations, and the uncertain economics of recycle. When construction began, it was not considered necessary to convert liquid plutonium nitrate to a solid oxide and to solidify the liquid waste stream for final disposal. At that time, plutonium nitrate transport was acceptable and waste resolidification was to occur sometime in the future. Regulatory recognition of the problems in both of these areas, however, caught up with the builders. In part as a result of these measures, the projected cost of the Barnwell facility has increased by at least a factor of ten over its original estimate of $70 million. Escalation led to an increase in the cost of the basic separation part of the plant to $250 million by late 1975. Estimates for the plutonium conversion and waste facilities are still uncertain. Even more uncertain is the impact of new occupational health standards and safeguards requirements. Allied General has petitioned the Energy Research and Development Administration (ERDA) to subsidize parts of the plant as a demonstration project.

Commercial-scale reprocessing plants are not operating elsewhere except in France, where the La Hague facility has been modified to reprocess LWR fuels to recover plutonium for possible use in breeders. France, the United Kingdom, and West Germany have formed United Reprocessors to coordinate technical development and delivery of reprocessing services on a cost-plus-fee basis. The United Kingdom, which has been reprocessing decomposing spent fuel from its Magnox (gas-cooled, graphite-moderated) reactors, has the largest commitment to reprocessing in Europe. The plant at Windscale, originally a military facility, was shut down in 1973 following an accident in reprocessing LWR fuel. Japan has only a small-scale plant under construction; the prospects for large-scale facilities are uncertain. Present Japanese plans are to send spent fuel to England or France for reprocessing.

The dominant motivation in advanced countries is varied. England sees a commercial advantage; Japan and West Germany apparently believe that reprocessing is essential in waste management; France is reclaiming plutonium for use in the commercial breeders it hopes to build and sell. Reprocessing is also of interest to some less advanced countries. However, these countries appear to be interested in inexpensive pilot or laboratory-scale plants and not in commercial-scale operations. None of these countries will have large enough nuclear deployments to justify commercial reprocessing.

ECONOMICS

The economic merits of reprocessing and recycle are difficult to assess. Many new operations are introduced, each of whose costs is uncertain. Moreover, the future prices of uranium and separative work, which determine the value of reprocessed uranium and plutonium, are uncertain. The most comprehensive analysis of the comparative economics of reprocessing and recycle has been performed by the Nuclear Regulatory Commission (NRC) in the *Final Generic Environmental Statement on the Use of Recycle Plutonium in Mixed Oxide Fuel in Light-Water Cooled Reactors* (GESMO, August 1976), issued in connection with the current hearings on whether to allow this extension of the commercial fuel cycle.

The cost/benefit analysis in the GESMO compares the total fuel cycle cost for fuel cycle options assuming that the nuclear industry grows to 507 GWe of light-water reactor capacity by the year 2000. We shall be interested in the comparison made there between an ordinary uranium fuel cycle and that in which reprocessing and recycle of both uranium and plutonium occur. The GESMO assumes that breeders are not introduced during the 1975-2000 period and expresses confidence that uranium is adequate even without recycle.

The GESMO predicts a net discounted benefit of $3.2 billion (in 1975 dollars, discounted at 10 percent) over the period 1976-2000, from uranium and plutonium recycle when this option is compared with that of disposing of spent reactor fuel without recovery of uranium and plutonium. This corresponds to a reduction by about 8 percent in the average nuclear fuel cycle cost. This is much smaller than the uncertainties in the cost estimates on which the analysis is based. Moreover, since fuel costs are only about 15 percent of bus bar electricity cost, and less than 10 percent of consumer electricity price, the possible reduction in electricity price would be less than 1 percent.

The narrow economic case made for recycle in the NRC analysis rests on cost factors derived from earlier industrial and governmental projections and upon certain assumptions. It is useful to examine sensitivities to changes in the dominant cost factors as they reflect present uncertainties and to see how changes in particular assumptions affect the analysis. The important economic factors and assumptions in comparing the option of reprocessing and recycle of uranium and plutonium with that of staying with a uranium fuel cycle and disposing of spent fuel appear to be: future uranium prices, separative work prices, enrichment tails assay, reprocessing costs, costs of fabricating mixed plutonium/uranium oxide fuels compared with uranium oxide fuel, costs of disposing of wastes from reprocessing compared to disposing of spent fuel, and physical security and safeguards costs made necessary by plutonium in the fuel cycle. Each of these is examined below.

Use of Uranium

Since any benefit from recycle comes largely from savings in uranium, the assumptions made about the way in which uranium is utilized affect the magnitude of the benefit. To provide a given quantity of enriched reactor fuel, natural uranium must be processed and enriched by separation into two components: the fuel material enriched in uranium-235, and rejected material (tails) containing lower concentrations of uranium-235. This process is variable, however, in that extra separative work can recover more of the uranium-235 from the tails. Thus, extra separative work stretches the supply of natural uranium. The incentive depends on the relative prices of natural uranium and separative work. If uranium is expensive and separative work less so, it pays to remove more uranium-235 from the tails and thereby use less uranium. There is an economically optimal balance between the amount of uranium used and the amount of separative work performed to produce a given amount of reactor fuel.

The GESMO computes total fuel cycle costs for the uranium fuel option and for the recycle option assuming 0.3 percent tails assay (recovery of 58 percent of the uranium-235 in natural uranium). But this is not the optimal tails, given the prices assumed in the GESMO. When the GESMO calculations are redone assuming optimal tails assay (0.2 percent tails, or 72 percent recovery of uranium-235), the benefit projected for recycle as compared to a uranium fuel cycle is nearly 25 percent lower than that projected with the uneconomic assumption. Reduced demand for uranium (about 20 percent less being required at 0.2 percent tails assay) would also mean lower average uranium prices and less advantage to recycle. Thus, the discounted benefit of recycle projected in the GESMO of $3.2 billion is reduced to about $2.5 billion simply by assuming that uranium is economically utilized in both of the fuel cycles being compared.

The reason GESMO gave for its enrichment assumption was that it is U.S. government policy, beginning in 1978, to enrich only to 0.3 percent tails assay. This policy is based on expectation that high enrichment demand might tax existing capacity. It is shown in Chapter 13 that this probably will not occur. Over the longer term, it would be reasonable to assume optimal uranium use in computing the long-term economic effect of recycle. According to the GESMO, the average constant-dollar uranium price over the period 1975–2000 would be about $28/pound of U_3O_8 with recycle and nearly $30 without. These are lower than current spot price quotations; however, the GESMO argues that such prices do not represent the long-term average prices which will actually occur. If the average price were increased to $40/pound, a representative current spot price, the discounted future benefit of recycle would be $2 billion larger than the GESMO result. There are, however, large uncertainties in the uranium price analysis (see Chapter 2). Taking all of these uncertainties into account, the average constant-dollar uranium price in 2000 may be closer to $35/pound than to the $28 in the GESMO analysis. The corresponding uranium savings would be about $1.2 billion larger, in favor of recycle, than indicated in the GESMO.

Enrichment

Technical advances in enrichment, such as laser isotope separation methods, could lead to a large reduction in enrichment price and a consequent reduction in the possible benefit of recycle. Moreover, a reduction in enrichment price would make more complete separation of uranium-235 desirable, thus extending low-cost uranium supplies and further reducing recycle benefits. Even without the latter effect, a factor of three reduction in enrichment price (from $75/SWU to an equivalent of $25/SWU) could reduce the discounted benefit of recycle relative to no recycle by about $1.2 billion. The effect on uranium price could result in an additional benefit reduction of about $0.5 billion. Conversely, failure of new technologies and a rise in diffusion enrichment price to $100/SWU would give an increase in benefit of about $0.4 billion. The commercial advantages of new enrichment technologies, however, may not become evident until about the time recycle could otherwise begin in the early 1980s.

Reprocessing

Reprocessing is the largest cost in the recovery of uranium and plutonium. In the GESMO analysis, a reprocessing cost estimate of $150/kilogram is used. This figure, which includes the resolidification of waste streams, preparation for disposal, and converting plutonium nitrate to a solid oxide, is that earlier cited by Allied General Nuclear Services (AGNS), owners of the partially completed Barnwell facility. However, most current estimates are in the range of $200–$400/kilogram,[b] with the great range reflecting the considerable uncertainties which still exist. Cost escalations in reprocessing have been very high in the past; there is no evidence that this process has stopped.

The impact of increased reprocessing cost on the economics of recycle is profound. A cost of $250/kilogram, instead of the $150 figure assumed in the GESMO, would reduce the discounted benefit of recycle by $2.4 billion, eliminating the benefit entirely if tails are optimized economically and uranium price does not rise above GESMO estimates.

Fuel Fabrication

There has been no commercial-scale experience with fabrication of mixed oxide fuels. However, some experience has been gained in fabricating experimental plutonium-bearing fuels, for LWRs and for the breeder program, using manual glove-box techniques and plutonium with less hazardous isotopic composition than that which would be involved in commercial recycle operations. An analysis of this experience by the Edison Electric Institute[c] suggests a fabrication penalty of about $8/gram for plutonium in mixed oxide fuels. This would imply a differential of about $260/kilogram for fabricating mixed oxide fuel compared to uranium fuel.

The GESMO states that uranium fuel fabrication costs $95/kilogram and

---

[b]This range is obtained from estimates of what would be required for profitable operation of the Allied General Nuclear Services (Barnwell) facility, from estimates of a European venture (which will operate on a cost-plus-fee basis), from ERDA estimates which range from $200 to $333 per kilogram, from estimates associated with a proposed Exxon reprocessing facility, and from estimates prepared for the Edison Electric Institute.

[c]*Nuclear Fuels Supply,* report by the Edison Electric Institute 76–17A, 7.5C-3176, March 1976.

projects a differential of $105 for mixed oxide fuel fabrication (or a $3.25/gram penalty for plutonium), contending that learning effects and technical advances will eventually reduce costs. During the early years of recycle, the GESMO observes, fabrication costs might be $350–$400/kilogram. It is quite likely that the present glove-box technology will give way to a complex remote fabrication technology. Since this technology does not yet exist, however, cost estimates are speculative. If fabrication costs remained as high as the EEI estimates suggest, the benefit of recycle would be reduced by about $1 billion.

Waste Management and Disposal

Reprocessing converts the radioactivity in spent fuel into several new categories of waste. There is a reduction in the volume occupied by the most radioactive elements, the fission products, when the liquid waste stream from the reprocessing plant is resolidified in a form suitable for disposal. However, large volumes of contaminated process materials also result from reprocessing and mixed oxide fuel fabrication. Since much of this material is contaminated with plutonium or other long-lived elements, it should be accorded the same care in long-term disposal as the much smaller volume of high-level waste. In the GESMO, the costs of putting the waste materials into a form suitable for disposal are included in the estimates for reprocessing and mixed oxide fuel fabrication, although no breakdown is given. The necessary commercial-scale waste processing steps have not been demonstrated.

Comparative estimates must be made for the costs of disposing of spent fuel without recycle and disposing of the variety of waste materials produced by reprocessing and recycle. Current estimates for disposing of spent fuel range from $70 to $150/kilogram. The GESMO assumes that assemblies would be sealed in stainless steel canisters and placed in a geological repository like that which would be used for reprocessing wastes. Noting that little attention has been paid to this option, the GESMO uses an estimate of $100/kilogram. Resolidified high-level waste from reprocessing occupies a smaller volume than the original spent fuel from which it is derived but generates about the same amount of heat. Since it is the heat load, and not volume, which is the major constraint in underground disposal, it is likely that high-level waste disposal costs will be based at least partially on heat output. In contrast, the GESMO assumes disposal costs for high-level waste will be based on volume rather than on heat output, thus giving a major advantage to reprocessing over disposal of spent fuel. Roughly $1 billion of the projected discounted benefit of recycle arises from this simple assumption.

Safeguards

The cost estimates in the GESMO analysis are based on present safeguards requirements. These requirements are being made much stricter and will impose higher costs on the presence of plutonium in the nuclear fuel cycle. This part of the GESMO analysis has been delayed, so it is difficult to estimate the impact of new standards; however, the discounted benefit will probably be reduced by less than $0.5 billion.

Net Economic Comparison

It is clear from the preceding discussion that there are numerous uncertainties in the economic comparison of the two fuel cycle options considered. For some items, notably uranium price, the GESMO analysis may have underestimated the benefit of recycle. For a larger number of factors, however, it may have chosen values for costs which are too optimistic. The net benefit of reprocessing and recycle appears to be consistent with zero, but with large uncertainties. However the uncertainties are resolved, recycle would have little effect on the cost of power: nuclear power costs in the United States would be at most 1 mill/kwh higher or lower with recycle.

Even with somewhat optimistic assumptions, the GESMO analysis finds little economic penalty in delaying recycle: a delay until 1986 reduces the discounted benefit in the GESMO analysis by $0.074 billion; a delay until 1991, with storage of spent fuel until then, gives a reduction of only $0.3 billion, or less than the cost of completing the Barnwell facility. With slightly more pessimistic cost assumptions for the interim period, there is a net savings from doing without recycle.

Economics of Recycle Abroad

The close calculations discussed above are not easily generalized to all situations in which reprocessing and recycle may be contemplated. The United Kingdom, France, and West Germany have made agreements for possible multinational ventures. England and France are offering toll reprocessing and several countries with small nuclear deployments are acquiring small reprocessing facilities. The cost comparison between the mixed oxide fuel cycle and the normal uranium fuel cycle may be different abroad if the scale of the operation is different or if the regulatory constraints on occupational and public health, safety, security and safeguards, and waste management and disposal are different.

Commercial-scale mixed oxide fuel cycle facilities appear to be characterized by high unit costs unless very large facilities can be built. For example, it is believed that the costs of building a commercial reprocessing plant do not change appreciably until a size of about 800–1,000 metric tons/year is reached. Several analyses have shown a plant capacity of about 1,500 metric tons/year to be preferable; such a plant would serve about fifty large reactors. With the same regulatory standards, the unit reprocessing cost for a small plant (say, 100 metric tons per year) might be as much as twice as high as for a large plant. In fabricating mixed oxide fuels manual fabrication is perhaps twice as costly as in the automated plants which the GESMO projects as being possible in the United States in the future. Similar statements apply to management of the several categories of waste from reprocessing and fabrication. Thus, unless the nuclear deployment to be served is larger than about 50,000 MWe, the economics of recycle may be even less attractive than in the United States. Cost reductions could probably only be achieved through lower regulatory standards and higher social costs. It should be noted, however, that high unit costs for reprocessing are not prohibitive for countries wishing to build pilot-scale facilities or even small commercial plants for noneconomic reasons.

NONECONOMIC FACTORS

A number of noneconomic factors are involved in decisions on whether to proceed with reprocessing to recover uranium and plutonium; these include increased fuel assurance for light-water reactors, closure of the back-end of the nuclear fuel cycle, technical experience preparatory to breeder reactors, stockpiles of plutonium for use in domestic LWRs or breeders or for sale abroad, and security or prestige attributed to plutonium technology and availability.

Fuel Supply Assurance

In some countries, spent reactor fuel may be regarded as a secure source of future fuel, not dependent on uranium availability or politically vulnerable. The value of this resource is not high, however. In the United States, full recycle of uranium and plutonium would result in a saving of about 22 percent in uranium needs through the year 2000 according to the GESMO.[d] In a growing reactor system, the reclaimed fissile resources from a previous year have less relative effect in displacing the larger fuel needs of subsequent years. Thus a country experiencing growth more rapid than in the United States will find even lower relative value in reprocessing.

In order to improve fuel supply assurance through recovery and use of plutonium, a country would have to invest in complete mixed oxide fuel cycle facilities or solve complex logistical problems. Indigenous fuel cycle facilities even for the simpler uranium fuel cycle are now limited to supplier nations because of the capital investments and high technology involved. Smaller countries unable to justify expensive mixed oxide facilities would have to arrange for fuel to be prepared in a supplier country; to do this, spent fuel would have to be reprocessed domestically or abroad and the products sent to the facility making fuel assemblies for the reactor in question. If this were done, despite what is very likely an economic disadvantage, it would involve further dependence on

---

[d]These figures are lower than others often quoted because of two effects. The first is that the plutonium and uranium recovered from spent fuel contain neutron-poisoning isotopes which reduce their value as new fuel. The second is that spent fuel from a growing reactor system can provide only a small fraction of the higher demand of larger future deployments.

the particular supplier countries performing the necessary fuel cycle services. It would also involve greater risk of diversion or seizure.

Fuel Cycle Closure

The expectation of reprocessing has conditioned thinking about waste management and disposal. For example, reprocessing in Germany and Japan is a legally required step in waste management. Little consideration has been given to longer-term storage or direct disposal of spent fuel, or to comparison of these options as they bear on risks in waste management and disposal. It is widely believed that reprocessing offers advantages in the long-term disposal of nuclear wastes through concentrating and reducing the volume occupied by radioactive fission products, through solidification in glass or ceramic form, and through separating long-lived plutonium from the other wastes. The necessity and efficacy of these measures are discussed in Chapter 8. It is our conclusion that the short-term risks of reprocessing and recycle outweigh the relatively small reduction in long-term risks which might be achieved under ideal circumstances.

Reprocessing reduces the volume occupied by the most radioactive isotopes in spent fuel by as much as a factor of ten. But a large volume of new waste materials, contaminated with plutonium or other long-lived isotopes, is also produced. The net result is that the volume and heat output of waste from reprocessing and recycle operations requiring permanent disposal is about the same as that of the original spent fuel. The amount of long-lived radioactivity in wastes is reduced by about a factor of ten by reprocessing and recycle. While about 99 percent of the original plutonium might be removed, recycle results in higher concentrations of other long-lived isotopes, notably americium, curium, and plutonium itself. Partitioning of long-lived species, if it could be accomplished in a commercial power program, might reduce long-term risks, but only at the expense of creating new short-term hazards in waste management and in fuel cycle activities, including partitioning itself. The possible reduction in long-term risk is small, if it exists at all.

We have concluded in Chapter 8 that permanent disposal in geological formations that are isolated from appreciable contact with ground water does not involve appreciable risk since the probabilities and consequences of postulated release mechanisms are both small. The risk has little dependence on the actual form or composition of the waste. The principal barrier to waste release is the geological formation chosen. Engineered barriers, whether stainless steel canisters for spent fuel or use of a glass form for reprocessed waste, cannot be relied upon beyond the first few hundred years, a period in which confidence in geological integrity is highest. Thus, the disposal risks from spent fuel and reprocessed wastes would appear to be similar.

The largest risks from nuclear wastes are in their interim management, before permanent disposal. The opportunities for failure in management are larger if wastes are reprocessed. This will be particularly true in countries with small nuclear deployments which do not justify investments in the more complex waste management facilities required by reprocessed wastes. U.S. experience in the handling of both military and commercial waste has shown the problems and eventual high costs of proceeding with reprocessing and using short-term solutions to waste problems, such as tank storage of neutralized wastes or shallow land burial of transuranics. Most countries are presently storing spent fuel in water-cooled facilities at reactor sites; decisions on what should follow this inexpensive interim measure will be made within a few years. Development of spent fuel storage and disposal options is seriously delayed and underemphasized.

Preparation for a Breeder Economy

Breeder reactors have been considered the next step in the evolution of fission power, offering greater efficiency in the use of uranium than present converter reactors. Reprocessing and recycle of plutonium are essential parts of breeder technology. Some countries lacking assured energy supplies presently regard breeders as a way of dealing with resource problems and thus look favorably on preparatory experience with reprocessing and on accumulation of stockpiles of plutonium as an energy resource for the future. The value to be attached to these now, however, depends on the timetable for breeder introduction and on the ultimate value of breeders in achieving greater energy independence.

It is our conclusion in Chapter 12 that breeders will not have appreciable advantage over conventional LWRs in the United States until at least well into the next century and that this conclusion is equally valid abroad if the expected world market in enriched uranium develops. The value of breeders in contributing to energy independence is also limited for most countries. The large scale of investments necessary for a completely independent breeder system will be beyond the reach of most countries, making them dependent upon a few supplier countries for technology and fuel cycle services. If breeders are delayed or offer only small benefits, the present value of experience in reprocessing and of plutonium stockpiles is low. For example, plutonium which might optimistically be worth $40 per gram in 2000 has a present worth, discounted at 10 percent, of $4 per gram, far less than the present cost of recovery. Plutonium is much more economically retained in spent fuel. The discounted cost of holding plutonium in spent fuel until 2000 is about $8 per gram, assuming a rather high spent fuel storage cost of $5 per kilogram per year. Major technical advances in reprocessing technology will be required for breeders, advances which would be possible only through the research programs of advanced countries. Present LWR reprocessing plants would not serve for breeder fuels without major modifications. There is thus no compelling reason, for perhaps several decades, to reprocess in anticipation of future breeders.

SOCIAL COSTS AND INTERNATIONAL IMPLICATIONS

In the absence of social costs, the preceding analysis would indicate that re-processing and use of plutonium has only small importance for contemporary nuclear power. On grounds of economics, fuel assurance, long-term risks from nuclear wastes, and technical development and bearing on possible future breeder economies, it is clearly a second-order decision. Projected benefits are either small or possibly negative. There are primary social considerations, however, which argue strongly against reprocessing and recycle. The largest are the inter-national dangers of proliferation of nuclear weapons and weapons capability and the potential for theft of plutonium. The potential health effects and accident risks associated with a widespread industry and commerce in plutonium may be significant but are difficult to access quantitatively.

Proliferation and Theft

The advent of separated plutonium in the nuclear fuel cycle would greatly increase the available technical opportunities for proliferation of nuclear weapons and of theft of weapons materials. This risk arises not only in the existence of the technology, a technology easily developed in most countries, but also in the existence of mixed oxide fuels which may be stolen or misused. Reprocessing and recycle thus present dangers larger than those presently involved in the LWR uranium fuel cycle where the low-enriched uranium fuel cannot be used for weapons.

Plutonium from power reactors is currently well protected against theft by the radioactivity of the spent fuel in which it is contained. National efforts to recover plutonium for weapons would require time for development or purchase of facilities. The contrast with a rapidly developing recycle fuel industry is striking. A worldwide plutonium economy reprocessing plutonium from about a thousand LWRs twenty-five years from now would involve a yearly commerce of more than 200,000 kilograms of plutonium. This amount would be separated each year from spent fuel in perhaps a dozen countries; thousands of shipments of separated plutonium would be made, from reprocessing to fabrication plants and as fresh mixed oxide fuel within and to as many as forty countries. Plutonium would be vulnerable to theft, diversion, and misuse at all stages following its separation at reprocessing plants. Shipments of fresh mixed oxide fuel for the yearly reloading of a single reactor would contain enough plutonium for fifty nuclear weapons. Separation of plutonium from fresh mixed oxide fuel would be much easier than from irradiated fuel, involving only simple chemical opera-tions and little radiation hazard.

Earlier, it was believed that reactor-grade plutonium would be unsuitable for weapons because of the presence of the isotope plutonium-240. This isotope, which can potentially lead to disruptive preinitiation of a weapon, is now not regarded as a prohibitive barrier to successful weapons construction. Its presence lowers the probability that less skilled terrorists would be able to make a high-yield weapon from stolen reactor plutonium. But even an inexpertly assembled terrorist weapon might well have a yield equivalent to a few hundred tons of chemical high explosive.

The process by which a country could move toward nuclear weapons through acquisition of reprocessing facilities need not involve conscious national decisions to pursue weapons capability. The political thresholds which stand in the way of nuclear weapons are lowered by reprocessing and by use of mixed oxide fuels. The timetable on which weapons could be developed is also shortened. A country without weapons may thus find itself in a situation in which the political, social, and economic costs of taking the final steps toward weapons are small at a time when external threats to national security are high. This possibility in one country may also induce it in another. Plutonium availability, as a result of reprocessing or domestic use of mixed oxide fuels, would thus amplify and destabilize conflicts.

A number of suggestions for reducing the proliferation and theft dangers while going ahead with reprocessing and plutonium recycle have been advanced. These include strengthened safeguards and security arrangements and multinational control of crucial steps in the fuel cycle. As discussed in Chapter 9, these measures, while their value should not be underestimated, do not yet provide assurance that risks can be reduced to a level commensurate with benefits which we have concluded range from small to nonexistent. Because it bears on the U.S. decision on whether to subsidize completion and operation of the Barnwell reprocessing facility, we are concerned here with proposals for multinational centers.

Multinational centers for the uranium fuel cycle would provide groups of countries with a way to take advantage of economies of scale and a stronger market position for fuel needs. Since the primary need is for uranium fuel, a multinational center limited to reprocessing and recycle would have very limited value in increasing fuel supply assurance or reducing costs. If countries which were otherwise going to build independent reprocessing facilities could be induced to participate in a multinational venture, this might help in applying safeguards or improving security. It would not, however, inhibit the diffusion of technology or restrict the presence of plutonium in the fuel cycle; indeed, international commerce and risks of theft might be greater. Most important, promotion of multinational reprocessing centers would only encourage the spread of reprocessing and the use of plutonium among countries that would otherwise not pursue them. Subsidy of the Barnwell facility as a demonstration, and particularly as a multinational center, would only provide a mistaken demonstration of the inevitability of separated plutonium in the fuel cycle and would subsidize its introduction. Until there are far more compelling reasons to begin reprocessing, U.S. efforts should focus on restraint and not on promotion.

Health and Safety

The potential health impacts of reprocessing and recycle are discussed in Chapter 5. Compared to the uranium fuel cycle, normal operations of the mixed oxide fuel cycle would result in additional occupational and public health effects. In part, these health effects are compensated by those avoided by mining and milling less uranium ore. However, the health effects from the uranium fuel cycle are probably more easily and inexpensively alleviated than those of the mixed oxide fuel cycle. Much of the public exposure from the present fuel cycle is due to inadequate treatment of uranium mill tailings, a situation which could be inexpensively remedied. The exposures in recycle could be reduced only by large capital investments at reprocessing and fabrication plants.

The health effects projections usually made assume normal operations satisfying stringent safety and exposure criteria and do not include exposures due to leaks, spills, and accidents, or to decontamination efforts following such occurrences. As yet, there is no analogue of the Reactor Safety Study on which to make realistic projections of accident risks from reprocessing and recycle. It is to be expected that these risks are larger than those associated with the uranium fuel cycle because of the large number of new operations and the nature of the material processed.

CONCLUSION

There is no compelling national interest to be served by reprocessing. There appears to be little, if any, economic incentive and it is unlikely that reprocessing and recycle could proceed without subsidy. The noneconomic benefits of reprocessing are small: fuel supply for LWRs would be little enhanced; present experience with reprocessing or plutonium stockpiles has little present value since the introduction of breeders is sufficiently far in the future and uncertain; and contemporary waste management risks with reprocessing are likely larger than possible reductions in long-term hazards from disposal. Health hazards and new accident risks argue against reprocessing. But the most severe risks from reprocessing and recycle are the increased opportunities for the proliferation of national weapons capabilities and the terrorist danger associated with plutonium in the fuel cycle.

In these circumstances, we believe that reprocessing should be deferred indefinitely by the United States and no effort should be made to subsidize the completion or operation of existing facilities. The United States should work to reduce the cost and improve the availability of alternatives to reprocessing worldwide and seek to restrain separation and use of plutonium.

✳ Chapter 12

Breeder Reactors

Fast breeder reactors have long been considered the next step in the evolution of commercial fission technology, since they can extract about seventy times more energy from uranium resources than do the ligh-water reactors (LWRs) now in use. There is little doubt of eventual technical success in accomplishing this goal, but the economic and social conditions under which breeders might be preferable to alternative energy sources are uncertain.

There is not an impending absolute shortage of energy necessitating early breeder introduction. Breeders should therefore be evaluated on their economic and social merits relative to other present and future energy technologies. Breeders must compete first with existing fission technologies, notably LWRs, and with other converter reactors that might prove competitive. At present, LWRs have a definite economic advantage over breeders. In the absence of plutonium recycle, LWRs also have a significant advantage on domestic and international security grounds. Projections show that breeders will have higher capital costs than LWRs and will be competitive only if and when LWR fuel costs rise so high that breeders can overcome their capital cost disadvantage through a lower fuel cycle cost. Our analysis indicates that, social costs aside, breeders are not likely to develop even a competitive economic edge over conventional fission technologies until well into the next century.

Plutonium breeders involve potentially serious dangers of proliferation, theft and diversion, and of accidents and hazards to health. These risks argue for delay in worldwide commitment to breeders until their economic advantages are decisive. Delay in commitment is not only possible, but probably would bring economic savings and opportunities to improve breeder technology for future deployment. If alternative sources of energy are developed during the early decades of the next century, there is some chance that a commitment to plutonium breeders will not be necessary. If domestic and international security dangers are perceived as very important, breeders could be delayed until at least 2025 without severe economic penalty.

The present U.S. program, like those of other countries, is aimed at commercial introduction of plutonium breeders in the 1990s. In the United States, efforts to establish a broad industrial base for breeder commercialization are now getting underway. We believe that this effort, which goes far beyond research and development, is premature, costly, and will not provide us with optimum choices for our long-term energy future. It is our conclusion that the U.S. program should be restructured to reflect longer-term goals.

Abroad, most notably in France, breeders are viewed as a means of providing for greater independence in energy supply. While this is an argument which bears less weight in the United States with its ample supplies of uranium and coal, the economic conditions for breeder success will not be different abroad if there is a world market in low-enriched uranium which provides assurance and diversity of supply. Autarky through the use of breeders would then be achieved only at higher economic cost. Moreover, independence of foreign energy resources through breeder systems would be achievable for only a very limited number of countries. The large scale of a breeder economy, which requires large numbers of breeders (perhaps fifty to one hundred) to justify full fuel cycle facilities, will be beyond the reach of most countries. Breeder fuel cycle facilities would, therefore, be located in a few advanced countries, involving a form of energy dependence for other countries comparable to that of oil, coal, or uranium. This also implies that the plutonium commerce associated with breeders would be highly international in scope.

This chapter examines the evolution of breeder technology, the nature of the current U.S. program, the comparative economics of breeders since this depends on date of introduction, the social costs associated with plutonium-based breeders, and alternative U.S. program strategies.

TECHNOLOGY, HISTORY, AND PRESENT PROGRAM

Very early in the U.S. nuclear energy program, it became clear that two basic lines of reactor development were possible. The first was based on fission by thermal neutrons (neutrons slowed down by a moderator material from the very high velocities with which they are emitted in the fission process). This line of development emerged from the plutonium production program, became ascendent when military applications (submarines) were found, and ultimately culminated in the commercial light-water reactors we know today. In other countries, thermal reactors utilizing heavy water or graphite as moderators were developed to commercial scale. The second possible development line was based on a chain reaction sustained directly by the fast neutrons emitted in the fission process. Reactors utilizing this principle are called fast reactors, in contrast to thermal reactors employing moderators which slow neutrons.

Fast reactor design has the great potential virtue of being able to produce more fissionable material than it consumes. There are, in principle, enough excess neutrons in the fission process to transmute large quantities of fertile material to new fissile material suitable for reactor fuel. Such transmutation also occurs in thermal light-water reactors, producing plutonium from uranium-238, but not at a rate high enough to replace the fissile fuel burned. The goal of fast breeder reactor design is to maximize the rate at which new fissile material is formed while at the same time ensuring good power production and high assurance of safe operation. Ingenuity is required since these goals may, at times, be in conflict. In principle, fast reactors can be designed to produce up to about 40 percent more fissile fuel than they consume; in practice, this has not yet been achieved. Since present thermal reactors produce much less fissile material than they consume, even a "breeder" which simply reproduced its own fuel supply would have a breeding gain of unity. Higher gains are of interest to provide excess fissile material if rapid growth in the number of reactors is required.

Like thermal reactors, fast breeders can be made which operate on either of two nuclear reaction cycles, using thorium as the fertile material and producing fissile uranium-233 or using uranium-238 to produce plutonium. Both require either the naturally occurring fissile material uranium-235 or man-made plutonium to begin operations. In part because of experience and knowledge gained in the weapons program, there has been an almost exclusive program emphasis on the plutonium fuel cycle for fast as well as thermal reactors, despite the fact that a thorium/uranium-233 cycle may offer some advantages in neutron efficiency,[a] in reduced social costs and in fuel availability.

[a]The neutronic characteristics of the thorium/uranium-233 cycle also make it possible, in principle, to build breeders which operate on thermal as opposed to fast neutrons. Such reactors might offer design advantages in the safety area. Two examples of this reactor type are of current interest: the light-water breeder reactor (LWBR) and the molten salt breeder reactor (MSBR). The LWBR, currently being developed in the naval reactor program, would utilize much of existing LWR technology (and perhaps even the reactor plants); however, these reactors would probably have a breeding gain close to unity, making them very efficient converters rather than true breeders, and they would involve large quantities of fissile uranium-233 in an external fuel cycle. Avoidance of large quantities of plutonium might reduce potential health risks and lead to some reduction in theft and proliferation dangers compared to plutonium cycle reactors. However, the problem is only modified since uranium-233 is suitable for weapons; addition of uranium-238, which can be separated only with isotope separation equipment, might provide a partial solution to the safeguards problem through denaturing and dilution, but some plutonium-239 would then be formed. The theft danger might also be reduced by the MSBR, which has its thorium and uranium-233 fuel dissolved in molten salt and which would employ a small-scale reprocessing plant to remove impurities right at the reactor. Only excess bred uranium-233 would be involved in transport. MSBRs would involve much smaller inventories of fissile material. However, there is not enough known to evaluate the comparative economic or safety aspects of this reactor type, which has recently been dropped from the U.S. program.

Because of the requirement to avoid slowing the neutrons that cause the chain reaction, fast reactors cannot be cooled by water. Instead, gases such as helium or liquid metals must be used. The inert gas helium is effective for cooling if highly pressurized. The liquid metal sodium, which is used in most present breeder programs, will boil only at temperatures well above those encountered in normal reactor operation and requires pressures only high enough to assure circulation. Sodium, however, corrodes some materials and reacts violently with water. The latter puts serious constraints on design and construction of the heat exchangers, which must transfer the heat in the sodium to water to make steam for turbines generating electricity.

The intense radiation encountered within a fast reactor core puts high demands on metals and other materials used, particularly when combined with the necessity of cycling the reactor through large temperature changes. The properties of materials presently available appear to limit the amount of time fuels can spend in high-radiation regions of the core, thus limiting the amount of energy which can be extracted (before reprocessing) below what would otherwise be economical. The swelling and creep of materials impose design constraints—room must be left for distortions to occur, and such distortions must not make continued operation less safe. This necessity appears to have reduced breeding gain below what might otherwise have been achieved and poses difficult problems for the analytical methods used to design reactors. Reducing reactor temperatures reduces the impact on materials at the expense of some loss in thermal efficiency. Future advances in materials may eliminate these constraints.

The benefits of breeding cannot be achieved without a fuel cycle more complex than that of present light-water or natural uranium reactors. High burnup rates require that fuel be removed from the reactor frequently. Spent breeder fuel must be reprocessed to recover fissile atoms (plutonium or uranium-233), and this material must be refabricated into new fuel. To do this requires sophisticated reprocessing and fabrication facilities, generally characterized by large economies of scale and probably requiring substantial and as yet undeveloped extrapolations from presently known technology. For these reasons, breeder systems would be advantageous, on economic or fuel supply assurance grounds, only in countries (or groups of countries) able to support a large reactor deployment and the complex of high-technology facilities required to support them.

Historical Evolution

The historical development of fast reactors is of interest for the light it sheds on the nature and directions of the present U.S. program. As noted above, fast reactors have been of interest since the beginning of the nuclear energy program.

In the early years, a rich variety of possible designs was examined, including gas-cooled, liquid mercury-cooled, and sodium- and potassium-cooled reactors. A number of small prototypes were also built. Virtually all work was based on the uranium/plutonium cycle, probably because highly enriched uranium and plutonium were the materials available and about which the most was known.

The beginnings of the line of fast reactor development dominant in the world today, the liquid metal fast breeder reactor (LMFBR), can be traced to the successful operation of the Experimental Breeder Reactor 1 (EBR-1) at the AEC laboratory in Idaho in 1951. This reactor, which was fueled with highly enriched uranium metal and cooled by liquid sodium and potassium, actually generated the first electricity from the fission process. The success of EBR-1 was marred, however, by a meltdown accident in 1955 when an operator failed to shut the reactor down during a test.

In 1963, a large (16.5 MWe) fast reactor, EBR-2, was brought to criticality at the Idaho laboratory. EBR-2 was also fueled with uranium metal but employed a pool-type sodium coolant system, rather than a loop-type as in EBR-1. The loop system has a relatively small core containment with cooling sodium circulated into this space and then out to heat exchangers; the pool design utilizes a large primary pool of sodium in which the core sits, with sodium coolant loops transporting heat away from this pool. The EBR-2 has been the workhorse of the U.S. experimental program, though its design limitations and relatively low radiation intensity make new experimental facilities necessary for work leading to advanced commercial plants.

The only fast reactor built so far in the United States for commercial power use was the Fermi breeder reactor. Under the leadership of Detroit Edison, the 200 MWt fast breeder reactor (FBR) was built near Detroit by a consortium of utilities with the assistance of the AEC. Planning for the reactor followed quickly upon the Atomic Energy Act of 1954 and the early success with the EBR-1. Its purpose was to demonstrate the benefits of nuclear power in a utility environment, with excess plutonium being sold to the government at an attractive price. The reactor went critical in 1963 after a long legal struggle in which construction of the plant near an urban area was opposed by political leaders, labor unions, and, initially, the AEC's own Advisory Committee on Reactor Safeguards (ACRS). Supreme Court action, upholding the right of the AEC to permit construction, was finally required.

Tests to bring the reactor up to full power were initiated in 1966. During these tests, the reactor suffered a partial core meltdown (before it had generated a large inventory of fission products and while it was still operating at a fraction of full power). The accident was initiated by a deflector plate which had come loose, blocking the flow of sodium to part of the core. The plate had originally been installed as a conservative design feature, to prevent critical reassemblies from occurring in the event of a meltdown accident. Parts of the core were ultimately replaced and efforts were made to resume the project. These efforts failed in late 1972 due to economic and licensing problems. Shortly thereafter, the AEC signed a memorandum of understanding with a new consortium of vendors, utilities, and other parties to build a new demonstration plant, the Clinch River breeder reactor (CRBR) near Oak Ridge, Tennessee. The initiation of the CRBR project marked the real beginning of the current U.S. breeder program.

Present U.S. Program

The U.S. breeder program has an almost exclusive emphasis upon a plutonium-cycle LMFBR and is dedicated to the rapid development of a competitive industrial base and utility deployments of LMFBRs by the early 1990s. To achieve this goal, the program envisions a rapid sequence of demonstration reactors and an extensive "base program" effort. The sequence of demonstration plants is such that plants are initiated before completion of the preceding reactor. The base program includes not only the research on physics, materials, chemistry, and fuels, but also a major component development and testing program. The latter is seen as necessary because of the sensitivity of demonstration plant operation to potential failure of major components such as sodium pumps or steam generators. Prototype and commercial-scale components would be developed and tested in base program facilities before installation in demonstration plants. The heavy emphasis (roughly half of the LMFBR budget) on the base program was the result of a 1974 joint AEC/industry review of the LMFBR program, which found a need for backup in the strenuous demonstration plant sequence. The U.S. program contrasts markedly with most foreign efforts, such as those in England, France, Germany, and Japan, which emphasize sequential construction of complete reactor prototypes on an experimental basis. In these foreign programs, design of a new plant follows only after experience with operation of its predecessor. Backup is provided by the ability to modify the succeeding plant.

The U.S. program also goes beyond foreign programs in its efforts to establish a competitive industrial base for breeder commercialization. Efforts are made to bring a wide spectrum of utilities and industrial participants (vendors, component manufacturers, fuel fabricators, architect/engineers, and so forth) into the program on a sustained basis. The problem of transferring technology from government research and development programs to the commercial sector is regarded as more difficult in the United States than in some other countries where only a few industrial and utility participants, often having close ties to the government, are involved.

The intensity and continuity of the U.S. program are regarded as essential to the commercialization goal. The concurrency of the demonstration plants, in which an industrial design team or component manufacturer may move directly from one project to the next, and the high level of industrial activity in the base program, which has its own natural continuity, prevent industrial development from faltering. Without this continuity and high level of activity, it is believed that industrial participation would not be possible on economic grounds and that creation of a competitive breeder industry would not result.

The rigorous timetable of the U.S. program is mandated primarily by the belief that the need for breeders is imminent. This assumption will be examined in the next section. It is true that a massive and rapidly moving program, once underway, has a better chance of sustaining political, as well as industrial, support than one which is stretched out in time and episodic. The demonstration plant sequence and the base program schedule are keyed to a decision by the ERDA Administrator on commercial acceptability in 1986 and commercial deployments of breeders beginning in 1993. The demonstration phase begins with the Clinch River breeder, now scheduled to become operational late in 1983, and continues with a full-sized Prototype Large Breeder Reactor (PLBR), now in design and scheduled for operation in 1988, and by one or more Commercial Breeder Reactors (CBRs). The CBR projects would begin design in 1983, begin construction upon an affirmative commercialization decision in 1986, and enter utility service in 1993. Since the broadened base program really began with the 1974 review, many of the facilities involved are only now getting underway or are still to be initiated.

Base Program. The base program has two parts: a traditional research and development effort in the basic areas—physics, chemistry, fuels, and materials—which underlie fast reactor design and operation, and a component development and test facility program directly supporting the demonstration plant sequence. The research part of the base program necessarily has a longer and more uncertain time horizon than the demonstration plants.

A sodium pump test facility is completed but will require modification to test commercial pumps. Most of the other facilities to test plant components such as steam generator heat transfer systems, to examine fuels, to fabricate fuels from LWR plutonium, and to investigate safety operations, will not be initiated until 1977 or later. Momentum in the base program, as in the demonstration plant sequence, is thus imminent but not yet established.

Fast Flux Test Facility. Since the United States does not presently have a reactor facility that provides a core environment comparable to that expected in commercial reactors, the U.S. program includes an experimental reactor with these characteristics. The Fast Flux Test Facility (FFTF) is a 400 MWt LMFBR with design parameters similar to those of the CRBR. It is intended to test fuels and materials properties, reactor design methods, basic sodium technology, and other aspects of LMFBR technology. The FFTF was authorized in 1967, the year in which the LMFBR was given first priority among fission alternatives, and was estimated to cost $87.5 million, with operation to commence in 1974. Current estimates are for a cost (including operation, in 1975 dollars) of $1.7

billion, with full power operation scheduled for 1980. Escalation appears to be due to inflation, changes in program definition, and management failures. Despite this, the FFTF will provide the first U.S. experience with a fast reactor prototypic of commercial plants. The FFTF will vent its heat directly to the atmosphere and has no provision for steam generators.

Demonstration Reactors. The Clinch River breeder project, which began in 1969, is the first large step toward the breeder commercialization goals.[b] The 380 MWe reactor was estimated by the AEC in 1972 to cost $699 million, with about $250 million to come from utility sources. Current estimates now exceed $2 billion, with nearly all of the cost increases to be borne by the government. The CRBR is a loop-type LMFBR with primary core design specifications similar to the FFTF (both designs are by Westinghouse) but with a uranium blanket to breed additional plutonium and steam generators and turbines to generate electricity. Ground has yet to be broken and there is little likelihood that operation could begin before 1983. While the rapid escalation of costs make it difficult for the CRBR to demonstrate economic success in utility operation, its construction would contribute to industrial experience and to licensability, a major goal of the commercialization program.

The CRBR project is still having difficulty in getting underway, but the next step in the demonstration phase has already begun. An agreement was reached in 1975 between ERDA and the Electric Power Research Institute (EPRI) to fund three independent design studies with $15 million each from ERDA and EPRI. Each study is being performed by a reactor manufacturer teamed with an architect/engineering firm (Atomics International with Burns and Roe; General Electric with Bechtel; Westinghouse with Stone & Webster). Each team will establish a design basis for a commercial plant, a Prototype Large Breeder Reactor. Each design will be accompanied by a Preliminary Safety Analysis Report (PSAR) and a cost analysis. When the comprehensive plant design work is completed in mid-1978, each vendor (and architect/engineer) will attempt to sell its design to potential utility customers. If buyers are found, present plans call for construction to begin in 1981 and operation in 1988, two years after the commercialization decision.

The PLBR designs will put a heavy emphasis on economic factors and will incorporate a number of advances over the CRBR design: fuel pins may be of larger diameter and, in at least some designs, cores that are not uniformly en-

---

[b]Proposals were solicited in 1972 and Westinghouse (assisted by Atomics International and General Electric) was named the lead reactor manufacturer; Burns and Roe is the architect/engineer. The application for a construction license was submitted to the AEC in 1974 and is still pending before the NRC. Though a large number of utilities are involved, leadership in the project has come from the TVA and Commonwealth Edison. Management, which involves such a large number of participants, has been a very complex arrangement. The increasingly large relative financial commitment by the government led in 1976 to a change in management structure that gave ERDA a larger role in the project.

riched may be used. The latter, in effect, means that unenriched uranium fuel elements will be put in the central core, along with those containing mixed oxide fuels, rather than only in the "blanket" surrounding the central core as in the CRBR. It is claimed that this development will enhance breeding gain and reduce the positive sodium void coefficient in the interior of the core with resulting safety improvement. The PLBR design teams were encouraged to explore both loop-type and pool-type designs. The latter is that favored by the French and English for large fast reactors; however, experience of vendors in designing loop-type machines (for example, the CRBR) and the assumed need to move rapidly to commercialization, probably mean that the United States will pursue a loop design for its commercial prototypes.

The demonstration program following PLBR(s) is to culminate in the first Commercial Breeder Reactor (CBR-1). Design of this reactor will be initiated in August 1983, according to the ERDA timetable. The design is expected to evolve from the design efforts for the PLBR and will be of about the same size. Construction initiation is intended to coincide with the ERDA Administrator's decision "to proceed with widespread LMFBR deployment" in August 1986, although it is clear that some expenditures and commitments will be necessary before this date. The plant(s) are scheduled to achieve criticality in 1993. The programmatic difference between CBR-1 and PLBR appears to be small, with PLBR intended to be a "target plant" prototypic of the commercial plant but intended for subsidized sale (in 1978) to utilities.

Fuel Cycle. Breeders will be economically successful only if fuel cycle technology is available at costs sufficiently low to overcome a capital cost disadvantage relative to light-water reactors. Research and development on the LMFBR fuel cycle is conducted in ERDA's Division of Waste Management, Production and Reprocessing and is thus separated from that being done on LMFBRs in the Division of Reactor Development and Demonstration. Considerable technical progress must be made in extending the aqueous reprocessing technology presently available for LWR fuels to LMFBR fuels. For this reason, and because commercial interest in this aspect of breeder technology cannot be stimulated until there is prospect of a sizable breeder industry to serve, the fuel recycle effort has less of the urgent demonstration emphasis apparent in the rest of the program. Only conceptual specifications for the recycle operations appear to have been defined, with laboratory development and test facilities still in the future. A hot-pilot plant capable of serving early breeders is projected for operation in 1988. LMFBR fuel cycle technology is currently budgeted at $10 million to $20 million per year.

Program Costs. The U.S. LMFBR program is expensive, particularly in comparison with foreign efforts. The requested authorization for fiscal year 1977 is $655.5 million with $343.3 million of this for the base program and $237.6

million for the CRBR. Much of the base program expenditure is for large component development and testing, including facilities; only a small fraction of base program expenditures go for traditional research and development. In addition, about $60 million is requested for safety research and less than $30 million for work on advanced fuels and the fuel cycle. In contrast, only about $80 million is requested for research and development on other reactor types; $50 million for the water-cooled breeder reactor program; and $30 million total for gas-cooled thermal converters (thorium/uranium-233 cycle) and gas-cooled fast breeders (uranium/plutonium cycle). Research on improvements in light-water reactor technology accounts for $12.5 million.

The large LMFBR program budget appears to be due to a perceived need to develop a broad base for commercialization and to provide insurance against failure of major components in the demonstration plants. The LMFBR program is thus in the process of developing a continuity and momentum. According to projections (for example, those in the draft LMFBR Program Plan, ERDA-67) future annual budgets will exceed $1 billion. Present total program estimates exceed $12 billion (current dollars but without escalation beyond 1977). This estimate assumes that subsidies to demonstration plants beyond CRBR will be about $350 million. This is undoubtedly an underestimate, given the impact of uncertainties on the willingness of utilities to make commitments to LMFBR plants in the 1980s. If introduction of commercial plants does occur in the 1990s, further subsidies may also be required if these plants are not competitive with conventional LWRs or with new converter reactors that may be developed, such as the HTGR. While the present U.S. program may produce breeders which are technically successful, the economic success of such reactors would depend on market conditions when the commercialization program peaks in the early 1990s. These conditions, and the prospective ability of breeders to succeed, are examined in the following sections.

ECONOMICS

The ability of the breeder to produce more fissile material than it consumes has long been thought to lead to a clear economic advantage in an era of increasingly expensive natural uranium. Though a cycle using thorium to breed uranium-233 is possible, emphasis in world breeder programs has been on relieving an anticipated strain on uranium supplies through use of plutonium-based breeders. The economic prospects for breeders at a given time depend upon the prospective costs of power from breeders as compared with LWR power costs and costs of alternative energy sources. There has been a widespread belief that fission in general would have a strong economic advantage over fossil fuels in a world characterized by high energy growth and that rapid growth of converter reactors would quickly deplete uranium resources, making breeders essential to continued supplies of low-cost power long before the end of the century. These beliefs are now being reexamined.

As discussed in Chapter 2, uranium resources will not put major constraints on LWR growth until well into the next century, since there are good reasons to believe that the growth of nuclear power will be slower and uranium supplies greater than previously thought. With increased energy prices, growth in energy consumption has fallen from its high historical rate. Under these circumstances, and probably well into the next century, the prospects for breeder introduction rest on close calculations of relative merit among breeders, LWRs, alternative converters, and coal. Beyond the early decades of the next century, breeders may have to compete with alternative energy sources whose economic merits are still very uncertain.

The relative economic positions of breeders and LWRs over the period 1993 to 2025 depend on the way LWR power costs change with uranium prices, on technical advances, and on breeder cost projections as this new technology evolves. LWR power costs are subject to uncertainties in uranium prices, particularly after the year 2000. However, the largest uncertainties are in the projection of breeder power costs, since the prospective capital and fuel cycle costs of breeders are still very uncertain.

Breeder capital costs will be higher than those of LWRs. Breeders will thus have to have a fuel cycle cost much lower than that for LWRs in order to overcome the capital cost disadvantage. The LWR fuel cycle will account for less than 20 percent of total power cost during the rest of this century, or about 6 out of 30 mills per kilowatt hour. It is thus difficult to establish a breeder competitive advantage on the basis of fuel cycle cost if its capital cost, which plays such a large role, is the dominant factor.

We first consider the uncertainties associated with these costs and then evaluate the possible benefits or losses attending breeder introduction on various timetables, using an illustrative model that allows examination of sensitivities.

Capital Costs

The comparison of capital costs of breeders with those of LWRs depends upon technical achievements in designing and building commercial breeders satisfying as yet undefined safety and environmental regulations and on improvements in LWR capital cost trends. Present breeder capital cost projections are based on design studies begun about ten years ago and on comparison of plant systems with those of light-water reactors. More up-to-date and direct estimates will follow from the EPRI/ERDA cooperative design projects now underway. Based on current technology, the economic analysis in the Environmental Impact Statement (EIS) for the U.S. LMFBR program (ERDA-1535, December 1975) projects a capital cost differential at introduction in 1993 of about 22 percent using an LWR capital cost of $460/kilowatt (1975 dollars) of installed capacity for a 1,000 MWe plant. This amounts to a capital cost differential of $100 per kilowatt.[c] In Chapter 3 an LWR capital cost of $667/kilowatt, in constant 1976 dollars, is projected. If breeder capital cost estimates are keyed to those of LWRs, one would project a capital cost differential of about $150/kilowatt using the EIS figures. In this section, we consider a range from 15 percent to 30 percent, or a differential of $100–$200/kilowatt in favor of LWRs. These figures do not include initial breeder core loads of plutonium, a cost we include in the fuel cycle cost.

The trend in capital cost differentials following introduction is also of importance. The EIS analysis assumes that breeder capital costs will decline to parity with those of LWRs within thirteen years of introduction. This optimism should be tempered by the observation that similar expectations for LWRs did not turn out to be correct. Indeed, it is perhaps more likely that maturation of the LWR industry, with larger plants, standardization, and improved capacity factors, will increase the capital cost differential with breeders during the early decades of breeder introduction. There appears to be no reason to believe that LWRs and LMFBRs will ultimately have the same capital requirements even if learning effects do occur.

To translate capital costs into power costs, it is necessary to make assumptions about the basis on which capital costs are assigned and about capacity factors for both kinds of plants. Capital charges are assigned here at a fixed annual rate of 12 percent and assuming that LWRs and breeders both operate at 70 percent capacity factor. Each $100 per kilowatt of capital cost differential between the two kinds of plants is then equivalent to a power cost differential of about 2 mills per kilowatt hour. The 12 percent fixed capital charge rate is appropriate to our later comparison of future benefits and losses on a discounted present value basis, using a discount rate of 10 percent. It is also approximately the rate which would be used by utilities in making technology choices if inflationary expectations were not a factor in interest rates; taking inflation into account, a utility would use a capital charge rate between 15 and 20 percent. Those favoring a social rather than a market decision on technology choice might use a lower capital charge rate in computing the relative merits of alternative technologies and a lower discount rate. We shall use the inflation-corrected utility rate, remembering that the effect of capital differentials on relative power costs may be reduced by about 30 percent if a social decision framework is preferred. This would not change our major conclusions.

The assumption that capacity factors for LWRs and breeders will be the same favors breeders if experience with LWRs is any guide. For example, one might expect LWRs to have achieved capacity factors of at least 70 percent

---

[c]The EIS assumes that larger LWR plants lead to reductions in LWR capital costs to $405/kilowatt in 1990, making the LMFBR differential $155/kilowatt at introduction.

twenty years from now while initial deployments of breeders might have difficulty in reaching the 70 percent figure. A difference of 10 percent in capacity would mean an additional power cost disadvantage for breeders compared to LWRs of more than 2 mills per kilowatt hour, comparable to an additional $100 per kilowatt capital differential.

Fuel Cycle Costs

Fuel costs for LWRs are well defined by the prices of uranium and separative work. This is not the case for breeder fuel cycle costs, which depend upon estimates for industrial operations still at the conceptual design state. Current estimates for the operations involved in the breeder fuel cycle, reprocessing, fabrication, waste management and disposal, and transportation, appear to be based on extrapolation from estimates made for an LWR recycle industry. Extrapolations are necessary because of basic physical differences between LWR and breeder spent fuels. For example, a reprocessing plant for breeders would have to deal with 12 percent plutonium concentrations as opposed to 0.6 percent for LWRs. Such higher concentrations are more difficult to dissolve and impose more severe criticality constraints, requiring a more expensive plant. Additional expenses also would arise from sodium contamination and radiation hardening of spent fuel assemblies, from higher heat and radiation output, and from higher unit waste handling charges. Similar differences affect other stages. Uncertainties in breeder fuel cost projections are made larger by the uncertainties and rapid escalation in LWR recycle cost estimates. Breeder fuel cycle costs also depend upon the fissile inventories required and upon breeding gain since carrying charges on plutonium in the fuel cycle will be high and perhaps only partly counterbalanced by the value of excess plutonium created.

In order to translate fuel costs into power costs, it is necessary to know how much power is generated, on the average, by each kilogram. For LWRs we assume a burnup of 33 thermal megawatt-days per kilogram at 33 percent thermal efficiency (these are the figures for a representative PWR). Breeders present a more uncertain picture. Breeder cores will have some elements with very little fissile material which undergo very low burnups and other elements which have high plutonium concentrations and undergo high burnups. The maximum burnup is presently limited by the properties of core materials which undergo degradation, swelling, and creep in the core environment. Some relief from these problems can be obtained by design, particularly through a reduction in core temperatures. For purposes of comparison we will use a burnup of 80 megawatt-days per kilogram, on the average over the core, at a thermal efficiency of 38 percent. This is better than has been achieved experimentally but below the ultimate design goal posed, some years ago, for the U.S. program of 110 megawatt-days per kilogram. The achievement of the 110 megawatt-day goal would require basic advances in materials properties. Using the 80 MWd/kilogram figure above, one finds that a kilogram of breeder fuel generates, on the average, nearly three times as much electrical energy as a kilogram of LWR fuel.

The fuel cycle contribution to the cost from LWRs, as a function of uranium price, is shown in Figure 12-1. The range of breeder fuel cycle costs considered is from 2 to 4 mills per kilowatt-hour.[d] These figures may still be low if experience with rapidly escalating predictions for LWR recycle are any measure. Fuel cycle costs are likely to be highest in the early years after introduction.

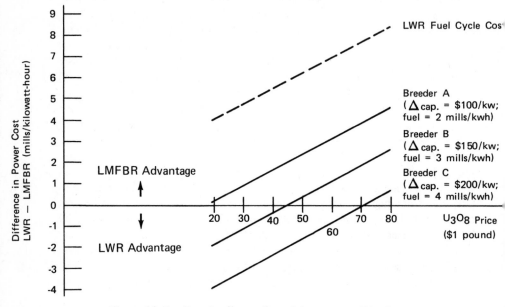

Figure 12-1. Breeder Power Cost Advantage or Disadvantage

Note: Power cost difference between LWR and LMFBR as a function of uranium price for breeders with different sets of capital and fuel cost assumptions. A capital cost differential of $100 per kilowatt of capacity is equivalent to about 2 mills per kilowatt hour in power cost. Thus breeder B could also be a breeder with a capital cost differential of $200/kilowatt, relative to LWRs, and a fuel cost of 2 mills/kilowatt hour. Also shown is the LWR fuel cycle cost (dotted line). Capital charges for the breeder are assessed at a 12 percent fixed charge rate in constant dollars. LWR fuel costs assume optimal tails assay at each uranium price, an enrichment cost of $75/SWU, a charge of 1 mill/kwh for conversion, fabrication, transportation, and waste management and disposal; interest on inventories of enriched and fabricated uranium is included.

[d]The precise cost depends not only on estimates for costs of currently nonexistent processes, but also on how plutonium inventory charges and plutonium credits are assigned. The fuel cycle cost here also includes the cost of the large initial inventory of plutonium needed to load the breeder reactor.

Comparative Power Costs

The power cost differentials between LWRs and breeders, as a function of uranium price, for various choices of breeder cost parameters are shown in Figure 12-1. The relative economic merit of breeders is critically dependent upon the cost factors which characterize the commercial version ultimately introduced and upon the trend of uranium prices. At low uranium price, say $40/lb U_3O_8, only a good breeder (A) would have a chance of competing with light-water reactors; at a uranium price of $60/lb, a less optimistic breeder (B) begins to compete; Breeder C begins to compete only when uranium prices reach about $100/ lb. Since average uranium prices are not expected to exceed about $40/lb by the end of this century, it appears that only breeder A, or one with even lower power costs, could be assured of successful introduction before 2000. Such a breeder would provide a 3 to 6 percent reduction in power cost during the first decade of the next century. Unfortunately, breeders deployable before 2000 may resemble breeder B or even C rather than breeder A, with some possibility of gradual improvements in performance as technical advances in fuels, materials, and design are made. This, and the necessity of competing with LWRs fueled with relatively low-priced uranium, militate against early economic success of LMFBRs.

Finally, the possible introduction of laser isotope separation technology having low unit costs and allowing more complete recovery of uranium-235 from the new uranium has a potential impact on relative costs. Success with this technology, which appears likely well before 2000, might reduce LWR power costs by 1 to 3 mills/kwh for uranium in the $40–$60/lb price range, depending upon the degree of separation feasible and upon ultimate cost of laser processes. Lasers might thus create an economic disadvantage for breeders comparable to that of an additional $100 capital cost penalty, greatly delaying the time at which breeders could be competitive.

An Illustrative Model

The uncertainties in breeder economics and the possibility that the breeder's advantage, if any, may be marginal, have important consequences for the U.S. program which is aimed at a decision on commercial introduction in 1986, with breeders coming on line in 1993. To evaluate the magnitude of gains or losses which might attend breeder introduction on this timetable, it is necessary to model the future growth of fission power and make projections about uranium prices. In the absence of hard information, all models are deficient. In what follows, we present a simple model which allows the reader to distinguish sensitivities to changes in major assumptions. It should be remembered that this model is not a prediction; like all similar models, it is of value only as a way of examining a range of potential future impacts of possible energy policy and R&D strategies to be pursued in the more immediate term.

Nuclear growth beyond the year 2000, including that of breeders, is conditioned partly by deployments of LWRs before that year. The rate at which LWRs are deployed has an effect on the price of uranium available at particular times in the future, with rapid growth leading to higher prices and an earlier potential competitive advantage for breeders. Conversely, low growth and lower uranium prices lead to delay in breeder opportunities.

Past comparative analyses of breeder benefits have used growth projections which now seem high. The Environmental Impact Statement for the current breeder program, prepared by ERDA in late 1975 (ERDA-1535), assumes as a base case 900 GWe of nuclear capacity in 2000 and 3,700 GWe in 2025.[e] Median estimates by ERDA for nuclear growth, only a year later, are now for 510 GWe of nuclear capacity in 2000. Since uranium prices are not likely to be high enough, nor uranium resources likely depleted appreciably by 2000,[f] it is unlikely that utilities will switch entirely to LMFBRs during the early part of the next century. By the year 2000, LWR deployments will have consumed less than 1.2 million tons of U_3O_8 (at 0.2 percent tails, less if lasers work) if only about 500 GWe are deployed. Commitments for the remaining lifetime of these reactors would be for an additional 2.8 million tons. This moderate LWR deployment would provide adequate plutonium to fuel a moderate growth of breeder capacity in subsequent decades. Whether plutonium would be a constraint on breeder growth depends upon the specific fissile inventory required (present estimates range from 3 to 6 kilograms of plutonium, in and out of reactor, per megawatt of capacity) and upon the availability of enriched uranium which could be used as an alternative to plutonium in initial breeder core loads.

Taking these factors into account, a large number of projections of nuclear growth through 2025, involving somewhat different mixes of LWRs and breeders, could be made. If an optimal breeder (e.g., breeder A) were introduced in 1993, the subsequent growth pattern might look something like that in Figure 12-2.[g] The LMFBR deployment rate shown would use most of the plutonium available after about 2010 if a specific fissile inventory of 4 kilograms per megawatt could be achieved by that time. Higher rates of growth would require use of enriched uranium. In order to compute benefits relative to using LWRs, it is necessary to assume a schedule of uranium prices. Uranium use in LWRs through 2025 would remain below about four million tons if a policy of optimum tails in gaseous diffusion enrichment is pursued; uranium use might remain below about three million tons if laser separation is introduced. Given

[e]A "low" nuclear growth case of 625 GWe in 2000, 1,730 GWe in 2025 and a "high" nuclear growth case of 1,250 GWe in 2000, 5,140 GWe in 2025 were also considered. The example considered here is very similar to the "low" growth case considered in the EIS; despite differences in capital, fuel cycle, and uranium price assumptions, the results of the EIS analysis are comparable to those presented here.

[f]See Chapter 2 for discussion.

[g]This scenario closely resembles the "low" growth case considered in the EIS (ERDA-1535).

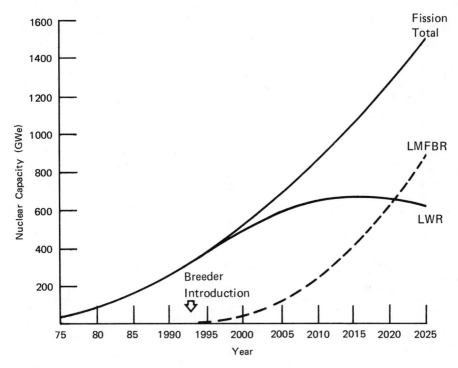

Figure 12-2. Illustrative Nuclear Growth Projection with Breeder Introduction in 1993

Note: Dashed line shows breeder deployments alone. Effects of alternative fission technologies are ignored.

this low demand and the likelihood that there are large quantities of uranium in moderate cost categories, uranium prices would not be expected to rise too rapidly. Again, the concreteness of these figures should be construed not as a prediction but only as a necessity in constructing a representative case.

The growth example in Figure 12-2 also assumes that some LWRS are deployed after 2000, reflecting the narrow competitive position of LMFBRs and the tendency of utilities to hedge against uncertainties which would be large for LMFBRs, by deploying a mix of power plant types. It should be noted that, at the maximum uranium prices assumed, LWR and coal power costs would be comparable and that LMFBRs would thus be competing with two readily interchangeable and economically similar power sources.

The annual discounted power cost savings associated with the deployment model of Figure 12-2 are shown in Figure 12-3. Introduction of breeder A

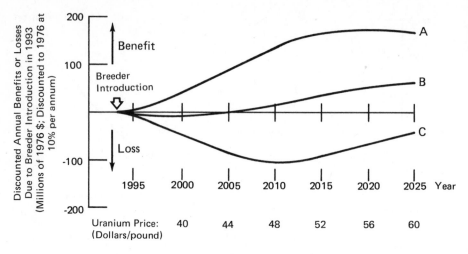

Figure 12-3. Yearly Savings or Losses Due to Breeder Introduction in 1993

Note: Annual savings or losses are discounted present values (discounted from the year incurred to 1976 at 10 percent per year) for three breeders with the illustrative cost parameters of Figure 12-1, the reactor growth pattern of Figure 12-2, and the schedule of uranium prices for LWRs shown at the bottom of the figure. All figures are in constant 1976 dollars.

in 1993 would give a total benefit of about $3.4 billion prior to 2025 when discounted to the present at 10 percent per year. By comparison, the U.S. development and demonstration program is estimated to cost at least $8 billion when discounted to the present at 10 percent. Most of the program costs are to be incurred in the next ten years; most of the potential benefit accrues after 2010. If the program costs are charged against potential benefits there is thus a net loss, in present values, of more than $4 billion. Also shown in Figure 12-3 are the losses and benefits resulting from introduction of less optimal breeders (B and C) in 1993. These less optimal breeders clearly present a disincentive for early deployment. Such breeders might be deployed, however, if their introduction were believed essential for other reasons. The public subsidy would not be prohibitive, perhaps on the order of $10 billion to $20 billion at 10 percent discount rate, including program costs (if these costs do not rise further). The changes in net power cost involved in all of these cases are relatively small, generally in the range of ±1-3 mills per kilowatt hour or between 3 and 9 percent of total power cost.

Breeder introduction may result in benefits or losses beyond those computed above, depending upon the effects on uranium prices for LWRs. If breeders free uranium supplies, leading to lower uranium prices, then there will be a benefit from lower LWR power costs. Using the elasticity implicit in the price schedule of our illustrative example, this induced savings effect may be as large as 0.2-1.0 mills per kilowatt hour, depending upon the period in question. If this effect should occur, it might be comparable to that due directly to power cost savings (or losses) for breeders as compared with LWRs. However, it is possible that breeder introduction could have the opposite effect. Uranium supplies and prices will undoubtedly reflect the enthusiasm with which new reserves are pursued; breeder introduction, or even the perception of its imminence, may reduce incentives to discover and develop new supplies, ultimately leading to higher prices for LWR fuels.

Economics of Delay

The preceding analysis, despite large uncertainties, suggests that prospective benefits from early breeder introduction are very small and that introduction on the present timetable may actually occasion a loss. Delay in introduction is thus an important policy alternative. As is discussed elsewhere in this report, social cost considerations provide substantial reasons for delay. What would the economic consequences actually be? If it were possible to deploy a good breeder (example A above), delaying it might imply some loss of benefits (as long as program costs are not too high). Conversely, delaying a less successful breeder until uranium prices have risen would avoid losses and the need for public subsidy. To estimate the economic magnitudes of these effects, it is useful to consider an example in which breeder introduction is delayed fifteen years beyond the 1993 date presently used as the basis for the U.S. program. Assuming that nuclear power growth continues at the rate used in the model above but with LWRs instead of LMFBRs until 2008, the simple model projection of Figure 12-2 becomes that in Figure 12-4. The additional LWR operations require an additional 750,000 tons of U_3O_8 (up to 40 percent less with lasers) before 2025. The rate of deployment of LMFBRs following introduction is slightly higher than with early introduction.

The changes in fission power costs in each year due to delay of breeder introduction, compared to those which would occur with early introduction, are shown in Figure 12-5. Delaying breeder A results in an increase in power costs; delaying the less advantageous breeder C results in a decrease. The total direct cost of delaying breeder A, should it actually be available in 1993, is $2 billion (present value discounted at 10 percent). If it is true that uranium prices would be higher without early breeder introduction, then there is a loss due to having to pay higher prices for power from LWRs; the annual magnitude of this loss is shown as the dashed line in Figure 12-5. The total loss due to this effect over the entire period would be $1.2 billion, assuming the uranium price increases shown. The greatest loss possible, under our model assumption, is thus $3.2 billion over the period 1976-2025.

These potential losses or benefits may be compared with changes in program costs which might be occasioned by delay. Reductions in program costs, or delay

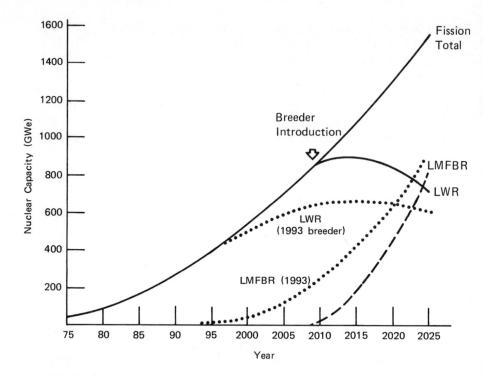

Figure 12-4. Illustrative Growth Pattern with Commercial Breeder Deployment Delayed by 15 Years

Note: Dotted lines show 1993 introduction deployments. Total fission growth is constrained to be the same as in Figure 12-2.

of such costs further into the future, might remove any possibility of loss, even from the delay of an advantageous breeder. For example, the direct power cost loss of delaying breeder A is comparable to the present cost estimate of the CRBR. The decision on whether to delay commercial breeders has very little impact on power costs: average fission power costs would be at most 1.5 mills/kwh (or less than 5 percent) higher or lower, during any of the years 1976–2025. If present program costs are charged against future benefits, as they should be, there is a net economic disincentive to introduction of breeders on the present timetable. This disincentive is even larger if breeders do not have capacity factors comparable to mature LWRs or if laser isotope separation technology reduces LWR power costs.

These cost-benefit figures may be compared with those yielded by the economic analysis in the Environmental Impact Statement (ERDA-1535). For a nuclear growth projection of 625 GWe in 2000 (20 percent higher than in our

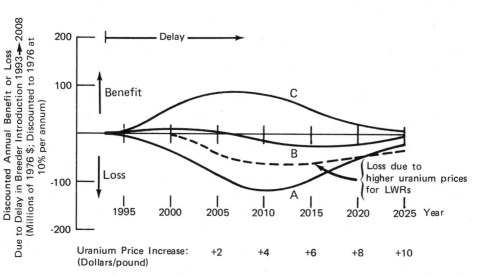

Figure 12-5. Benefits or Losses Due to Delayed Breeder Introduction

Note: Yearly savings or losses due to delaying breeder introduction from 1993 to 2008, as shown in Figure 12-4. Solid lines are direct costs (or benefits) of substituting LWRs for breeders in given years. Thus, there is always a gain from delaying breeder C and a loss from delaying breeder A. If uranium prices rise due to breeder delay (see text for discussion), LWR energy costs will be higher, occasioning the additional loss shown by the dashed line. All amounts are discounted present values (discounted to 1976 at 10 percent per year).

case), with four million tons of uranium available at prices below $60/lb and LMFBR capital cost declining to parity with LWRs, the discounted total power cost savings due to 1993 introduction is given as $8 billion (discounted at 10 percent) up until 2025, before program costs are subtracted. Subtracting program costs, the net benefit is zero. Delay of introduction until 2000 reduces total power cost savings from $8 billion to $5 billion; unless program costs were also reduced by delay, there would thus be a small net loss due to delay. With a capital cost differential of $100 per kilowatt, the discounted power cost saving of 1993 introduction is only $2 billion; for introduction in 2000 the saving is $1 billion. Neither of these is large enough to offset the current program costs. Despite optimism about both LWR and LMFBR power costs (which appear to be about half of those projected in this report), the EIS thus projects results similar to those given by the very simple model in this chapter.

Analyses done earlier have generally projected larger potential breeder benefits than those suggested here. Earlier analyses usually assumed rapid growth of LWR capacity, early depletion of low-cost uranium reserves, and favorable breeder capital and fuel cycle costs. Thus the EIS for the breeder program projected a breeder benefit for the period prior to 2025, of $52 billion (in 1975 dollars, discounted at 10 percent) for a "reference" growth case envisioning 900 GWe of fission power in 2000. The factors responsible for our specific conclusions, which are more optimistic for LWRs and less so for breeders, are lower fission growth, greater availability of uranium, and higher breeder capital and fuel cycle costs. These are by no means extreme assumptions, as we have seen from comparison with the EIS. Our general conclusion that breeder delay is economically acceptable, however, would hold even with assumptions much more favorable to breeders. For example, the high nuclear growth case considered in Chapter 1, which assumed high uranium prices, a rapid displacement of LWRs and coal plants by breeders and favorable capital and fuel costs for breeders, projected a loss in delaying breeders from introduction in 2000 to 2020 of $28 billion, discounted at 10 percent. The changes in total energy costs, even projected under assumptions optimistic for breeders, are a very small fraction of annual GNP and would have little bearing on future economic health or growth. Since this conclusion is virtually independent of what happens outside the fission energy sector, our analysis may be generalized to situations outside the United States. As long as low-enriched uranium is available in the world market at prices comparable to those projected for the United States, there appears to be little or no economic reason for any country to install breeders in preference to LWRs or comparable converter reactors until well into the next century. However, some countries may find noneconomic reasons to do so.

One is led by the analysis above, and by that in Chapter 1, to the conclusion that, under most reasonable circumstances delayed introduction of breeders has little economic consequence. Delay of an economically superior breeder by fifteen years occasions a loss over the period 1993-2025 comparable to potential savings in LMFBR program costs, most of which costs would otherwise be incurred before 1986. In the more likely event that breeders deployable in 1993 will not offer such large advantages but will improve slowly, if at all, compared with LWRs, delay is advantageous. Unless new, more competitive, sources of energy are developed, the real opportunities for breeders are beyond 2025; our ability to claim these benefits will be little different whether commercial introduction of an early breeder design occurs in 1993 or a more advanced design in 2008 or even later.

SOCIAL COSTS

The risks and potential costs associated with light-water reactors have been discussed extensively in this report. The potential risks presented by LMFBRs are qualitatively similar to those of LWRs. Waste management and disposal would not be very different, and both technologies present risk of very serious accidents. The differences arise in the implications of a massive and irreversible commitment to a plutonium fuel cycle.

The Plutonium Economy

Unlike LWRs, where use of plutonium recycle is optional, breeders as they are presently envisioned require commitment to a plutonium fuel cycle involving extensive transportation and industrial processing of this material. The scale of this undertaking would be very large. If domestic LMFBR deployments reach 1,000 GWe (as in the example of the last section), there would be an active inventory of three to six million kilograms of plutonium. About half of this amount would be removed in spent fuel each year from reactors, reprocessed, fabricated into new fuel and transported in relatively pure form several times. This annual civilian commerce would be more than ten times the total amount of plutonium produced in the U.S. military program over the past several decades. The rate of growth of the plutonium economy would also be high. A yearly increment of eighty breeders would require at least one new reprocessing plant, up to ten fabrication plants, ancillary facilities, and new transportation links to handle more than two hundred thousand kilograms of new plutonium (comparable to total U.S. military plutonium inventory). The risks and social costs of the breeder and the resulting plutonium economy include an increase in the threat of nuclear proliferation and in the possibility of thefts of materials suitable for weapons by terrorist groups, and, unless ideal conditions can be achieved, potential increase in the long-term hazards to human health.

Proliferation. Introduction of breeder reactors would certainly complicate efforts to control nuclear proliferation. The associated fuel cycle will inevitably result in the spread of facilities that provide all but the last step in the production of nuclear weapons. Anticipation of early introduction of plutonium breeders has in fact led some countries to value more highly early experience with plutonium technology, such as reprocessing of LWR spent fuels, or even to stockpile accumulations of plutonium, as spent fuel or in separated form. As discussed in Chapter 9, these actions lower thresholds and shorten timetables for weapons acquisition.

The U.S. view of its own need for plutonium-based breeders and its timetable thus have a bearing on the proliferation problem. The United States has played a major role in shaping world expectations regarding breeders and, indeed, inspired many of the foreign breeder programs now underway. Reexamination of U.S. needs and estimates of what it is possible to do at economic advantage or low cost in delaying commitment to a plutonium economy can still have an impact on decision processes in other countries. Rapidly rising program costs and estimates of commercial breeder capital costs have recently stimulated reexamination in several other countries. While a delay in the U.S. breeder program and reconsideration of alternative nuclear options may have only modest impact on the proliferation problem, a continuation of the present urgent U.S. program would probably guarantee the early introduction of a plutonium economy without significant economic benefit to the United States or to other countries.

Theft. The large quantities of plutonium in a worldwide breeder economy also pose increased risk of theft. Annual plutonium commerce might provide enough plutonium for several hundred thousand nuclear weapons. An annual core reload for a single large LMFBR would provide enough for 200 to 400 weapons. Plutonium from LMFBRs having lower concentrations of plutonium-240 would be at least as suitable for this use as that from LWRs. It is likely that breeder introduction, if it occurs, will be attended by a rapid growth in international plutonium commerce, despite the prevalent belief that breeders provide a way to achieve energy self-sufficiency. With many shipments of spent fuel, separated plutonium, and fresh mixed oxide fuel, it would be difficult to provide physical security and safeguards on the scale required, particularly in the present world.

Health Effects. The health effects associated with plutonium are discussed in Chapter 5. While there is considerable uncertainty about the health consequences of exposure to plutonium, the largest uncertainties involve society's ability to manage the large quantities of plutonium involved. Risk assessments should take into account the cumulative effects of small spills, leaks, and mishaps, as well as accidents, rather than relying only on design objectives. Such assessments are at present embryonic at best. The health risks of the breeder plutonium fuel cycle should be compared with fuel cycles that do not require separation and use of plutonium, such as LWRs and thorium cycle reactors.

Safety. Reactor safety considerations play an essential role in the prospects for success of the breeder. As discussed in detail in Chapter 7, the LMFBR safety issues differ substantially from those associated with LWRs. Research on LMFBR safety is focused primarily on five areas: whether an abnormal condition in one part of the core can propagate to involve a larger fraction of the core; whether positive reactivity effects from sodium voids can be controlled; whether recriticality can occur in a core meltdown accident; whether the energetics of criticality and recriticality accidents are such that containment failure is likely; and what quantities and types of radioisotopes might be released subsequent to containment failure.

The present U.S. program dealing with these basic safety issues sets target dates for resolution which extend beyond final conceptual design for the commercial demonstration plants. For example, study of the sodium void effect is scheduled for completion in 1981 and that of the analysis of the consequences of core disruption for 1986. It is difficult to say how the results of these investigations will be integrated into commercial plant design.

The present schedule implies a certain degree of concurrency of safety research and design with engineering of a commercial LMFBR. Accordingly, the LMFBR program will face a similar situation as that encountered with LWRs, where conservatism of design and multiple safety barriers have to be substituted for fully adequate information. Such conservatism in turn increases costs and thereby delays the date by which the LMFBR can successfully compete with LWRs on economic terms.

Conservative design is frequently of uncertain efficacy and some of the costs engendered are other than financial. For instance, lack of knowledge on the reliability of materials at high temperatures forces reduction in operating temperatures with consequent loss in efficiency. Circumstances can arise in which strengthening a reactor in one way makes failure in another more likely. This problem is illustrated by the accident initiated by the deflector plate in the Fermi reactor which was intended to be a conservative design feature. Once conservative design has been adopted, it tends to be frozen into regulatory requirements. Therefore, economies which would be possible through future relaxation of conservatism in view of acquired superior knowledge are difficult to achieve in practice.

One should recognize that the identified problems could be alleviated but never eliminated if the LMFBR schedule was lengthened. There will always be some residual doubts on safety issues in systems as complex as those in an LMFBR. In this situation, public acceptance will be enhanced by the judgment of the technical community, both inside and outside the breeder program, that there exists a convincing basis for confidence in breeder safety. Although we do not believe that a schedule delay of the LMFBR is required for safety reasons, we conclude that such a delay would make it possible for safety research to contribute more effectively to LMFBR designs.

U.S. PROGRAM STRATEGY

While there are good reasons to delay commitment to plutonium breeders and such a delay would not be economically consequential, breeders are a major energy resource. They provide high confidence insurance against failure of other energy sources in the future. The breeder option should be preserved since the cost of uranium will eventually rise, coal may eventually be found to have unacceptable adverse consequences that cannot be avoided, and other alternative energy sources may prove to be very expensive. The nature of the U.S. program strategy most appropriate to the goal of long-term insurance is a difficult issue. Although a shift away from the present orientation, which promotes early commercialization, is clearly indicated, the task of designing a new approach is beyond the scope of this study.

The costs of delayed breeder introduction lie primarily in the loss of program momentum that might result from the removal of an early commercial goal. The present schedule of commercial prototype reactors is seen by those working in the breeder program as the culmination of several decades of effort and as the basis on which past research and development will be given the test of commercial viability, licensability, and public acceptance. Proponents of the program would argue that such a focus is essential to further research and development and to resolving questions, such as safety, which require research and engineering development to meet standards set by an independent regulatory process.

The present urgent program schedule requires the nearly concurrent development of a sequence of increasingly large reactor prototypes with design and construction steps of each reactor following closely upon those of the preceding. The higher risk entailed by the lack of usable operating experience in the U.S. program is backstopped by extensive and expensive component development and testing efforts in the base program. The concurrent approach offers the advantage of continuity; the sequential approach favored in some foreign programs offers that of usable experience. A more relaxed timetable would allow either a stretchout of a prototype sequence or a delay in initiating a concurrent sequence. In either case, there would be larger opportunities to integrate important new research efforts into the reactor development process.

The emphasis on early commercialization requires that choices be made now on design features such as a loop rather than a pool design and on conservative design features where knowledge of better materials or components are lacking. These choices narrow the scope of future alternatives and may have adverse consequences for the ultimate success of breeders. If the range of economic and socially acceptable breeders is narrowly circumscribed, as it appears to be, an early commitment to a less than optimal breeder could compromise the availability of a better breeder later.

A broader program might involve thorium technologies, gas-cooled designs, molten salt reactors, and other possibilities. Some of these reactors and fuel cycle systems may offer advantages in having lower proliferation and theft dangers than the present LMFBR. Some of the research presently being done on physics, materials, chemistry, and so forth could be extended to lay a basis for alternate reactor concepts at a cost small compared with that of the present program. Construction of experimental prototypes could also be inexpensive when compared with commercial-scale demonstration projects like the Clinch River breeder. It is at the commercial development and demonstration phase that large program budgets are involved. Less than $1.8 billion were spent on LMFBR research and development through fiscal year 1974 while current projections are for more than $10 billion additional federal funds, primarily for demonstration plants and commercial-scale component development and testing. This figure assumes substantial utility contributions for plants beyond CRBR and is likely to be a low estimate, given the history of cost escalation in other reactor projects.

In part, the high cost of the present U.S. program, in contrast, say, to that of France, may be traced to a program philosophy which goes beyond technical demonstration to include commercial demonstration and the creation of a "competitive" industrial base for breeders. The latter requires continuity in order that embryonic industries (design teams, component manufacturers, fuel fabricators, etc.) not falter and this can only be provided by a concurrent plan. This emphasis applies at present only to the reactor part of the program. The lag in developing LMFBR fuel cycle technology on a commercial-scale basis may, in part, be due to recognition that it will not have a commercial incentive until there are adequate numbers of breeders to service. The necessity of involving large numbers of participants (government, many different utilities, and potential vendors and others) in the demonstration phase is also awkward, as the CRBR project has shown, and probably more expensive. The problem of transfer of technology from government laboratories to industrial practice in the United States is not a new problem, nor is it an easy one to solve. However, it should be easier to solve when there is a more compelling economic incentive than is likely to be the case with the present program schedule.

The risk of delaying the creation of a breeder industrial base in the 1980s, or even the 1990s, is not that the United States will run out of energy or that its economy will falter but that electricity prices in the early decades of the next century will be a mill or two higher, or lower. Since the timetable for the breeder can be relaxed without large economic risks, it is natural to reopen the question of whether alternative converter reactors can still play a role. The present U.S. program does not pay much attention to new converters such as High Temperature Gas-Cooled Reactors, natural uranium reactors like the Canadian CANDU, molten salt reactors, or other technically feasible machines. Such technologies, rich in their variety in the early years of the U.S. program, have been driven out by the success of the LWR and by the presumed imminence of the LMFBR. Light-water reactors are very inefficient in their use of uranium. Natural uranium reactors, like the CANDU reactor for example, can already obtain 20-40 percent more energy from unenriched uranium. The commercial use of more efficient converters is largely a question of economics, a situation which will change if uranium prices rise and breeders are less urgent. The CANDU appears to be close to being competitive; it has a higher capital cost than an LWR but could have a compensatingly better average capacity factor. Use of a thorium cycle converter would also tap a new resource base. There are regulatory and other hurdles to introduction of such technologies, but these can hardly be more substantial than those associated with the LMFBR.

The connections of the U.S. breeder program with foreign programs are of particular interest in considering revisions in the U.S. course. Plutonium cycle breeders are being pursued in a number of countries, notably France, West Germany, the United Kingdom, Japan, and the Soviet Union. While many of these foreign efforts were stimulated by earlier U.S. programs and technical exchanges, they are now independent and in some cases more advanced than U.S. efforts. This apparent technical success abroad is often cited as an imperative to more rapid U.S. development efforts, lest preeminence and future markets be lost. It has also been suggested that the United States could not afford to be dependent on foreign technology since foreign reactors could not satisfy U.S. regulatory standards.

Being first with a technological development program does not always mean success in economic or other terms since technical success must be matched to real needs. The supersonic transport provides a contemporary example. The examples of France, Germany, and the United Kingdom in adapting U.S. LWR technology to their own needs, often with improvements, on a short timetable and with little economic penalty refute the argument that such transfers are impossible or necessarily costly. The largest economic losses in some of these countries have been associated with the abandonment of domestic reactor technologies in favor of the more satisfactory U.S. versions. Foreign breeder technologies could, in most cases, be matched with U.S. regulatory standards through a process of mutual adaptation similar to that occurring for a domestic technology: a process that would be more difficult if a regulatory framework for a domestic breeder had already been detailed. However, this is not yet the case, and there is indeed evidence that some U.S. vendors are quietly exploring the possibility of importing and adapting foreign breeder technologies. Despite the urgent schedule of the U.S. program, the first information on the competitive merits of breeders may well come from foreign programs. Motivated by rapid escalations in prototype cost estimates, Japan and some European countries are apparently reexamining the economics of the breeder.

Because it is the first in the commercialization sequence, the Clinch River breeder project deserves immediate consideration in any reexamination of the U.S. program strategy. The stated purpose of the CRBR is "to demonstrate technical performance, reliability, maintainability, safety, environmental acceptability and economic feasibility of an LMFBR central station electric power plant in a utility environment. . . . It is further intended that this project will establish the licensability of the LMFBR in today's regulatory environment." The CRBR is thus not intended as a research facility; the design similarity and far greater flexibility of the Fast Flux Test Facility, and its much more immediate availability, make it a superior engineering laboratory. Our analysis implies that the construction of a commercial demonstration plant is premature. However, even if this were not the case, the CRBR would probably not provide very convincing demonstrations of the goals stated for it. Begun in 1969, the CRBR project still has not begun construction and criticality is now predicted for late 1983 at the earliest. In the interim, cost estimates have escalated to nearly $2 billion, or roughly $5,000/kilowatt. The utility industry's lack of enthusiasm for the project may be reflected in its unwillingness to commit additional funds much beyond the $250 million originally pledged. Finally, the design for the CRBR, based on design studies conducted in the late 1960s and early 1970s, is about to be superseded by the new ERDA/EPRI cooperative design studies for the PLBR, the scheduled successor to the CRBR. It is thus difficult to escape the conclusion that the CRBR does not have a value consistent with the likely cost, even under the assumption, which we believe incorrect, that there is need for an urgent LMFBR commercialization program.

CONCLUSION

There is no question that the present version of the breeder stressed in the U.S. program, the plutonium-cycle LMFBR, will successfully generate power. It thus provides insurance against very high energy costs in the future. The hard question concerns the timetable on which a breeder might have sufficient economic merit, compared with other sources, to outweigh the potentially large social costs associated with a plutonium fuel cycle.

On the basis of our analysis, we conclude that there is little advantage in terms of economics or energy supply assurance in early commercial introduction of LMFBRs. Introduction of the breeder may be deferred for ten, twenty, or more years without seriously affecting the economic health or energy security of the United States. As long as there is a world market in low-enriched uranium, a similar conclusion appears to apply to other countries. The social costs associated with breeder introduction argue strongly for deferral.

The relaxation in the breeder timetable recommended here has implications for the nature of the U.S. strategy in research and development on fission alternatives. There is time to pursue a broader research and development program on the breeders and on more efficient converter reactors. Delay will permit the development of more successful breeders should they be needed. In such a program, the Clinch River project, as presently conceived, is not necessary and could be canceled without harming the long-term prospects of breeders. In fact, a premature demonstration could even be detrimental to these prospects.

It is important to continue work on the breeder, with a longer time horizon and an emphasis on its role as insurance. The goal should be to provide a range of more attractive choices at a series of decision points extending into the early decades of the next century.

Chapter 13

Uranium Enrichment

Approximately 90 percent of the world's present and planned nuclear generating capacity requires slightly enriched uranium as fuel. The questions of assurance of enrichment services will thus be critical for virtually all nations with a substantial commitment to nuclear power. The major exception to this is Canada, whose reactors use natural uranium. In the past, the United States has provided such services for all countries outside of the Communist world, using enrichment plants built in connection with its weapons program. The Soviet Union provides similar services for its reactors and those in Eastern Europe, and has contracted to provide some enrichment for Western European countries.

At present there is a surplus of enrichment capacity. There is a question, however, whether there will be adequate capacity in the future. The answer to this question depends on the growth of nuclear power and on expansion of enrichment capacity. Because lead times for construction of both reactors and enrichment plants are long—seven to ten years—it is now possible to estimate upper limits to growth in nuclear reactor deployments and enrichment capacity until about 1985, but the uncertainties in both are large for the post-1985 period. Because of slippage or cancellation of nuclear plants, the demand for enrichment services may be well below current estimates for the period 1980–85.

As recently as mid-1974, it appeared that the United States would not be able to meet all future demands for enrichment, even with the completion of an upgrading plan that would expand the capacity of its three gaseous diffusion plants from 17.2 million separative work units (SWUs) to 27.7 million SWUs.[a]

---

[a]Separative work is measured in units of mass, and throughout this chapter when the term SWU is used it is to be construed as meaning kilograms of separative work. In the literature

There was a consequent widespread belief that U.S. enrichment capacity should be expanded and, considering the lead times involved, that a decision on expansion should be made quickly. The issue was complicated by dispute about whether additional capacity should be built by the Energy Research and Development Administration (ERDA) or by the private sector, and by the possibility that centrifuges might perform enrichment at costs lower than those of gaseous diffusion plants. Implicit in expansion plans was the assumption that it was desirable that the United States continue to be the major supplier of enrichment services for the non-Communist world.

The fear of a possible shortage of enrichment services was exacerbated by a change in Atomic Energy Commission (AEC) contracting policy from "requirements" contracts to "fixed commitment" contracts, a change which would make private investment in enrichment more attractive. Under the old policy, the AEC agreed to provide service as required for reactors, but with fixed commitment contracts, utilities must pay for and accept enrichment even if a cancellation or slippage in the commercial service date for a reactor makes the delivery of enriched fuel on schedule unnecessary. The effect of this was to commit AEC/ERDA capacity fully and thus to make future availability uncertain without early expansion. The situation was relieved somewhat by an ERDA offer to permit amendment of fixed commitment contracts, on a one-time-only basis, until August 18, 1975.

Since mid-1974 the situation with respect to expected enrichment service demand and supply has changed dramatically because of deferral and slippage of nuclear power plant construction; while there may be distributional problems, it seems unlikely that there will be a world shortage of enrichment capacity in the mid-1980s, even if no additional American plants are built. Still, there are questions about when new capacity will be needed; which technology should be used in expanding enrichment capacity; how large a role the United States should attempt to play in meeting demand for enrichment services in the non-Communist world, and whether it should attempt to continue to dominate this market; and whether additional capacity in the United States should be built by ERDA or by the private sector, and if by the latter, with what kinds of government guarantees.

From the perspective of the world community it is desirable that proliferation of enrichment plants be limited since they can be used, if need be with modification, to produce highly enriched uranium (uranium with 90 percent or more uranium-235), which can be used to manufacture weapons. It is particularly

---

the term metric tons of separative work is sometimes used. For light-water reactor fuel it is usual to enrich the concentration of uranium-235 from the natural composition, 0.71 percent, to about 3 percent. This requires 3.425 SWUs per kilogram of product if the depleted uranium, the tails, contains 0.3 percent uranium-235 or 4.306 SWUs per kilogram of product if the tails assay is 0.2 percent uranium-235. A 1,000 MWe reactor operating at 65 percent capacity will require about 90,000 SWUs per year, assuming tails assay of 0.3 percent, or 110,000 SWUs per year if tails assay is 0.2 percent.

troublesome that centrifuges, which are now becoming operational, can be more easily adapted than gaseous diffusion plants to produce highly enriched uranium, and that lasers, if development is successful, will probably be even more adaptable.

SUPPLY AND DEMAND FOR ENRICHMENT SERVICES

Projections of enrichment capacity available to the future non-Communist world are given in Table 13-1. The growth in ERDA capacity projected is due solely to improvements at the three existing gaseous diffusion plants. The projections for Western Europe include small amounts from the existing diffusion plants during the next several years; for the post-1982 period, they include only the URENCO[b] centrifuge capacity (2 million SWUs) and that of the Eurodif[c] diffusion plant (10.7 million SWUs). If some of the proposals that have been made for additional non-American plants are actually implemented, capacity could be substantially greater. These include the possibility of an expansion of URENCO capacity, possibly to 10 million SWUs per year by 1985; production of perhaps 5 million SWUs per year by UCOR (South Africa) by 1986; and 9–10 million SWUs per year by COREDIF[d] in the late 1980s. Thus, even aside from possible further expansion in American capacity, the supply projections in Table 13-1 should probably be regarded as reasonably conservative.

Demand for enrichment services will depend on the growth in nuclear power, on the tails assay at which enrichment plants operate, and on the extent to which plutonium and uranium from spent fuel is used in fueling light-water reactors. Recycle of plutonium and uranium is not likely to be very important. The Nuclear Regulatory Commission (NRC) has estimated that recycle would reduce separative work requirements in the United States by 14 percent over the period 1976–2000. Our estimates of demand for separative work may prove somewhat high since they are based on the assumption that uranium and plutonium from spent fuel will not be reprocessed and recycled. This is thus a conservative assumption in assessing the need for additional enrichment capacity.

The economically optimum percentage of uranium-235 in the tails from an enrichment plant increases with the cost of separative work and diminishes with an increase in the cost of uranium. At present uranium and enrichment prices, the optimum tails assay is about 0.2 percent. When uranium prices were lower, a few years ago, operating at a higher tails assay (0.3 percent) made more sense. High tails assay also makes sense if there is a shortage of separative work capacity; this prospect was responsible for ERDA's current policy of gradually increasing

---

[b]URENCO is a joint British, Dutch, West German organization with plants in The Netherlands and the United Kingdom.
[c]Eurodif is a joint venture involving France (52 percent), Italy, Belgium, Spain, and Iran, which is building a gaseous diffusion plant in France.
[d]COREDIF will involve the same countries as Eurodif.

Table 13-1. Projected Schedule of World SWU Supply* (millions of SWUs)

Year	ERDA	Western Europe	Soviet Union	Total	Cumulative Total
1975	13.6	0.6	0.4	14.6	32.6
1976	16.1	0.6	0.8	17.5	50.1
1977	17.1	0.6	1.6	19.3	69.4
1978	18.4	0.8	2.5	21.7	91.4
1979	21.6	2.6	3.0	27.2	118.3
1980	24.6	6.1	3.0	33.6	152.0
1981	25.3	9.4	3.0	37.7	189.7
1982	25.3	12.5	3.0	41.2	230.5
1983	25.5	12.7	3.0	41.2	271.7
1984	26.7	12.7	3.0	42.4	314.1
1985	27.7	12.7	3.0	43.4	357.5
1986	27.7	12.7	3.0	43.4	400.9
1987	27.7	12.7	3.0	43.4	444.3
1988	27.7	12.7	3.0	43.4	487.7

*Based on estimates given in "Report of the Edison Electric Institute on Nuclear Fuels Supply." Cumulative total is based on an ERDA stockpile of 18 million SWUs as of January 1, 1975. The projections for Western Europe include only small amounts from existing plants for the next few years. After 1982, only the firmly committed URENCO capacity (2 million SWUs) and Eurodif capacity (10.7 million SWUs) are entered. Proposed expansions and new plants would greatly increase the figures in this table in the later years.

tails assay to 0.3 percent. Our projections of demand are made on the basis of both 0.2 percent and 0.3 percent tails. Separative work requirements are about 20 percent less if tails are stripped to 0.3 percent rather than to 0.2 percent uranium-235, but uranium ore requirements are about 26 percent greater.

Estimates of the nuclear generating capacity for the non-Communist world (excluding reactors fueled with natural uranium and possible breeder reactors) given in an Edison Electric Institute study released in March 1976[e] are shown in Table 13-2. More recent projections suggest that these estimates are much too high. Table 13-3 shows a recent ERDA projection for the United States.

ERDA's low projection now seems more probable than the other cases. In establishing a range of requirements for separative work, we have used ERDA's low projection and ERDA's mid-case (which is hereafter referred to as the high estimate) in this analysis. Noting that the growth for these two cases is only about 55 percent and 72 percent respectively of that projected in Table 13-2, we have scaled the projections for foreign growth using these factors to produce what we believe are more reasonable projections. This results in a low to high range of non-Communist world generating capacity of 123 to 143 GWe in 1980; 280-339 GWe in 1985; and 463-594 GWe in 1990.

In calculating requirements for separative work, we have used these projec-

[e]"Report of the Edison Electric Institute on Nuclear Fuels Supply."

Table 13–2. **Non-Communist World Nuclear Generating Capacity, GWe Edison Electric Institute, March 1976 (Excluding Natural Uranium and Breeder Reactors)**

	1975	1980	1985	1990
U.S.	39.5	77.3	185	340
Other	21.9	97.3	261	469
Total	61.4	174.6	446	809

Table 13–3.   U.S. Nuclear Generating Capacity, GWe

	1975	1980	1985	1990
Low case	39	60	127	195
Mid case	39	67	145	250
High case	39	71	166	290

Source: Edward J. Hanrahan, Richard H. Williamson, and Robert W. Brown; Office of Planning, Analysis and Evaluation, U.S. Energy Research and Development Administration. Presented at the Atomic Industrial Forum, "International Conference on Uranium," Geneva, Switzerland, September 14, 1976.

tions, assuming a 65 percent capacity factor, and 0.2 and 0.3 percent tails assay. Figure 13-1 shows the cumulative surplus in separative work occurring for each demand projection, assuming the supply estimates of Table 13-1.

It is apparent from Figure 13-1 that it is very unlikely that there will be a world shortage of enrichment capacity before about 1988, and in fact unlikely that there will be one before the early 1990s if uranium is stripped to a tails assay of 0.3 percent or more uranium-235. The United States alone will be able to meet all non-Communist world demands through about 1983 at 0.2 percent tails and through 1986 at 0.3 percent tails if our "high" projections of demand are not exceeded. One should not infer that surplus stockpiles of the size suggested by Figure 13-1 will necessarily develop. They could if plants were operated at capacity as a matter of policy. In the absence of such policy, it is likely that they would operate at reduced capacity considering the carrying charges involved in maintaining a large stockpile.

It should be noted that in addition to the supply of separative work projected in Table 13-1, there is a large amount of separative work in the weapons stockpile. This is of course in highly enriched uranium and would have to be mixed with natural or depleted uranium to produce fuel for light-water reactors. It takes 30 to 40 percent more separative work to produce 3 percent fuel this way than if one starts with natural uranium and enriches to only 3 percent. But some of the large amount already produced for weapons may be surplus.

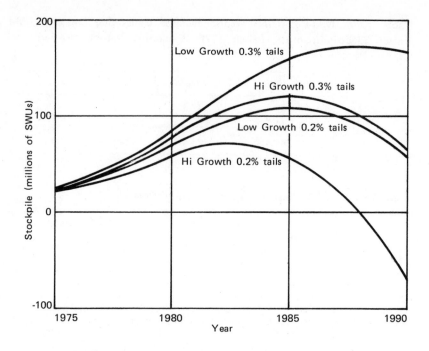

Figure 13–1. Cumulative Stockpile of Separative Work

Note: The curves are based on an assumption that one-third of LWRs are BWRs and two-thirds are PWRs; that the thermal efficiencies are 0.33 and 0.32 respectively; and that the irradiation levels of the fuel are 24,000 MWD_{th}/MTU and 30,000 MWD_{th}/MTU respectively. Greater thermal efficiency, higher irradiation level, or lower capacity factor would reduce demand for enrichment services. Proposed additions to enrichment capacity, beyond the committed capacity shown in Table 13–1, would greatly increase stockpiles in later years.

ENRICHMENT TECHNOLOGIES

Virtually all enrichment so far has been carried out using gaseous diffusion, a process that involves pumping uranium hexafluoride gas through a porous barrier. The lighter uranium-235 hexafluoride molecules move through the barrier more readily than the heavier uranium-238 hexafluoride molecules, but the degree of enrichment in a single pass is very slight. To obtain 3 percent enriched fuel, the process must be repeated more than 1000 times, and this requires an enormous amount of energy. One of the problems in upgrading programs now underway, and in construction of any additional gaseous diffusion capacity, is arranging for sufficient power.

Separation using gas centrifuges is, in principle, more attractive. Since a much higher degree of enrichment is possible in each stage, the process need be repeated only about a dozen times to produce 3 percent fuel, and the power requirement is only about 10 percent of that for gaseous diffusion. The centrifuge may be especially attractive as a source of enrichment for smaller power programs. While 9-10 million SWUs per year appears to be the preferred size for gaseous diffusion plants, 2-3 million SWUs per year is a reasonable size for centrifuge plants. URENCO is already producing some separative work using centrifuges, but the price is about 50 percent more than that for gaseous diffusion enrichment plants. However, American centrifuges have at least ten times the capacity of the European ones, and it is likely that enrichment costs will be as low, and possibly lower, than in diffusion plants. Lower unit costs, and the likelihood that small centrifuge plants will be feasible for meeting each new increment of demand, may make new diffusion plants noncompetitive.

The aerodynamic processes—the Becker nozzle process and the South African process—may prove attractive in special situations. The technology, at least in the case of the Becker process, appears to be less demanding than for gaseous diffusion or gas centrifuge plants. The power requirements, however, are greater than for either. A Becker nozzle plant is planned for Brazil. Details of the technology are in the public domain, while crucial aspects of the technologies for gaseous diffusion, centrifuges, and the South African process are not.

Finally, there is the possibility of laser enrichment: processes in which either uranium-235 atoms or uranium-235 hexafluoride molecules are made to absorb radiation preferentially to uranium-238 atoms or uranium-238 hexafluoride molecules in such a way that it is possible to separate the valued isotope. Laser separation has been accomplished on a laboratory scale with uranium atoms; the excited uranium-235 atoms are ionized and then separated from the uranium-238 atoms using electrical and magnetic fields. No separation has yet been reported in the case of the molecular process. Although there are difficult problems to be solved for both processes, we believe it likely that laser enrichment will be demonstrated on a commercial scale some time during the 1980s. In principle, there are two great advantages over the other processes: the possibility that power requirements will be very low and the possibility that a high degree of enrichment will be reached in a single stage. It is probable that if laser enrichment is practical at all, it will separate nearly all the uranium-235 from natural uranium. This will mean that approximately 30 percent more fuel will be obtainable from a given amount of natural uranium (assuming the alternative is stripping to 0.2 percent tails) and that it will be feasible to strip uranium-235 from the tails accumulated from other enrichment processes. By extending uranium supplies and reducing enrichment costs, lasers would make plutonium recycle less attractive and delay the time at which breeders might be competitive. While both of these results would serve nonproliferation goals, the laser process itself may make highly enriched uranium for weapons more readily available.

While there appears to be little need for great concern about availability of enrichment, there are many uncertainties about the next stage of technological evolution. Investment in gaseous diffusion, or the aerodynamic processes, may be undercut by centrifuge development; and it in turn may be undercut by laser development. Some of the uncertainties are likely to be resolved in the next few years. There is, therefore, a strong argument for delaying going ahead with construction of new plants, particularly gaseous diffusion plants. This argument is particularly compelling in light of our conclusion that existing plants and those under construction can meet demands through the mid- or late 1980s.

ASSURED SUPPLY AND THE U.S. ROLE

During the debate of the last few years about whether the next enrichment plants in the United States should be owned privately or by government, there was broad support for the position that the United States should retain as much of the export market as possible. This was argued on the basis of a favorable effect on the balance of trade and on the export prospects for U.S. reactors. It was also argued, questionably in our view, that an expansion in U.S. capacity and even a continued U.S. dominance would be desirable in limiting nuclear proliferation.

There are two components to the usual nonproliferation argument. The first is that U.S. ability and willingness to provide service, with credible guarantees, and perhaps subsidized prices, would reduce incentives for other nations to build enrichment plants. The second is that U.S. dominance and the threat of not providing service or cutting it off could be used to induce others to refrain from taking actions judged incompatible with efforts to limit nuclear weapons proliferation. The United States could also use its dominance to exert pressure in situations unrelated to nuclear proliferation.

The two components of the argument, however, are mutually contradictory. The United States can either provide assurance of enrichment service or try to maintain a situation of dependence where the possibility of termination of supply can be used for political purposes; it cannot do both. We believe that the second approach is unrealistic and counterproductive. Enrichment is a vital but economically small component of most nuclear power programs; nations are thus likely to construct their own plants rather than submit to precarious dependence on a politically motivated supplier. Any use by the United States of a dominant position in enrichment supply for political purposes would tend to encourage independent capabilities. Nations willing to accept the economic disadvantages of a small enrichment plant, perhaps based on less than optimal technology, in exchange for security of supply are likely to be able to obtain such a facility in due course.

Assurance of supply remains an effective nonproliferation argument. It is important not only in reducing incentives to acquire indigenous enrichment plants but also in some measure in reducing incentives to acquire spent fuel reprocessing plants or premature breeder systems. The first requirement in assuring enrichment services is that there be adequate capacity to meet world needs. It now seems probable that this will be the case even if no new American plants are constructed.

Some surplus U.S. capacity would be desirable from a nonproliferation perspective, particularly if means could be found to make excess enrichment available on an assured basis to the countries needing it. Since large blocks of power are only available on the basis of advanced commitments, ERDA must take steps to insure that power will be available when needed as the existing U.S. diffusion plants are upgraded to 27.7 million SWUs per year capacity. It is by no means clear, however, that U.S. capacity should be expanded beyond this level in an attempt to keep as much of the market as possible. Countries about which there is concern from a nuclear proliferation perspective may feel more comfortable about assurance of supply if there are several independent suppliers of enrichment service than if they are dependent on a single source. Indeed, from a nonproliferation perspective, one would like to see the development of a buyer's market, as long as new facilities are limited to those nations that are not regarded as likely candidates for new weapons programs.

Guarantees will help provide supply assurance but may not be sufficient since they may not survive political difficulties between the states involved. The United States' announcement in 1974 that it would not accept new foreign contracts, and its unilateral determination that it would regard some of its existing commitments for future enrichment services as conditional on a U.S. plutonium recycle decision, hardly helped in this respect. It may have been a factor in Brazil's interest in acquiring an indigenous enrichment capacity. The credibility of guarantees may be enhanced somewhat by part ownership of the facilities, as is the case with Eurodif.

One means of providing absolute assurance is stockpiling. The United States or other suppliers could offer to supply fuel in advance sufficient to operate a nation's reactors for seven to ten years, the time required to construct a plant if there should be a cutoff in delivery. It is unlikely that many nations not having enrichment plants would accept such an offer, considering the carrying charges and the likelihood that the cost of enrichment may drop as the new technologies are developed. However, for those nations really concerned about security of supply, such an offer would be an attractive alternative to building an indigenous plant. This is especially true if technological change would make investment in a present technology commercially risky. According to the projections of the world stockpile shown in Figure 13-1, it would probably be possible to implement such a proposal. If all plants were run at capacity, the surplus, assuming our low projection and 0.3 percent tails, would be sufficient to provide all nations outside the Soviet Bloc, France, and the United States with a seven-year forward stockpile of separative work during the period 1980 to 1985. Since not all countries would want to invest in such a stockpile, it would probably be possible to meet the demand even with 0.2 percent tails or, under our high projection, with

0.3 percent tails. The negative aspect of this approach is that such a stockpile would shield a nation launching a nuclear weapons program from the effects of a nuclear fuel cutoff.

PRIVATE OWNERSHIP OF URANIUM ENRICHMENT FACILITIES

During the past several years, there have been proposals by industrial organizations to build and operate four new enrichment plants in the United States: a 9 million SWU gaseous diffusion plant to be built by Uranium Enrichment Associates (UEA), a consortium headed by Bechtel; and three separate 3 million SWU centrifuge plants, to be built by Garrett, Exxon, and CENTAR, a joint venture involving Atlantic Richfield and Electro-Nucleonics. In all cases, guarantees by the government have been sought. To meet this demand, the Ford Administration proposed the Nuclear Fuels Assurance Act. It would have:

- Authorized ERDA to enter into cooperative arrangements with as many private firms wishing to build, own, and operate enriching plants as the ERDA Administrator believed necessary to develop a competitive industry.
- Authorized ERDA to provide various forms of assistance and assurances under such arrangements.
- Limited the government's total potential liability to $8 billion in the event that the private ventures failed and the government had to take them over.
- Authorized ERDA to start construction planning and design activities for expanding one of the government's existing enrichment facilities as a contingency measure.
- Provided for Congressional review of the basis for the cooperative arrangements by the Joint Committee on Atomic Energy.

During consideration by the Congress, there were extensive negotiations between ERDA and Uranium Enrichment Associates on the provisions of a contract which would meet UEA's conditions for going ahead with its proposed plant. UEA asked that the government:

- Supply essential components to UEA.
- Provide technical assistance and know-how on the installation and operation of the gaseous diffusion process.
- Assure that the plant would operate successfully.
- Assure domestic partners that the government would assume all liabilities and obligations, if UEA could not successfully complete the plant.
- Permit UEA access to the ERDA stockpiles up to 9 million SWUs, decreasing to 0 after five years.
- Purchase up to 6 million SWUs during the first five years.

In the early negotiations, UEA asked that the government terminate some of its long-term contracts to supply enrichment services so that UEA could sell all of its product, but this provision was dropped. ERDA also discussed guarantee and assistance provisions with the three organizations proposing centrifuge plants.

The Act was widely seen as a device leading to government taking most of the risk in these ventures with the entrepreneurs realizing profits if the ventures were successful. It narrowly failed of passage in the Congress. The question of whether additional enrichment capacity should be built by ERDA or by the private sector will come up again. If this occurs soon, the demands for guarantees are likely to be at least as strong since even without such plants it now seems likely that there will be a surplus in capacity during the late 1980s. In 1974 when the proposals above were first made, it seemed likely that there would be a shortage.

There have been two major arguments for turning to the private sector for additional enrichment capacity: to reduce the federal budget by removing large enrichment plant investments with long payback timetables from it; and to realize advantages in efficiency from competition. Opponents of private ownership have argued that the costs of private ownership would be greater, because of the greater costs of raising capital and of taxes. They have also argued that government control is preferred because provision of service for export should be tightly coupled to foreign policy, and particularly to nonproliferation policy. There have, for example, been expressions of concern about the extent of foreign investment in the proposed UEA venture.

The arguments relating to taxes, the cost of money, the effect on the federal budget can be dismissed if one looks at the question from the viewpoint of society as a whole. If private or government plants are built and run with equal efficiency, the total costs and benefits will not differ. There may be merit in the argument that privatization leads to competition and lower prices. However, the potential benefits of cost reductions that might follow from privatization would be small at best, considering that enrichment costs will be only a few percent of the cost of generating electricity. The major argument against privatization is that it would loosen government control of enrichment technology. Considering the importance of nonproliferation objectives, we believe that any additional enrichment capacity, whether using gaseous diffusion or centrifuge technology, should be built by the government. Heavily subsidized or guaranteed private ownership would seem to offer the worst of both worlds: failure to capture the benefits of competition and looser control of the technology and its products.

In addition to the issue of ownership of new enrichment capacity, there is the question of the role of private industry in the development of new enrichment technology. Development of laser enrichment technology is being carried out on a proprietary basis by Exxon and Avco, as well as in ERDA laboratories. Although we cannot be overly optimistic about preventing or greatly delaying the spread of technology, once it is developed, we believe that an effort should be made to delay the spread of laser enrichment techniques for as long as possible because of its potentially serious impact on proliferation. We believe therefore that the government should maintain strict classification and export controls on this technology.

Chapter 14

Nuclear Export Policy

Nuclear facilities, technology, and materials are already significant items of international trade. For most countries, the only access to nuclear power is through imports from a small group of supplier states. The terms on which such trade is carried out can contribute to nonproliferation objectives. Although the United States is no longer in a position to impose its views unilaterally, its export policies will affect decisions of other countries on fuel cycle issues.

Almost any nuclear facility acquired by a country contributes to the skills and experience needed to develop a bomb. Some items, such as reprocessing and enrichment facilities, contribute directly to nuclear explosive capability. For this reason, the United States has consistently refused to permit the export of reprocessing or enrichment equipment or technology.

Nuclear exports offer a potential market for U.S. industry and for other supplier states. The Energy Research and Development Administration (ERDA) has estimated that U.S. export revenues from reactors and fuel cycle services might reach $2.3 billion annually by 1980, and amount cumulatively to over $100 billion by the end of the century. These figures appear to be substantially overstated, since they are based on inflated estimates of nuclear power deployment, particularly in developing countries, and on an underestimation of the competitive roles of European and Japanese suppliers. Nevertheless, even if these figures are high by a factor of two, the potential is large enough not to be ignored in U.S. decision-making.

Exports of reactors and fuel can help meet the energy needs of the importing country. Thus, U.S. regulation of its nuclear exports affects the energy policies and decisions of other countries. Further, it raises the issue of access by developing countries to modern technologies.

U.S. nuclear export policy has tried to accommodate these contrasting aspects of the problem. The Atomic Energy Act of 1954, as amended, embodied President Eisenhower's Atoms for Peace program, marking the end of the postwar U.S. effort to maintain a nuclear monopoly. The new policy called for widespread sharing of nuclear technology for peaceful puposes, combined with safeguards against military use. On the export side, the Act provided a two-tier structure of control. No exports or information exchange could take place except within the framework of an Agreement for Cooperation between the United States and the foreign state or international organization involved. Then, each transfer of nuclear materials or facilities to the foreign partner would require an individual license from the Atomic Energy Commission (AEC).

In practice, until recently, the primary control mechanism has been the Agreement for Cooperation, with licenses granted routinely for individual exports falling within the terms of an Agreement. The Act provides that the Agreement must contain an undertaking by the foreign government to maintain "security standards and safeguards as set forth in the agreement" and guarantees against use of the transferred material for weapons or military purposes and against transfer to third parties. Each Agreement for Cooperation must be approved by the President with a determination that it "will promote and will not constitute an unreasonable risk to the common defense and security." The Agreement is then submitted to the Joint Committee on Atomic Energy. Since 1974, Congress has had the power to veto, by concurrent resolution, any Agreement authorizing a reactor of more than five thermal megawatts, which means, in effect, all reactor agreements.

Agreements for Cooperation exist with almost thirty foreign countries. While a few are for research only, most authorize exports of power reactors and fuel. Agreements also are in effect with the International Atomic Energy Agency (IAEA) and Euratom. The Agreements are similar in form, but the obligations of the foreign country have become increasingly stringent and explicit in recent versions, reflecting a more sober attitude toward the dissemination of technology. For example, in the past few years the United States has demanded an explicit understanding against the use of exports covered by the Agreement in the development of peaceful nuclear explosives, as is required by its commitment under the Non-Proliferation Treaty (NPT).

The NPT reflects the same conflict between the goal of nonproliferation and the desire for nuclear exports that has confronted U.S. policy. Article I of the NPT pledges the United States and all other signatories, "not in any way to assist, encourage, or induce any nonnuclear weapon State to manufacture or otherwise acquire nuclear weapons or other nuclear explosive devices." At the same time, under Article IV, "All parties to the Treaty undertake to facilitate, and have the right to participate in, the fullest possible exchange of equipment, materials and scientific and technological information for the peaceful uses of nuclear energy."

In dollar volume, the bulk of exports consists of light-water reactors (LWRs) and slightly enriched uranium fuel, which under existing safeguards create small risk of weapons proliferation. More severe risks are associated with exports of reprocessing plants, or of weapons-grade uranium and separated plutonium. The amount of money involved in these exports is and would remain small compared with reactors, but in the past they have affected the export picture as inducements for larger reactor sales. Controversies have therefore focused on fuel cycle facilities and materials.

EXPORTS OF REPROCESSING AND ENRICHMENT FACILITIES

In February 1976, West Germany and Brazil concluded an "Agreement Concerning Cooperation in the Field of Peaceful Uses of Nuclear Energy." The areas of proposed cooperation include "uranium enrichment and enrichment services" and "reprocessing of irradiated fuels." Shortly thereafter, South Korea and Pakistan contracted for reprocessing plants from France. South Korea canceled its contract under pressure from the United States, and the status of the Pakistan contract was uncertain at the end of 1976. These events have been troublesome because of the sensitivity of reprocessing and enrichment for weapons development and because of the absence of compelling economic justification for the purchases at this time.

In assessing these events, it must be recognized that the construction of a small-scale plant is within the reach of any industrial country and, with some additional lead time and effort, of advanced developing countries as well. India, for example, and probably Israel also, have developed reprocessing facilities with indigenous resources. Large-scale commercial reprocessing of LWR fuel is another matter. There is no commercial reprocessing plant operating in the United States; abroad, only the plant in La Hague, France, which came on line in May of 1976, is in operation but has not yet attained its rated capacity or engaged in foreign commercial activities.

The West German-Brazilian deal includes an experimental reprocessing plant. As a follow-on, the Germans have agreed to provide a larger plant—perhaps on the order of 100 kg per day—subject to approval by their partners in United Reprocessors (a consortium of France, West Germany, and the United Kingdom), which controls the relevant technology. Delivery dates are unknown, perhaps not yet specified, and in any case some years hence. Not even this much detail is available on the French-Pakistani agreement. Apparently, it was exempted from the requirement of United Reprocessors approval in the negotiations leading to the establishment of the consortium.

The current technology for the enrichment of uranium by gaseous diffusion remains classified. It could be replicated only by countries with a high technological base, and then only at great cost. Gaseous diffusion plants could not be built and operated clandestinely. The situation has been changed somewhat by centrifuge technology and may change even more if laser separation is perfected since these technologies allow the construction of smaller plants with fewer capital requirements. However, both of these separation techniques are very high technology and are still classified. Developing countries will for some time be forced to purchase the critical equipment for these facilities from outside suppliers. The aerodynamic nozzle technology, sold by West Germany to Brazil and under development by South Africa, is not classified. However, it involves high capital and especially high operational costs and is not likely to interest most countries.

It is generally thought that at least in Brazil and Pakistan the sensitive facilities were not offered for their own sakes but as inducements to secure lucrative reactor contracts. Intense export pressure among European suppliers derives from the failure of their own domestic orders to absorb available manufacturing capacity. Without exports, the condition of European reactor manufacturers is precarious. The charge has been made that some companies and their governments are prepared to shade antiproliferation considerations to increase their market.

In the closing weeks of 1976, there were indications of a major shift in attitude by both France and West Germany. The French government has announced that in the future it will no longer export sensitive fuel cycle facilities, and it appears that West Germany is prepared to follow suit. Moreover, France has said that it would not object if Pakistan should wish to cancel the contract for purchase of a reprocessing plant. Again, there are indications that West Germany is moving toward a similar stance on the fuel cycle elements of its contract with Brazil. Canada has announced that in the future it will not export nuclear equipment or materials to a country unless it accepts safeguards on all its nuclear facilities, and an official Australian report proposes a similar policy on uranium exports. These are welcome developments, and they illustrate the results that can be expected from vigorous and sensitive diplomacy in combination with a growing international appreciation of the economic doubts and the security risks of a plutonium economy.

These events seem to have overtaken some earlier suggestions in the United States about an export cartel among nuclear suppliers enforced by U.S. control over reactor fuel supplies. Such an arrangement would be difficult to negotiate and would be seen as coercive by suppliers and users alike. The United States now has an opportunity to build on the developments of recent months.

U.S. policy should seek to encourage a broad consensus on the desirability of avoiding a plutonium economy. This is a fundamental element of a nonproliferation strategy. Although no one can guarantee the success of such an effort, the international climate now seems propitious. The report of the French Commission on Nuclear Energy Policy, the British Royal Commission on Environmental Pollution (the Flowers Commission), the Swedish election, all suggest some change in mood on nuclear matters throughout the world.

Any U.S. program for developing a common appreciation of fuel cycle problems must have the personal involvement of the President and his principal foreign policy officers. One way of mounting such an effort would be for the President, after suitable diplomatic preparation, to call for a two- or three-year moratorium on exports of enrichment and reprocessing technology. During this period, there could be an intensive reexamination by suppliers and importers alike of the risks and problems of reprocessing and recycle of plutonium. There are many ways such a reexamination could be carried out. Without dictating the process, the United States should make sure that there is broad understanding of the technical difficulties and economic uncertainties of plutonium recycle and breeder reactors and of the potential dangers of the plutonium economy for nuclear proliferation, terrorism, health, and safety. In general, it would appear most effective to approach the problem by considering the energy and security problems of specific countries rather than through large international conferences.

Any U.S. proposal for international reexamination of the fuel cycle could hardly be credible if the United States were forging ahead with its own plans for reprocessing fuel for LWRs and with its program for early commercialization of the breeder reactor. The recommendations in this study for deferral and slowdown of these programs take on added significance in the context of export policy. Other countries may not follow the U.S. lead but most would at least reassess their own plans in the face of U.S. restraint.

In the context of clear U.S. restraint, we believe the proposed evaluation could lead to a broad consensus, including many developing countries, on the dangers that are inherent in widespread diffusion of national plutonium reprocessing and enrichment facilities. In such an environment a more formal agreement embargoing exports to those countries—hopefully few—that did not voluntarily renounce national reprocessing and enrichment could be pursued with some prospects for success.

Meanwhile, the United States should continue prohibiting its own exports of the sensitive technologies. If, despite the recent shifts in French and West German policy, new contracts for reprocessing or enrichment seem to be in process, the United States may still be able at a high level to dissuade the supplier or the customer. Continuing consultations among the suppliers in London would provide a means for coordinating policy and adjusting the burdens of restraint.

The concept of a multinational fuel cycle center presents more formidable complexities of negotiation, but there are other basic problems with the concept. In a plutonium economy, there is some prospect that multinational facilities would have some nonproliferation advantages. Until an international plutonium economy is inevitable, however, it would be a mistake to encourage commercial reprocessing facilities in any form, national or multinational. Such facilities would result in an international commerce in plutonium and would develop international experience in the separation and handling of plutonium.

Skepticism about multinational reprocessing does not necessarily extend to other multinational activities. New enrichment plants and spent fuel storage are promising candidates for international cooperation. The IAEA's current approach of encouraging joint analysis and assessment of regional fuel cycle problems by countries in a region is sound and deserves U.S. support.

FULL FUEL CYCLE SAFEGUARDS

Although safeguards cannot prevent a nation from developing a nuclear capability, they can help to deter it and provide assurances to others that it has not done so. The value of safeguards is obviously greatly weakened if they do not apply to all of the facilities in a country, in particular to indigenous facilities for separating plutonium or enriching uranium. While the NPT obligates non-nuclear weapon states that are members to place all their nuclear facilities under safeguards, the United States and other suppliers have interpreted their treaty obligations on exports to non-members as requiring only that the non-members place the exported facilities or materials under safeguards and not *all* their nuclear facilities. The consequence is that, for non-parties, indigenously developed nuclear activities not using imported materials or equipment (such as in India and South Africa) or facilities acquired without a bilateral safeguards agreement (such as Israel's Dimona reactor) are not subject to any international safeguards.

The question arises whether the United States should condition exports of materials and facilities to non-NPT countries on the acceptance of "full fuel cycle safeguards"—safeguards on all peaceful nuclear activities, just as though the importing countries were parties to the NPT. The argument is straightforward: if importers accept this condition, safeguards are extended; the policy would eliminate a situation that discriminates against parties to the NPT in favor of non-parties.

On the negative side, the analysis is complicated, particularly in the case of exports of fuel for reactors in place, as distinguished from exports of new reactors and other facilities. The threat to cut off fuel supplies for existing reactors because of a refusal to extend safeguards beyond the scope of commitments given when the reactor was purchased would be resented by countries that felt that they had fulfilled their part of a formal agreement. Since most, if not all, of these countries would presumably not be in the process of actually developing nuclear weapons, the damage to the perception of the United States as a reliable supplier of nuclear fuel could outweigh the benefits of an extension of the safeguards regime.

In the case of new reactor exports, a distinction may be made between transactions under an existing Agreement for Cooperation and the negotiation of new Agreements (or the renewal of expired ones) establishing the terms and conditions for the future. As to new Agreements, the foreign partner would have no basis in principle for objection to full safeguards, although it may be unwilling to agree to the terms and may forego the cooperation.

As to reactor exports under existing Agreements, the importing country may argue that the requirement of full fuel cycle safeguards changes the rules in the middle of the game. Although the argument has a certain plausibility, it is not justification for resistance to broader safeguards. That must rest on equally abstract notions of sovereignty or, in some cases, on an explicit desire to protect a weapons potential if not a weapons program. Neither of these provide compelling grounds for the United States as supplier not to demand full fuel cycle safeguards.

On the other hand, if the United States were to insist on full safeguards, the countries of greatest concern such as India, South Africa, and Israel might forego exports from the United States, and possibly even rely entirely on indigenous resources. The tradeoff for the United States is between the uncertain effect of a cutoff and the practical possibilities for influencing the other country's nuclear program if contact is maintained. This tradeoff will vary from case to case. The balance is close enough so that an unyielding stance is not justified. Nevertheless, it would be desirable to maintain pressure for full fuel cycle safeguards in the negotiation and renegotiation of Agreements for Cooperation. This objective should be pursued in concert with other suppliers.

PLUTONIUM RECAPTURE

A proposal that has received some attention is that the United States should buy back spent fuel elements from reactor cores it has supplied to foreign operators. This would eliminate the possibility that the plutonium might subsequently be separated for weapons purposes. A different form of the same concept is that the reactor fuel would not be "sold" in the first place, but only "leased," presumably facilitating the legalities of recapture. A number of the earlier U.S. bilateral Agreements for Cooperation contain more or less explicit provisions covering repurchase of spent fuel, but the option has never been exercised and the considerations bearing on issues have not been systematically analyzed.

Spent fuel disposal will be an increasingly difficult problem for many countries. Temporary storage is not difficult or expensive but, eventually, permanent arrangements are needed. It has been widely assumed that such permanent disposal would require reprocessing, which would also separate plutonium. However, as discussed in Chapter 8, permanent or retrievable storage does not require reprocessing, a process which may in fact involve a significant economic penalty, especially for a country with a small program. Adequate sites for permanent disposal, whether or not reprocessing has occurred, are probably not available in most countries, and such arrangements could be costly for a small program.

The disposal problem could be solved and the proliferation dangers mitigated if countries with suitable geologic endowment would accept this spent fuel and store it on a permanent or retrievable basis. There may be resistance in the receiving country; in the United Kingdom, for example, earlier willingness to accept responsibility for the high-level wastes from reprocessed Japanese fuel has given way to a policy requiring an option to return the wastes to Japan. But such resistance would be misconceived in countries like the United States, the Soviet Union, and Canada, where there are geological formations suitable for long-term waste storage. The countries of concern from a proliferation standpoint will have produced only small increments of fuel compared to U.S. waste generation.

Another factor complicating the recapture issue has been the expectation that the plutonium and uranium in the spent fuel will have a large value. If so, reactor operators would be justified in demanding payment for their spent fuel. It is doubtful, however, whether recycle will prove economical. To cover these uncertainties, a recapture scheme should provide a credit for the value of any plutonium or uranium when it is ultimately recovered.

These considerations suggest that it would be desirable for the United States to include a recapture provision in future Agreements of Cooperation and to exercise existing options that are legally or politically available. However, a unilateral recovery program has drawbacks since there might be suspicion of U.S. motives, and determination of costs would be difficult.

More promising than unilateral U.S. repurchase of spent fuel would be development of some form of multinational spent fuel storage. The objections to full-scale multinational fuel cycle centers do not apply to this more modest venture. If the effort to develop a more realistic appreciation of the outlook for reprocessing, as suggested above, were to prove successful, a considerable number of countries might well be willing to participate in such a scheme. The incentive would be improved if there were a provision for return of the plutonium (or its equivalent) in the form of safeguarded fabricated fuel if and when recycle became economical. If not, the multinational organization could assume responsibility for ultimate waste disposal. Finding a host country for storage of spent fuel might be troublesome, but some geologically favored countries should be willing to provide suitable locations—perhaps for a fee. A multinational spent fuel enterprise might be cast in the form of a depository under the supervision and management of the IAEA, which is expressly empowered by its statute to perform such functions. This would assure participants that the arrangements would be fair and would minimize the risk of the host country's expropriating the assets. U.S. financing for such a scheme might be more acceptable than its unilateral repurchase of spent fuel.

EXPORT LICENSING

Each U.S. export of nuclear facilities, fuel, or technology must be individually licensed. Until recently, such licenses were the responsibility of the AEC and were granted almost automatically. No hearings were conducted, and licensing was *pro forma*. Under the Export Control Act, a Commerce Department license is required for certain nonnuclear components of reactor exports, but these also have been issued routinely.

In the past few years, these arrangements have begun to change. The Energy Reorganization Act of 1974 lodged the licensing function in the Nuclear Regulatory Commission (NRC). The NRC announced a policy of intensified consideration of export license applications, with review of the most important ones by the Commissioners themselves. Shortly thereafter, the Natural Resources Defense Council and other environmental groups sought to intervene and to require public hearings in a proceeding for an export license for nuclear fuel for the Tarapur reactor in India. In May 1976, the NRC denied the petition to intervene but agreed to hold a "legislative-type" hearing at which intervenors could make their case. This was the first public hearing on an export license application.

The actions of the NRC and the increasing political sensitivity of the issues have already had an impact on nuclear exports. Although one lot of fuel for Tarapur was licensed, a second and larger shipment was delayed, pending a hearing on India's safeguards. A proposed sale of two reactors to South Africa by General Electric was canceled in anticipation of a sharply contested hearing. The order has since been placed with a French exporter. An application for 23 kg of 93 percent enriched uranium for South Africa's research reactor was also held up in the licensing process. A license for the sale of a Westinghouse reactor to Spain was approved, but for the first time one of the Commissioners filed a dissenting opinion.

Despite this increased attention to licensing problems, the licensing process continues to be fragmented. The NRC's jurisdiction extends only to exports of hardware and special nuclear material. Transfers of technology and know-how require express approval in advance from ERDA. And the Commerce Department retains its Export Control Act responsibilities. These two latter agencies act without a hearing and without any formal requirement for inter-agency coordination.

Under the Atomic Energy Act, in order to issue a license, the NRC must find that the proposed export is in accordance with the terms of the applicable Agreement for Cooperation and "is not inimical to the common defense and security and the health and safety of the public." The NRC interprets this as requiring a finding that the safeguards of the importing country are adequate and that its peaceful uses assurances are reliable. The Commission imposes no explicit condition with regard to the foreign health and safety standards. The license determination is probably subject to judicial review, although the scope of review may be more limited than in the case of domestic licenses.

A basic question is whether the NRC—a regulatory commission, independent of direct Presidential control and acting in a quasi-judicial manner—should be involved in export licensing. Under present law, the Executive Branch probably has authority to prevent a transaction it *disapproves* even if the NRC should grant a license. But if the NRC refuses a license, the export cannot go forward even if the Executive Branch favors it. New legislation, proposed during the 94th Congress and narrowly defeated, would have consolidated licensing authority in the NRC, with a Presidential veto over licenses and authority in Congress to review licenses backed by the Executive Branch but disapproved by the NRC. The tendency toward more open, judicial-type proceedings on NRC export license applications would have been reinforced. Formal participation would have been mandated for ACDA in both the Agreement and the licensing process.

The main difficulty with consolidating licensing authority in the NRC is that the process requires the Commission to make foreign policy judgments for which it is not equipped. The reliability of peaceful assurances on the part of the foreign state is hardly the kind of question that should be decided after an open hearing and made subject to judicial review.

On the other hand, it is hard to make a case that the State Department is the proper agency to make determinations on safeguards and physical security. It would be dependent on external sources for information and technical advice. It is quite plausible, moreover, that these determinations would benefit from public review and scrutiny.

As a middle position, the reliability determination could be made by the State Department, while the safeguards and security issues would be left to the NRC. The difficulty with such a compromise is that it makes it more difficult for the United States to give positive assurances to foreign governments about future fuel supply.

These considerations lead us to suggest a procedure like that used in granting international air routes. There, the Civil Aeronautics Board makes an initial determination, based on economic and other "objective" factors. Its decision is then subject to approval, disapproval, or modification by the President in his unreviewable discretion, thus allowing for the integration of foreign policy elements. A comparable procedure for nuclear exports would provide for an initial determination of the safeguards and security issues on the open record by the NRC, with the final decision on whether to issue the license left to the President. The procedure would ensure systematic evaluation and public scrutiny. The decision to override an NRC determination would have to be made at a level high enough to overcome narrow political or bureaucratic perspectives, and the President would have to be prepared to take the political consequences of such a decision. The disadvantages of such a system are delay and public review of a foreign state's conduct in a domestic forum, but we believe that these problems are overridden by the advantages of such a procedure.

A second issue is the proper role for Congress in export controls. Existing law gives Congress, acting by concurrent resolution, veto power over Agreements for Cooperation, and this authority should be retained. The recent legislative proposals would have added similar authority over export licenses in cases where the Executive Branch wished to go forward but the NRC refused to issue a license.

If the suggested approach to licensing were adopted, it would eliminate the occasion for Congressional participation, since there would be no deadlock between the Executive Branch and the NRC. On the other hand, there is a strong case for Congressional involvement at the Agreement stage, where it is most appropriate. The statutory provisions governing Agreements for Cooperation should be more detailed and explicit. At present, the requirements are few and the language is vague. Congress should specify the undertakings and obligations that should be included in the Agreement on such subjects as peaceful uses, safeguards (including full fuel cycle safeguards), plutonium recapture, fuel supply, and duration.

A third issue concerns exports of reactor fuel. Assuming that the NRC were to retain its present role in the licensing of reactor exports, there is still a question whether the Commission should have the power to deny an export license for fuel for an existing reactor on any ground other than that the importing government is in substantial breach of an agreement or undertaking with respect to its nuclear program. A subsidiary question is whether the NRC should make an independent determination that there has been a breach and that it is substantial enough to warrant denial of the export license.

The argument for a broader NRC role on both counts boils down to mistrust of the Executive Branch. In this view, the Executive Branch has given nonproliferation low priority in the past and will tend to subordinate it to other political objectives when fuel exports are being considered. Moreover, these questions may be decided at a level where the existence of the tradeoff is not even perceived. Only if an independent agency has a statutory mandate to act without regard to "irrelevant" political considerations will the nonproliferation objective be adequately subserved. While this argument may have some plausibility, the argument against giving the NRC final authority in these decisions is that the United States needs to be able to make credible guarantees for the supply of reactor fuel. Such assurances may be crucial in forestalling the spread of other nations' enrichment and reprocessing capabilities. It is hard to get foreign governments to understand the U.S. system of independent regulatory agencies. It is even harder to ask them to submit to the quasi-judicial determination of such an agency.

CONCLUSION

U.S. policy toward nuclear exports can play a significant supporting role in pursuit of broader nonproliferation objectives. However, the ability of the United States to influence events unilaterally by its export policy is limited.

The most important U.S. objective should be, therefore, to develop an international consensus on the problems and risks of a plutonium economy, and to encourage a more cautious approach to plutonium reprocessing and recycle and breeder reactors. Within such a consensus it will be possible to pursue export policies that will inhibit the spread of technologies carrying the most serious proliferation risks, while at the same time ensuring the availability of nuclear power to meet world energy needs.

�֍ *Appendix*

Nuclear Power Technology

Many of the policy questions relating to nuclear power are closely related to the details of nuclear power technology. This appendix presents a simple description of the fission process, current reactor types, and fuel cycles.

NUCLEAR FISSION AND REACTOR OPERATION

Nuclear fission is the process by which atoms of certain heavy elements, notably uranium or plutonium, split into two lighter atoms after absorbing a neutron. The lighter atoms are called fission products and are accompanied by two or three energetic neutrons. A chain reaction is sustained when, on the average, at least one of the neutrons emitted in a fission reaction goes on to initiate a subsequent fissioning. The fission process generates a great deal of energy, most of it in the form of kinetic energy of the neutrons and fission products. The fission products are usually unstable and undergo radioactive decay, a process which also produces heat. The fissioning of 3 grams of uranium-235 in a reactor produces enough energy to generate about a megawatt-day of electricity. Not all of the neutrons emitted in a fission reaction go on to initiate subsequent fissionings: some escape from the region of the reaction, some are absorbed, and some are captured by heavy nuclei without occasioning fissioning. The latter is important because it begins the process by which plutonium and other elements are made in a reactor.

A nuclear reactor is a device containing quantities of fissionable materials in which a chain reaction can be sustained on a constant basis. To avoid having the chain reaction either stop or get out of hand, precisely one neutron, on the average, from each fissioning must go on to initiate a new one. If too few neutrons do this, the device is said to be subcritical; if many, it is supercritical. A nuclear explosive is designed to operate, very briefly, in a highly supercritical condition, in which the neutron population may double each hundredth of a microsecond. Nuclear reactors have the fuel much less compact than do nuclear explosives, so that a neutron may live on the average about a millisecond (for reactors with large amounts of water present) or a microsecond (for so-called fast reactors). The reaction is controllable because about 1 percent of the neutrons do not emerge promptly with the fission process, but about a second later. Reactors normally operate so that they are critical or supercritical only if these delayed neutrons are included, giving some seconds as the time available for control of the reaction.

Chain reactions and hence reactors are made possible by the existence of fissile[a] nuclei. The naturally occurring fissile isotope, uranium-235, is the basis for all current power programs. Neutrons from the fissioning of this isotope can be used, however, to make new fissile material. When neutrons are captured by the fertile nuclei uranium-238 or thorium-232, reactions are initiated which result in the formation of plutonium-239 or uranium-233, both of which are fissile. These may be used in new reactor fuel. A reactor is usually loaded with fertile as well as fissile material. The amount of new fissile material formed in the reactor depends upon the way in which the reactor operates. The conversion ratio is the ratio of the number of fissile nuclei formed to the number consumed. A reactor in which there is little conversion, and hence one with a low conversion ratio, is termed a burner; one in which the conversion ratio approaches 1 is called a converter; any reactor in which the conversion ratio exceeds 1 is called a breeder.

Ultimately, a nuclear reactor is a device to generate heat which may be used,

---

[a]The terminology of nuclear power is complicated but reflects simple physical distinctions. Heavy nuclei, such as uranium and plutonium, are fissionable in that they can be fissioned by a neutron of some energy. Most fissionable nuclei require very energetic neutrons to fission. Those which can be fissioned by a neutron of low energy (a slow neutron) are termed fissile. Each heavy element exists in several subspecies or isotopes. All the isotopes of a given element have the same number of protons in their nuclei and the same atomic structure; since it is atomic structure which determines chemical activity, isotopes are chemically indistinguishable. Isotopes are distinguishable at the nuclear level because they have different numbers of neutrons in their nuclei. Thus uranium-235 has 92 protons and 143 neutrons in its nucleus while uranium-238 has 92 protons and 146 neutrons. This nuclear difference is what makes some isotopes fissile and others not. Uranium-235 is the only fissile nucleus which appears in nature in quantities large enough to use as a starting point for nuclear power. Some nonfissile isotopes of heavy elements can capture a neutron and, rather than being fissioned by it, initiate a sequence of radioactive decays which culminates in it being transformed, or converted, into a new, and fissile, isotope. An example is uranium-238, which can capture a neutron, decay, and be converted by this process to fissile plutonium-239. A fissile isotope may also capture a neutron without fissioning; in nuclear power production this is undesirable since it simultaneously removes a neutron which could otherwise be causing a fission and a fissile nucleus which could otherwise be fissioned. Some fission products also parasitically capture neutrons. Fuel is removed from a reactor every year or so to prevent these products from building up in it.

directly or indirectly, to make steam, which can then be used to generate electricity, just as is done in a fossil fuel power plant.

Nuclear Reactor Components

There are several basic components in most power reactors: fuel, moderator, coolant, control elements, reflector, and vessel.

Fuel. Every reactor must include fuel material containing fissile nuclei. The fuel may consist of natural uranium containing 0.7 percent fissile uranium-235 and 99.3 percent nonfissile uranium-238, or "enriched" uranium (which has been enhanced in its concentration of uranium-235), or uranium-233 (produced earlier by conversion of thorium-232), or plutonium-239 (produced earlier by conversion of uranium-238), or some combination of these.

Moderator. Reactors in which most of the fission reactions are caused by the fast, and hence very energetic, neutrons emitted in earlier fissionings are called fast reactors. In general, however, a slow neutron is more likely to cause fission than a fast one. To take advantage of this, all commercial power reactors now operating contain a moderating material which slows the speed of neutrons released in fission, by a factor of about 10,000, to the energy at which they approach thermal equilibrium with their surroundings. The moderating materials most commonly used are ordinary or light water (H_2O), heavy water (D_2O), which is like water but with deuterium, a heavy isotope of hydrogen, taking the place of ordinary hydrogen in the water, and graphite (a form of carbon). Reactors in which most of the fissions are caused by such low energy neutrons are called thermal reactors. Thermal reactors operating on the uranium/plutonium cycle are generally burners or inefficient converters; those operating on the thorium/uranium-233 cycle can be efficient converters or inefficient breeders. Fast reactors operating on either cycle are generally breeders.

Coolant. Heat released by fission is removed from the fuel by circulating a fluid coolant through the reactor core. Water is most commonly used to cool thermal reactors and it simultaneously provides the moderator. In some designs the coolant water boils within the core; in others, it remains a liquid. Other coolant fluids of practical interest are heavy water, liquid metals (especially sodium), and the gases carbon dioxide and helium. Heavy water is both a coolant and a very efficient moderator, in that it does not capture neutrons. Liquid metals do not moderate, though they may change the energy distribution of fast neutrons. Gases may be used as coolants in fast reactors and in graphite-moderated thermal reactors.

Control Elements. Control of a reactor is achieved by varying the fates of neutrons within the core. The primary means for accomplishing this is to insert or withdraw neutron-absorbing materials such as boron or hafnium into the core, never allowing the reactor to become critical on prompt neutrons alone.

Reflector. The core of a reactor is surrounded by a reflector, the purpose of which is to redirect back into the core some of the neutrons which might otherwise escape. For thermal reactors, good moderating materials also make good reflectors. Fast reactor reflectors usually are made of natural uranium or the depleted uranium left from the enrichment process. A reflector made of fertile material, such as a fast reactor reflector made of natural or depleted uranium, is usually referred to as a blanket. In some recent fast reactor designs the blanket is woven into the main part of the core; in older designs, it only surrounded the core.

Vessel. The nuclear portion of the reactor is contained in a reactor vessel. If the coolant is maintained at a high pressure, the reactor vessel is referred to as a pressure vessel.

POWER REACTOR SYSTEMS

The types of nuclear power reactors currently operational on a significant commercial scale are the light-water reactor (LWR)—of which there are two kinds: the pressurized-water reactor (PWR) and the boiling-water reactor (BWR)—and the heavy-water reactor (HWR). Other types of reactors of current interest include the high temperature gas-cooled reactor (HTGR) and the liquid metal fast breeder reactor (LMFBR). Also of possible interest are the gas-cooled fast breeder reactor (GCFBR), the light-water breeder reactor (LWBR), the steam-generating heavy-water reactor (SGHWR), and the molten salt breeder reactor (MSBR). The important characteristics of each type of power reactor are listed in Table A-1.

Power reactor systems most familiar in commercial practice or currently stressed in reactor development programs are described below.

Light-water Reactors

More than three-fourths of the power reactors operational and under construction worldwide are light-water reactors, and LWRs are expected to dominate the industry through this century. The LWR was developed by the United States; and, with the exception of a few experimental systems, all current U.S. power reactors are LWRs.

A schematic diagram of a boiling-water reactor (BWR) power system is shown in Figure A-1. The reactor fuel consists of stacks of pellets of slightly enriched uranium (about 3 percent uranium-235) in the form of uranium dioxide, encased in tubular metal cladding. Some of the fuel pellets could also be made of mixed oxide fuel (a mixture of plutonium and uranium dioxide) if it is decided to reprocess spent fuel and recycle plutonium. Each fuel rod is between three-eighths

Table A-1. Characteristics of Nuclear Power Reactors

Reactor Type	Neutron Energy	Fuel*	Moderator	Coolant	Converter or Breeder
LWR/PWR	Thermal	Enriched U (3.2% U-235) and possibly recycled Pu	Water	Water	Converter
LWR/BWR	Thermal	Enriched U (2.8% U-235) and possibly recycled Pu	Water	Water	Converter
HWR	Thermal	Natural U	Heavy Water	Heavy Water	Converter
HTGR	Thermal	Enriched U (~90% U-235), recycled U-233, and Th-232	Graphite	Helium	Converter
LMFBR	Fast	Recycled Pu and U-238	None	Liquid Sodium	Breeder
GCFBR	Fast	Recycled Pu and U-238	None	Helium	Breeder
LWBR	Thermal	Recycled U-233 and Th-232	Water	Water	Breeder
SGHWR	Thermal	Enriched U (~3% U-235) and possibly recycled Pu	Heavy Water	Water	Converter
MSBR	Thermal	Molten fluorides of U-233 and Th-232	Graphite	Molten Salt	Breeder

*Reactors using Pu-239 or U-233 may be started with enriched U-235 substituting for the other fissile materials.

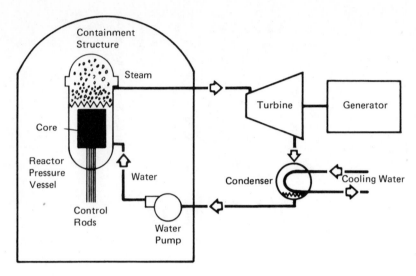

Figure A-1. Schematic Diagram of a Boiling-Water Reactor Power System

and one-half inch in diameter and about 12 feet long, and the core comprises an open array of several tens of thousands of these rods. The heat generated by fission in the fuel rods is transferred through the cladding to the water flowing past. The water boils, and a mixture of steam and water flows out of the top of the core. After passing through steam separators (to remove the entrained liquid water) the steam leaves the reactor vessel and is piped to the turbine. The turbine drives a generator which, in turn, produces electricity. The exhaust from the turbine then passes through a condenser and is pumped back to the reactor inlet. The BWR system is a direct cycle: the coolant which passes through the reactor also flows through the turbine. BWRs typically operate at pressures of about 1,000 pounds per square inch (psi). At this pressure, water boils and forms steam at about 545° F.

A schematic diagram of a pressurized-water reactor (PWR) power system is shown in Figure A-2. The main difference between the PWR and BWR systems is that the PWR system employs an indirect cycle: the cooling water which passes through the PWR does not flow through the turbine. The primary coolant water passes through the reactor core (which is similar in composition and configuration to that of a BWR) and is heated to about 600° F. Boiling does not occur, however, because the pressure in a PWR is maintained at about 2,250 psi (at which pressure water only boils at about 650° F). The primary coolant then leaves the reactor vessel, is piped through two or more steam generators, and is pumped back to the reactor inlet.

In the steam generators, the primary coolant water circulates inside tubes, the outsides of which are in contact with a second stream of water at lower pressure,

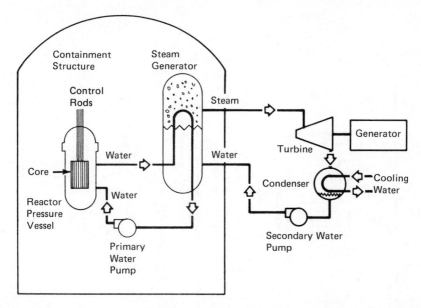

Figure A-2. Schematic Diagram of a Pressurized-Water Reactor Power System

about 1,000 psi. Heat is transferred through the tube walls from the hotter primary coolant to the cooler secondary stream, and the water in the secondary stream boils, providing steam at about 500° F for the turbine. The exhaust from the turbine then passes through a condenser and is pumped back to the steam generators.

Heavy-Water Reactors

The principal advantage of the HWR over the LWR stems from the fact that heavy water is sufficiently more effective than light water as a moderator to permit the use of natural unenriched uranium as fuel. The commercially available HWR is of Canadian design and is known as CANDU, a name derived from Canadian-deuterium-uranium. CANDU reactors employ natural uranium in the form of uranium dioxide as fuel, with heavy water as moderator and coolant. Unlike the LWR, in which the coolant flows through one large pressure vessel containing all of the fuel elements, in CANDU the coolant flows through several hundred individual pressure tubes containing the fuel. These pressure tubes are immersed in unpressurized heavy water (at room temperature) which serves as moderator, and the core is surrounded by a large cylindrical vessel. This configuration makes possible an important design feature of CANDU: fuel elements can be inserted and removed automatically while the reactor operates at full power, unlike LWRs, which must be shut down for refueling.

A schematic diagram of a HWR CANDU power system is shown in Figure A-3.

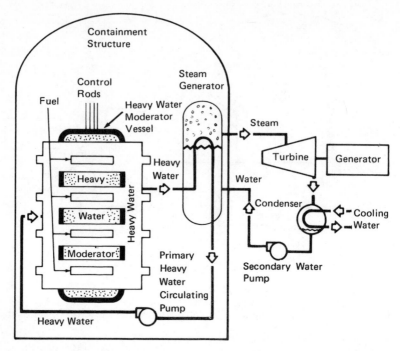

Figure A–3. Schematic Diagram of a Heavy-Water Reactor Power System

The CANDU system employs an indirect cycle in which the primary coolant, heavy water at a pressure of about 1,450 psi, passes through the pressure tubes and is heated without boiling to about 590° F. It then circulates through the tubes in the steam generators, and is pumped back into the reactor vessel. Ordinary light water, serving as the secondary coolant, flows past the outsides of the steam generator tubes, is converted into steam at about 570 psi and 480° F, and is piped to a turbine. The turbine exhaust then passes through a condenser and is pumped back to the steam generator.

High Temperature Gas-Cooled Reactors

The high temperature gas-cooled reactor (HTGR) has been the subject of considerable development effort in the United States but its commercial future in the United States has been clouded significantly since late 1975 when Gulf General Atomic, the developer and vendor of HTGRs, canceled all orders and suspended sales. HTGRs are being developed in a government research and development program in West Germany.

The HTGR makes use of three fuels: highly enriched uranium (over 90 percent uranium-235), uranium-233, and thorium-232 (for conversion to uranium-233), with the mix of fuels changing over the reactor lifetime. The initial loading consists of highly enriched uranium and thorium-232; in subsequent loadings recycled uranium-233 would replace highly enriched uranium. The fuel materials, in the form of graphite-coated particles, are stacked in holes in blocks of graphite (which serves as moderator), and are cooled by a stream of gaseous helium. The ability of the graphite core of the HTGR to absorb a great deal of heat without melting may give it a safety advantage like that of the HWR with its large bath of heavy water at room temperature.

An HTGR power system is shown schematically in Figure A-4. The primary coolant, gaseous helium, leaves the reactor at 1,430° F and 700 psi, circulates through the tubes in the steam generator, and then returns to the reactor. Water in the steam generator is converted into steam at 1,000° F and 2,500 psi, passes through the turbine and condenser, and returns as liquid to the steam generator. The relatively high steam temperature of HTGR power systems (1,000° F steam compared to the 480°-500° F steam produced in HWR and LWR power systems) permits more efficient conversion of the nuclear fission heat into electricity.

Liquid Metal Fast Breeder Reactors

The liquid metal fast breeder reactor (LMFBR) is expected to be a true breeder reactor; that is, it will convert fertile uranium-238 into fissile plutonium at a rate faster than its consumption of fissile fuel. Experimental LMFBR plants are operating in several countries.

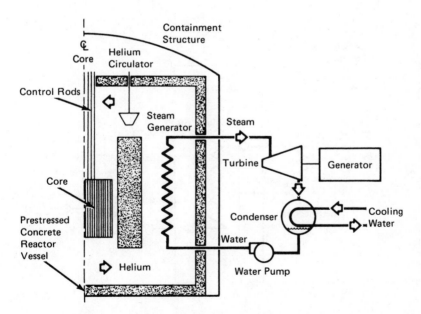

Figure A-4. Schematic Diagram of a High Temperature Gas-Cooled Reactor Power System.

The LMFBR fuel rods contain a mixture of plutonium dioxide and depleted uranium dioxide, and a blanket of rods containing depleted uranium dioxide surrounds the core. The initial loading could use either plutonium recovered from spent LWR fuel or enriched uranium; subsequent loadings would employ plutonium bred in the LMFBR itself.

An LMFBR power system is shown schematically in Figure A-5. The fission heat is transferred to liquid sodium in the primary loop. This sodium becomes radioactive in passing through the reactor core, and a heat exchanger is therefore used to transfer its heat to nonradioactive sodium in a secondary loop. The secondary loop sodium then passes through a steam generator in which the heat is transferred to water, creating steam at temperatures and pressures up to 900° F and 2,400 psi. The steam is piped to a turbine, passes through a condenser, and is pumped as liquid water back into the steam generator.

Light-Water Breeder Reactors

The LWBR would utilize much of existing LWR technology and is being developed in the same U.S. program which gave rise to the LWR. It is of interest because it would use an additional resource, thorium, and because it would be a very efficient converter or low-efficiency breeder. Given an initial load of fissile material (probably uranium-235), the LWBR would produce enough fissile

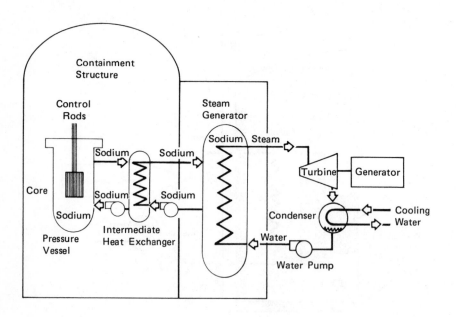

Figure A-5. Schematic Diagram of a Liquid Metal Fast Breeder Reactor Power System

uranium-233 from thorium to refuel itself repeatedly. A thorium/uranium-233 cycle avoids potential health problems from plutonium and the transplutonium actinides but perhaps involves different risks; proliferation and theft dangers would not be appreciably different since uranium-233 is suitable for weapons. The fact that the LWBR is a water-cooled thermal reactor might be found to imply safety or engineering advantages over sodium-cooled fast reactors when engineering of both systems is completed.

NUCLEAR FUEL CYCLES

The processes undergone by materials in their use as fuel in nuclear reactors comprise the nuclear fuel cycle. Each of these processes potentially affects human health, safety, the environment, nuclear proliferation, and theft.

Fuel Cycle Activities

The major fuel cycle activities, not all of which may be involved for a particular type of reactor, are described below.

Mining. Uranium mining techniques are similar to those used in coal mining. In the United States, about 70 percent of the uranium sources being worked are in underground mines, about 20 percent are in open pit mines, and the remaining 10 percent are in low-grade stockpiles or by-products. Some thorium is available as a by-product of uranium and titanium mining.

Milling. Following its removal from the mine, uranium ore is mechanically and chemically processed, or milled, to produce yellowcake, a concentrate containing about 80 percent uranium oxide (U_3O_8). The ore typically contains only about 0.1 percent U_3O_8; the rest is rejected as tailings. The tailings constitute a health hazard since they release radioactive radon gas to the atmosphere. Thorium milling converts the thorium-bearing ore into concentrated ThO_2.

Conversion to Uranium Hexafluoride (UF_6). In preparation for uranium enrichment, yellowcake is converted to hexafluoride (UF_6), which is a solid at room temperature but vaporizes at 135° F.

Enrichment. To be used as fuel in an LWR or HTGR, uranium must be enriched; that is, the concentration of the fissile isotope uranium-235 must be increased beyond the 0.7 percent fraction characteristic of natural uranium. Isotopes of a given element (uranium-235 and uranium-238, for example) exhibit identical chemical behavior and therefore cannot be separated by chemical means. The slight differences in the masses of isotopes, however, can be exploited to achieve separation. There are four methods of current interest: gaseous diffusion, gas centrifuge, aerodynamic, and laser techniques. Each has the potential for producing highly enriched uranium and thus nuclear weapons, though the ease with which an enrichment plant may be adapted to this purpose depends on the technology employed.

The gaseous diffusion process has been used for almost all of the uranium enrichment which has been performed for both weapons and reactor fuel. When gaseous uranium hexafluoride (UF_6) passes through a porous barrier, the lighter UF_6 molecules containing uranium-235 pass through at a faster rate than do the heavier UF_6 molecules containing uranium-238. The amount of separation accomplished by a single barrier is slight, and the process must be repeated many times to achieve enrichments of practical interest. Starting with natural UF_6, enrichment to reactor fuel levels, 3 percent uranium-235, requires about 1,250 stages, and enrichment to weapons standards, 90 percent uranium-235, requires about 4,000 stages. The energy required to convert natural feed material into enriched product and depleted tails is measured in terms of separative work units (SWUs). For a tails assay of 0.2 percent uranium-235, the production of one kilogram of 3 percent enriched uranium requires 5.5 kilograms of natural uranium feed and 4.3 SWU of separative work. Decreasing the concentration of uranium-235 in the tails requires more separative work but reduces the mass of feed material required to produce a given amount of product.

The gas centrifuge enrichment process is just reaching commercialization. URENCO—a consortium of the United Kingdom, West Germany, and The Netherlands—has two pilot plants in operation and has begun construction of commercial facilities. This process, which makes use of centrifugal forces to separate lighter UF_6 molecules containing uranium-235 from the heavier UF_6 molecules containing uranium-238, involves higher capital costs than does gaseous diffusion, but requires fewer stages and uses less energy to produce enriched uranium. While there are very large economy-of-scale penalties for small gaseous diffusion plants, economics is not a severe constraint on the size of a centrifuge plant.

Several aerodynamic methods have been used to separate isotopes of uranium. Best known among these is the jet nozzle process, in which a gaseous mixture of UF_6 and hydrogen is forced to flow at high velocity in a semicircular path, establishing centrifugal forces which tend to separate the heavier molecules from the lighter ones. Capital costs are expected to be lower than for the gaseous diffusion process, but energy requirements will be higher. The jet nozzle process has been demonstrated in West Germany, and a larger plant is to be sold to Brazil. South Africa has a pilot plant based on an aerodynamic process resembling the jet nozzle process.

Laser techniques for uranium enrichment, now at the laboratory research stage, could be of great importance. These processes make use of finely tuned lasers to exploit the slight differences in excitation energies of uranium-235 and uranium-238 atoms or of molecules containing them. In principle, a laser process could achieve a large degree of separation in a single stage while consuming relatively little energy. This would virtually eliminate the waste of uranium-235 in enrichment tails; indeed, the tails remaining from years of operation of gaseous diffusion plants might well serve as feed for laser enrichment plants. On the other hand, the highly enriched uranium-235 resulting from single-stage separation could be suitable for use directly in nuclear weapons.

Fabrication. The final step in the portion of the fuel cycle preceding reactor operation, the front end of the fuel cycle, is the fabrication of the fuel elements. This includes conversion of enriched UF_6 (or, in the case of HWR fuel, U_3O_8) into uranium dioxide (UO_2) or uranium metal; conversion of recovered fissile material, if any, into a suitable form; conversion of fertile materials, if any, into a suitable form; fabrication of small ceramic pellets containing the fuel material(s); encapsulation in fuel rods; and assembly.

Reactor Operation. The initial fuel loading of a typical 1000 MWe LWR power plant contains about 80 metric tons of slightly enriched (about 3 percent uranium-235) uranium. These fuel elements must be removed from the reactor before all the fissile material (uranium-235 and plutonium produced by conversion of uranium-238) is consumed. The buildup of fission products also constitute parasitic absorbers of the neutrons and threaten the mechanical integrity of the fuel elements. An LWR is shut down annually for a period of several weeks, during which time about one-third of the fuel elements are replaced with fresh ones. CANDU reactors are refueled continuously without interrupting their operation. The necessity of interrupting LWR operation is an economic penalty while the fact that HWRs can be refueled continuously would make it more difficult to safeguard against diversion of spent nuclear fuel.

Spent Fuel Storage. The "back end" of the fuel cycle begins with the temporary storage of the highly radioactive spent fuel in a special water-filled pool at the reactor site. After the spent fuel has cooled for several months, it can be moved to an interim storage or permanent disposal facility.

Reprocessing. After a year or so, during which the most intense radioactivity decays, spent fuel could be reprocessed to separate residual uranium and plutonium. After the mechanical and chemical steps involved in reprocessing, the recovered uranium could be converted into uranium hexafluoride for subsequent reenrichment, the recovered plutonium would be converted into plutonium dioxide for subsequent use in mixed oxide (a blend of uranium and plutonium dioxides fuel elements, and the radioactive waste would be converted into forms suitable for permanent disposal. The separated plutonium would also be usable in the manufacture of weapons. Only the French reprocessing plant at La Hague is now reprocessing LWR fuel; however, several others are planned, under construction, or nearing operation, and a number of countries have laboratory-scale and pilot plant reprocessing facilities planned or in operation. Reprocessing is the subject of Chapter 11.

Waste Disposal. The final steps in the nuclear fuel cycle are the management and disposal of the radioactive wastes, including fission products and materials which, through capture of neutrons or through contamination by radioactive materials become radioactive themselves. Some of these wastes have radioactive half-lives extending to tens of thousands of years; the problems of waste management and the extent to which disposal is safe over these time periods are discussed in detail in Chapter 8.

LWR Fuel Cycle

The fuel cycle currently associated with LWRs is shown in Figure A-6. This fuel cycle involves no reprocessing of spent fuel and no recycle of uranium or plutonium. The fuel cycle operations involved in this cycle are: mining of natural uranium ore; milling to separate U_3O_8, conversion to uranium hexafluoride; enrichment to about 3 percent uranium-235; conversion of the slightly enriched hexafluoride to dioxide; fabrication of fuel elements; reactor operation; temporary storage of spent fuel at the reactor site; and (in principle, but not yet in practice) storage or disposal in special facilities.

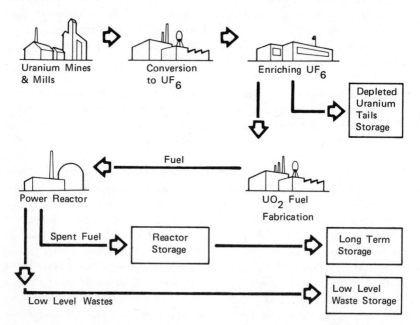

Figure A-6. Light-Water Reactor Fuel Cycle—No Uranium or Plutonium Recycle

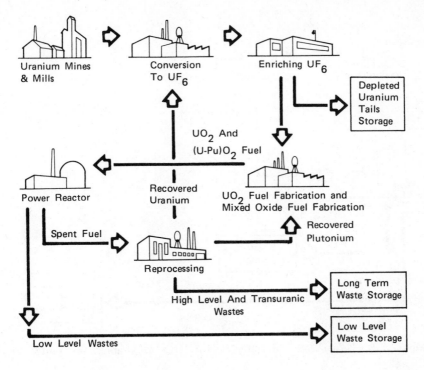

Figure A-7. Light-Water Reactor Fuel Cycle—Uranium and Plutonium Recycle

The extension of the LWR fuel cycle to reprocessing and recycle is shown in Figure A-7. The fuel cycle operations involved are the same as above except spent fuel is reprocessed to separate and recover uranium (in the form of uranium hexafluoride) and plutonium (in the form of plutonium dioxide) from radioactive wastes. The recovered uranium hexafluoride is then reenriched. The plutonium dioxide is combined with natural uranium dioxide to make mixed oxide fuel pellets which are fabricated to make mixed oxide fuel elements. The mixed oxide fuel elements are assembled with enriched uranium elements to make reactor core loadings.

HWR Fuel Cycle

The HWR fuel cycle is shown in Figure A-8. Because there is no need for uranium enrichment, the HWR fuel cycle is simpler than the LWR fuel cycle. The operations involved are: mining and milling of uranium ore; conversion to uranium dioxide; manufacture of fuel elements; reactor operation; and storage or disposal of spent fuel. Modification of this cycle to provide for plutonium recycle or use of fertile thorium-232 to produce fissile uranium-233 is technically feasible, though not now employed.

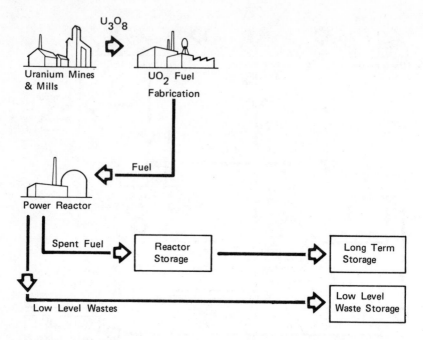

Figure A–8. Heavy-Water Reactor Fuel Cycle—No Uranium or Plutonium Recycle

HTGR Fuel Cycle

The feasibility of the HTGR is dependent upon the inclusion of reprocessing and recycle of uranium-233. The steps in the HTGR fuel cycle may be summarized as follows. For initial HTGR fuel loads, uranium ore is mined, milled, converted to UF_6, enriched to over 90 percent uranium-235, and converted to a carbide form. Thorium ores also are mined, milled, and converted to carbide. These materials are assembled in carbon blocks for loading into the reactor. Following its use in the reactor, the spent fuel is reprocessed to recover unburned material and uranium-233 bred from the thorium. Uranium-233 is used in subsequent fuelings. The wastes from reprocessing are converted to forms suitable for permanent disposal.

LMFBR Fuel Cycle

Reprocessing and recycle are essential elements of a breeder reactor fuel cycle. Present plans are for development of a plutonium-cycle LMFBR although a thorium/uranium-235 cycle is feasible. While initial loadings of early LMFBRs would contain highly enriched uranium or plutonium recovered from spent LWR fuel, LMFBRs would eventually depend for fuel upon the plutonium produced in the breeding process. The steps in this steady-state fuel cycle may be summarized as follows. Depleted uranium from enrichment tails (or natural uranium which has been mined and milled) is converted to uranium dioxide and combined with recovered plutonium dioxide and recovered uranium dioxide to make mixed oxide fuel. After exposure in the reactor, the spent fuel is reprocessed to separate the uranium and plutonium and the radioactive waste is converted to forms suitable for permanent disposal. Unlike LWRs, where reprocessing and recycle are optional, LMFBRs (as presently planned) involve a commitment to a fuel cycle in which separated plutonium is present at many stages.

Glossary

ACTINIDES: A group name for the series of radioactive elements from element 89 (actinium) through element 103 (lawrencium).

BASE LOAD UNIT: An electric generating facility designed to operate at constant output with little hourly or daily fluctuation.

BARNWELL: The site in South Carolina of a large unfinished plutonium reprocessing plant owned by Allied General Nuclear Services.

BOILING-WATER REACTOR (BWR): A light-water reactor that employs a direct cycle; the water coolant that passes through the reactor is converted to high pressure steam that flows through the turbine.

BREEDER REACTOR: A nuclear reactor that produces more fissile material than it consumes.

BRITISH THERMAL UNIT (BTU): The amount of energy necessary to raise the temperature of one pound of water by one degree Fahrenheit, from 39.2 to 40.2 degrees Fahrenheit.

CANDU: A nuclear reactor of Canadian design, which uses natural uranium as a fuel and heavy water as a moderator and coolant.

CAPACITY FACTOR: The ratio of energy actually produced to that which would have been produced in the same period (usually a year) had the unit operated continuously at rated capacity.

CAPITAL CHARGES: The annualized costs of borrowing plus the amortization of investment and allowance for taxes.

CENTRIFUGE ISOTOPE SEPARATION: A new isotope enrichment process that separates lighter molecules containing uranium-235 from heavier molecules containing uranium-238 by means of ultra high speed centrifuges. The process is now being successfully exploited by the URENCO consortium in Western Europe.

CLINCH RIVER BREEDER REACTOR (CRBR): A proposed demonstration of the liquid metal fast breeder reactor. The CRBR, which would operate on a plutonium fuel cycle and be cooled by molten sodium, would have an electrical output of about 350 MWe.

COGENERATION: The generation of electricity with direct use of the residual heat for industrial process heat or for space heating.

COMBINED CYCLE: The combination, for instance, of a gas turbine followed by a steam turbine in an electrical generating plant.

CONSTANT DOLLARS: Dollar estimates for which the effects of inflation or deflation have been removed, reported in terms of a base year value and assumed to have constant purchasing power.

CONVERSION EFFICIENCY: The percentage of total thermal energy that is actually converted into electricity by an electric generating plant.

COOLING TOWER: A heat exchange device that transfers rejected heat from circulating water to the atmosphere.

CURIE (CI): A measure of the radioactivity level of a substance; one curie equals the disintegration of 3.7×10^{10} nuclei per second and is equal to the radioactivity of one gram of radium-226.

CURRENT DOLLARS: Dollar values that have not been corrected for inflation or deflation.

DEGREE-DAY: A unit of mean daily outdoor temperature duration representing one degree of difference from a standard temperature, usually 65°F. for a day.

DISCOUNT RATE: A rate used to reflect the time value of money. The discount rate is used to adjust future costs and benefits to their present day value. The effect of the discount rate (r) on the present value of a cost or benefit at time (t) in the future (C_t) is given by the expression $C_{(t)}[\frac{1}{(1+r)^t}]$. The selection of discount rates appropriate to particular situations is a matter of debate among economists, although 10 per cent has been used in government calculations.

EMERGENCY CORE COOLING SYSTEM (ECCS): A safety system in a nuclear reactor, the function of which is to prevent the fuel in the reactor from melting should a sudden loss of normal coolant occur.

ENRICHMENT: The process by which the percentage of the fissionable isotope uranium-235 is increased above that contained in natural uranium.

ENVIRONMENTAL IMPACT STATEMENT (EIS): A statement required by the National Environmental Policy Act (NEPA) of 1969 for any federal action significantly affecting the environment.

EURODIF: A joint venture, involving France (52 percent), Italy, Belgium, Spain, and Iran, which is building a gaseous diffusion plant in France.

FAST FLUX TEST FACILITY (FFTF): An experimental liquid metal fast breeder reactor, currently under construction, which will be used to test various fuels and reactor components.

FISSILE MATERIAL: Atoms such as uranium-233, uranium-235, or plutonium-239 that fission upon the absorption of a low energy neutron.

FISSION: The splitting of an atomic nucleus with the release of energy.

FIXED CHARGE: Expenses that are independent of the volume of business activity. Items included in this category are a company's interest on funds or other borrowings, some taxes levied by the government, insurance payments, and depreciation due to obsolescence.

FLUIDIZED BED COMBUSTION: The combustion of coal using the fluidized bed technique with the solids composed of finely crushed coal and limestone or dolomite and a fluidizing gas. This method burns coal efficiently with reduced emissions of sulfur and other pollutants.

FUSION: The combining of certain light atomic nuclei to form heavier nuclei with the release of energy.

GASEOUS DIFFUSION: A process used to enrich uranium in the isotope uranium-235. Uranium in the form of a gas (UF_6) is forced through a thin porous barrier. Since the lighter gas molecules, containing uranium-235, move at a higher velocity than the heavy molecules, containing uranium-238, the lighter molecules pass through the barrier more frequently than do the heavy ones, producing a slight enrichment in the lighter isotope. Many stages are required to produce material sufficiently enriched for use in a light water reactor.

HALF-LIFE: The period required for the disintegration of half of the atoms in a given amount of a specific radioactive substance.

HIGH TEMPERATURE GAS-COOLED REACTOR (HTGR): A graphite moderated, helium cooled reactor using highly enriched uranium as initial fuel and thorium as a source of new fuel.

INCOME ELASTICITY OF DEMAND: A measure of the change of consumer demand for a particular good with change in income. It is defined as the ratio of the percentage change in demand to the percentage change in consumer income. If income elasticity of demand for a commodity is low, there will be little change in consumer demand for that commodity in response to changes in income.

IRRADIATED FUEL: Nuclear fuel that has been used in a nuclear reactor.

ISOTOPE: One of perhaps several different species of a given chemical element, distinguished by variations in the number of neutrons in the atomic nucleus but indistinguishable by chemical means.

KRYPTON-85: A fission product and radioactive isotope of the element krypton, an inert gas, with a half-life of about 10 years.

LASER ISOTOPE SEPARATION: A new isotope enrichment process now in the development stage. Atoms of uranium-235, or molecules containing them, would be selectively ionized or excited by lasers, allowing physical or chemical separation of one of the isotopes.

LIGHT-WATER REACTOR (LWR): A nuclear reactor that uses ordinary water as a coolant to transfer heat from the fissioning uranium to a steam turbine and employs slightly enriched uranium-235 as fuel.

LINEAR ENERGY TRANSFER (LET): The density of ionization events along the path of a nuclear particle. Beta and gamma radiations have low LET whereas alpha particles and neutrons have high LET.

LOSS OF COOLANT ACCIDENT (LOCA): A reactor accident in which the primary coolant is lost from the reactor core.

LOSS OF FLUID TEST (LOFT): An experimental device, one-sixtieth the size of a commercial pressurized-water reactor, which will be used by ERDA to simulate loss of coolant accidents.

MAN-REM: A measure of human radiation exposure; it is the total dose in rem received by the individuals in a given population.

MATERIAL UNACCOUNTED FOR (MUF): The difference between the amount of plutonium or uranium entering a facility and that coming out; it is an indicator of the uncertainty inherent in inventory measurements.

MEGAWATT: The unit by which the rate of production of electricity is usually measured. A megawatt is a million watts or a thousand kilowatts.

MILLING: A process in the uranium fuel cycle by which ore containing only a very small percentage of uranium oxide (U_3O_8) is converted into material containing a high percentage of U_3O_8, sometimes referred to as yellowcake.

MILLIREM (MREM): A unit used to measure radiation dose; one one-thousandth of a rem.

NANOCURIE: One billionth of a curie.

NUCLEAR WASTE: The radioactive products formed by fission and other nuclear processes in a reactor. Most nuclear waste is initially in the form of spent fuel. If this material is reprocessed, new categories of waste result: high-level, transuranic, low-level wastes and others.

PARTICULATES: Fine solid particles that remain individually dispersed in emissions from fossil fuel plants.

PEAK LOAD: The maximum power demand on a power supply system.

PRESENT VALUE: The current value of a future stream of costs or benefits calculated by discounting these costs or benefits to the present time. (See discount rate.)

PRESSURIZED-WATER REACTOR (PWR): A light-water reactor that employs an indirect cycle; the cooling water that passes through the reactor is kept under high pressure to keep it from boiling, but it heats water in a secondary loop that produces steam that drives the turbine.

PRICE ELASTICITY OF DEMAND: The responsiveness of demand to changes of price. It is defined as the ratio of the percentage change in the quantity demanded to the percentage change in the price of the commodity: If a given change in price results in a large change in demand, then demand is elastic; if the change in price has only a slight effect on demand, then demand is inelastic.

QUAD: A measure of energy equal to 10^{15} British thermal units.

RAD: A measure of the radiation absorbed in tissue, corresponding to an energy absorption of 100 ergs per gram.

REACTOR YEAR: A year's operation of one reactor.

REM (ROENTGEN EQUIVALENT MAN): The unit of biological dose given by the product of the absorbed dose in rads and the relative biological efficiency of the radiation.

REPROCESSING: The chemical and mechanical processes by which plutonium-239 and the unused uranium-235 are recovered from spent reactor fuel.

RESERVES: For uranium, those resources that are known in location, quantity, and quality and that are recoverable below a specified cost using currently available technologies.

RESOURCES: Deposits that may be known to exist but not in such quantity or state as to be economically recoverable by present technologies, or those that are unidentified but suspected or probable on the basis of indirect evidence.

SCRUBBER: A device for removing pollutants, such as sulfur dioxides or particulate matter, from stack gas emissions.

SEPARATIVE WORK UNITS (SWUs): A measure of the work required to separate uranium isotopes in the enrichment process. It is used to measure the capacity of an enrichment plant independent of a particular product and tails. To put this unit of measurement in perspective, it takes about 100,000 SWUs per year to keep a 1,000 MWe LWR operating and 2,500 SWUs to make a nuclear weapon.

SPECIAL NUCLEAR MATERIALS (SNMs): Plutonium, uranium-233, enriched uranium-235, and any other material that the NRC determines to be SNMs.

SPENT FUEL: The fuel elements removed from a reactor after several years of generating power. Spent fuel contains radioactive waste materials, unburned uranium and plutonium.

STRATEGIC SPECIAL NUCLEAR MATERIALS (SSNMs): SNMs that have been enriched to a concentration of at least 20 percent.

TAILINGS: Rejected material after uranium ore is processed. Since uranium ore contains less than 1 percent uranium, essentially all of the processed ore is left as tailings near uranium mills.

TAILS: Uranium, depleted in the isotope uranium-235, remaining after production of enriched uranium.

TAILS ASSAY: The percentage of uranium-235 in tails. Natural uranium contains 0.71 percent uranium-235; the tails assay may be 0.3 percent or lower.

URANIUM HEXAFLUORIDE (UF_6): A gaseous compound of uranium used in various isotope separation processes.

URANIUM OXIDE (U_3O_8): The most common oxide of uranium that is found in typical ores.

URENCO: A joint British, Dutch, and West German organization that operates centrifuge isotope separation plants.

Abbreviations and Acronyms

ACDA	Arms Control and Disarmament Agency
AEC	Atomic Energy Commission
AIF	Atomic Industrial Forum
BEIR	The Advisory Committee on the Biological Effects of Ionizing Radiations
BTU	British thermal unit
BWR	boiling-water reactor
CANDU	Canadian deuterium-uranium reactor
CBR	commercial breeder reactor
CBR-1	first commercial breeder reactor
ci	curie
CRBR	Clinch River breeder reactor
DBoA	delayed breeder or alternative
EBR-1	Experimental Breeder Reactor 1
EBR-2	Experimental Breeder Reactor 2
ECCS	emergency core cooling system
EIS	Environmental Impact Statement
EPA	Environmental Protection Agency
EPRI	Electric Power Research Institute
ERDA	Energy Research and Development Administration
EURATOM	European Atomic Energy Community
FBC	fluidized bed combustion
FBR	fast breeder reactor
FFTF	fast flux test facility

GESMO	Final Generic Environmental Statement on the Use of Recycle Plutonium in Mixed Oxide Fuel in Light Water Cooled Reactors (Nuclear Regulatory Commission Report, August 1976)
GNP	Gross National Product
GWe	gigawatt(s) (electric) = 1,000 MWe
GWt	gigawatt(s) (thermal) = 1,000 MWt
HTGR	high temperature gas reactor
IAEA	International Atomic Energy Agency
ICRP	International Commission on Radiological Protection
KeV	kiloelectron volt(s)
kg	kilogram(s)
kw	kilowatt(s)
kwe	kilowatt(s) (electric)
kwh	kilowatt-hour(s)
kwt	kilowatt(s) (thermal)
LEMUF	limits of error on material unaccounted for
LET	linear energy transfer
LMFBR	liquid metal fast breeder reactor
LOCA	loss of coolant accident
LOFT	loss of fluid test
LWR	light-water reactor
MB	millions of barrels of oil
MBD	millions of barrels of oil per day
MBTU	million BTU
MPC	maximum permissible concentration
mrem	millirem(s)
MT	metric ton(s)
MTU	metric tons of uranium
MUF	material unaccounted for
MWd	megawatt day(s)
MWe	megawatt(s) (electric) = 1,000 KWe
MWt	megawatt(s) (thermal) = 1,000 KWt
NCRP	National Council on Radiation Protection and Measurements
NFS	Nuclear Fuel Services
NPT	Non-Proliferation Treaty
NRC	Nuclear Regulatory Commission
NSC	National Security Council
NURE	National Uranium Resource Evaluation Program
OECD	Organization for Economic Cooperation and Development
OMB	Office of Management and Budget

OPEC	Organization of Petroleum Exporting Countries
OSHA	Occupational Safety and Health Administration
PLBR	prototype large breeder reactor
PWR	pressurized-water reactor
quad	quadrillion (10^{15}) British thermal units
rem	roetgen equivalent man
SALT	Strategic Arms Limitation Talks
SNM	special nuclear materials
SSNM	strategic special nuclear materials
SWU	separative work unit
TCF	trillion cubic feet
TRU	transuranics-contaminated
TVA	Tennessee Valley Authority
UNSCEAR	United Nations Scientific Committee on the Effects of Atomic Radiation
USGS	United States Geological Survey

NUCLEAR ENERGY POLICY
STUDY GROUP
List of Members

Spurgeon M. Keeny, Jr. (Chairman), Director, Policy and Program Development, The MITRE Corporation, Washington Operations

Seymour Abrahamson, Professor of Genetics, University of Wisconsin

Kenneth Arrow, James Bryant Conant University Professor, Harvard University

Harold Brown, President, California Institute of Technology

Albert Carnesale, Associate Director, Program for Science and International Affairs, Harvard University

Abram Chayes, Felix Frankfurter Professor of Law, Harvard Law School

Hollis B. Chenery, Vice President, Development Policy, International Bank for Reconstruction and Development

Paul Doty, Director, Program for Science and International Affairs, Harvard University

Philip Farley, Senior Fellow, The Brookings Institution

Richard L. Garwin, IBM Fellow, IBM Corporation, Thomas J. Watson Research Center

Marvin Goldberger, Eugene Higgins Professor of Physics, Princeton University

Carl Kaysen, David W. Skinner Visiting Professor of Political Science, School of Humanities and Social Sciences, Massachusetts Institute of Technology

Hans H. Landsberg, Co-Director, Energy and Materials Division, Resources for the Future

Gordon J. MacDonald, Henry R. Luce Third Century Professor of Environmental Studies and Policy, Dartmouth College

Joseph S. Nye, Jr., Professor of Government, Center for International Affairs, Harvard University

Wolfgang K.H. Panofsky, Director, Stanford Linear Accelerator Center

Howard Raiffa, Frank P. Ramsey Professor of Managerial Economics, John F. Kennedy School of Government, Harvard University

George Rathjens, Professor of Political Science, Massachusetts Institute of Technology

John C. Sawhill, President, New York University

Thomas C. Schelling, Lucius N. Littauer Professor of Political Economy, Harvard University

Arthur Upton, Professor of Pathology, State University of New York at Stony Brook